HUMAN PHYSIOLOGY AND THE
ENVIRONMENT IN HEALTH AND DISEASE

Readings from
**SCIENTIFIC
AMERICAN**

HUMAN PHYSIOLOGY AND THE ENVIRONMENT IN HEALTH AND DISEASE

With Introductions by
Arthur J. Vander
University of Michigan

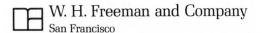 W. H. Freeman and Company
San Francisco

Most of the SCIENTIFIC AMERICAN articles in
HUMAN PHYSIOLOGY AND THE
ENVIRONMENT IN HEALTH AND DISEASE
are available as separate Offprints. For a complete list
of more than 950 articles now available as Offprints,
write to W. H. Freeman and Company, 660 Market
Street, San Francisco, California 94104.

Library of Congress Cataloging in Publication Data

Main entry under title:

Human physiology and the environment in health and
   disease.

   Includes bibliographies and index.
   1.   Adaptation (Physiology) — Addresses, essays,
lectures.   2.   Environmentally induced diseases —
Addresses, essays, lectures.   3.   Human physiology —
Addresses, essays, lectures.   I.   Vander, Arthur J.,
1933-   II.   Scientific American.   [DNLM: 1.
Physiology — Collected works.   2.   Environmental
health — Collected works.   WA30 H918]
QP82. H93   612   76-1923
ISBN 0-7167-0527-3
ISBN 0-7167-0526-5

Printed in the United States of America

9 8 7 6 5 4 3 2 1

PREFACE

This anthology of articles from *Scientific American* is designed to supplement the standard texts for courses in human physiology, environmental physiology, pathobiology, and environmental medicine. Its major purpose is to illustrate the profound and ubiquitous influences of the external environment on the body, the physiological responses to environmental challenges, and the ways in which these responses contribute to either health or disease.

The scope of physiology is extremely broad and is certainly not limited to the work of those who bear the label "physiologist." Thus, some of these articles are written by physiologists, but others have arisen from the research of other specialists: psychologists, biochemists, toxicologists, pathologists, clinicians, etc. This breadth is all to the good, for this anthology should, I think, emphasize the central importance of physiology in human biology, and should present a panorama of the extraordinary possibilities for future physiology-related research. At a time when the external environment is changing rapidly because of human intervention, physiologists are also expanding their research focus, so that the effects of noise, synthetic chemicals, radiation, crowding, psychosocial stress, and many other components are now considered part of the field studied by environmental physiology.

I have several rationales for presenting this anthology. First, the closely related fields of physiological responses to the environment and of pathophysiology (the physiological basis of disease) are barely touched upon in most physiology textbooks. Yet their content has great inherent interest, nicely illustrates basic physiology, and is of enormous practical importance for human health and disease. Second, there is little opportunity in textbooks to present experiments and research data; these *Scientific American* articles clearly and accurately report experiments, with a minimum of technical detail, so that the reader gets a feel for how physiology is done and for the complexity of the problems. Finally, and perhaps most important, it has been my experience, in using most of these articles in my courses, that students find them highly interesting and enjoyable, and receive an additional strong stimulus from them to continue pursuing their study of basic or applied physiology.

The articles have been organized into six sections. They are preceded by a general introduction which is an essential prelude to the articles. The first four sections deal with the physiological effects of (and physiological responses to) the nutritional, chemical, physical, and psychosocial environments. Section V, "Immune Defenses of the Body," is concerned not only with our responses to the microbial environment, but also with reactions to cancer cells and transplants. Section VI, "Aging," discusses current concepts of the aging process, and, in many respects, serves to bring together information and concepts from the earlier sections. Accordingly, sections VI and VII are best read last, and the General Introduction should definitely be read first. The order of all the other sections is logical but arbitrary, and need not be followed.

October 1975   *Arthur J. Vander*

CONTENTS

General Introduction: Adaptation and Disease 1

V   IMMUNE DEFENSES OF THE BODY

VI   AGING

VII·   EPILOGUE

Note on cross-references: References to articles included in this book are noted by the title of the article and the page on which it begins; references to articles that are available as Offprints, but are not included here, are noted by the article's title and Offprint number; references to articles published by SCIENTIFIC AMERICAN, but which are not available as Offprints, are noted by the title of the article and the month and year of its publication.

GENERAL INTRODUCTION: ADAPTATION AND DISEASE

Two key concepts in environmental physiology are *homeostasis* and *adaptation*. Homeostasis is the total process by which the internal environment of an individual is maintained relatively constant in the face of changes in the external environment. This stability is achieved by regulation and integration of the various organs and tissues of the body in such a way that any significant change in the internal environment automatically initiates a reaction that minimizes the change. These homeostatic mechanisms, which are the focus of study in basic physiology, are a subset of the more general category known as *adaptations*. Unfortunately, this last term has come to have a variety of meanings, but I shall use it to mean any structural or functional property of an organism which favors survival in a specific environment. An adaptation permits maintenance of physiological activity and favors survival when the environment changes.

Adaptation and the Environment

Biological adaptations may be classified, depending on how they originate, as *genetic adaptations* or as *acclimatizations* (sometimes also called *physiological adaptations*). The origin of genetic adaptations is described by basic evolutionary theory: the adaptive property *arises* from a mutation, and the environment then *selects* for it if it is *adaptive* for that particular environment (by means of the fact that those individuals who possess it will be more likely to survive and to reproduce). In contrast to genetic adaptations, an acclimatization is an adaptive change induced during the individual's lifetime by an environmental stress, and does not involve an alteration in his genetic endowment. Accordingly, it follows that an acclimatization is not passed on genetically to the offspring.

As an example to clarify the distinction between the origins of these two categories, let us consider what occurs when we expose identical twins to heat stress. On day 1, we shall expose them for 30 minutes to a temperature of 37°C and ask them to do a standardized exercise task. They will soon begin to sweat, and will probably sweat very similar amounts during the exposure period. Sweat glands and sweating are examples of genetic adaptations, in this case for survival in a hot environment; they are present more or less in all persons (except for a tiny number of people who cannot survive for long in heat), because the genes which code for them are passed on from generation to generation. There is no reason to suspect that exposure of man's ancestors to heat was the trigger which caused these genes to form; rather, we presume that these genes were formed as a result of mutations that occurred by chance, and were then selected for by the hot environment of early man (or his ancestors).

Now we will perform a second experiment with our identical twins: for one week twin A will enter the 37°C heat chamber for one or two hours per day and will exercise, but twin B will be excused for the entire week. On day 8, both twins will again be brought into the heat chamber and their sweating rates measured. Now we observe that, in contrast to day1, when their sweating rates were approximately the same, twin A begins to sweat earlier and more profusely than twin B. Twin A is said to have *acclimatized* to the heat, that is, he has undergone an adaptive change induced by the environmental stress, to which he is now better adapted, but with *no* change in his gene structure. Thus, acclimatization may be viewed as an environmentally induced *improvement* in the functioning of an already existing, genetically based homeostatic system.

In the preceding example, the heat acclimatization of twin A is completely reversible, in that, if the daily exposures are discontinued, then within a relatively short time his sweating rate upon exposure to 37°C will have reverted to its preacclimatized value. However, not all acclimatization is reversible. If the acclimatization is induced very early in the individual's life, it may be irreversible. Thus, the barrel-shaped chests of Andean natives seem not to represent a genetic difference between them and their lowland relatives, but rather an irreversible acclimatization which was induced during the first few years of their lives by their exposure to high altitude. These so-called *developmental acclimatizations* appear to be more common than was previously thought, and make the phenomenon of reversibility a less valuable criterion for deciding whether variation in a given trait is due to genetic differences or to acclimatization.

How does acclimatization occur? Usually, if not always, it reflects environmentally induced changes in the functioning of certain of the body's macromolecules, ultimately the enzymes and other proteins. Thus, in our previous example, the increased sweat production which characterized the acclimatized state reflected, at least in part, changes in the activities or concentrations of the enzymes responsible for catalyzing the transport of the sweat electrolytes, the formation of structural proteins and neurotransmitters, etc. There are many biochemical mechanisms by means of which the environment brings about these changes in enzymes and reaction rates, but one of the most important is that of gene induction and repression. The rate at which any single gene operates (i.e., codes for RNA and ultimately for synthesis of a single protein) is highly variable, and is susceptible to influence by environmental factors, which can increase (induce) or decrease (repress) it. In addition to influencing the rate at which a protein is formed by acting at the level of the genetic apparatus, environmental factors can alter the activity of an already formed enzyme. Finally, the environment can influence reaction rates by altering the supply of substrates, which are the substances on which the enzyme works.

So far we have focused on the distinction between acclimatization and adaptations that arise from changes in genetic endowment. Given any genotype (that is, a particular genetic endowment), an individual has remarkable *phenotypic plasticity*, i.e., the ability to manifest different traits and different magnitudes of any particular trait or physiological response, depending on the environment. However, I should now like to emphasize that acclimatization is, itself, limited by the genes. Phenotypic plasticity in an individual is not an infinite continuum; one's genes set the possible range of plasticity, and the environment, by influencing the expression of the genes, determines whether (and how much) any given trait within this range will be manifested. Human beings are extraordinarily adaptable because of their great phenotypic plasticity (especially in their behavior). This adaptability provides a considerable safety factor against environmental change, but there are limits to it. A key question for modern man is whether he possesses enough adaptive capacity to meet the environmental challenges he himself is creating.

Finally, it should be stressed that man's plasticity is not acted upon only intermittently, that is, only when the environmental stress is very great. Rather, the entire development and everyday functioning of the individual is conditioned by the environment. For example, several profound physical changes have occurred in the peoples of the Western world during the past 100 years, and these changes can be attributed to environmental influences (since this time span is much too short for genetic change to have played a role), although the relative contributions of the various influences are not clear. Most visible is people's increased rate of growth, but even more remarkable has been the change in the rate of sexual maturation; in Norway, the average age at which menstruation begins has dropped from 17 to 13 during the past 110 years.

To reiterate, at all times environment shapes our physical functioning and behavior. The most profound and lasting—often irreversible—effects occur during prenatal and early postnatal periods. This fact is expressed by the concept of *critical periods*, which states that developing organ systems are most vulnerable to environmental influence during their periods of maximal growth. For example, we shall see in this anthology how early malnutrition produces permanent stunting of growth.

Because of the complexity of the interactions between the genes and the environment, it has proven extremely difficult to determine whether differences in physiological responses between populations are due to differences in their gene pools or to differences in their environments. To take a single example, the blood flow to the hands during exposure to cold is greater in Eskimos than in other populations. One might suppose it would be easy to discover whether this fact results from an acclimatization in Eskimos or whether it represents a genetic difference, but the evidence is far from conclusive on this subject. If scientists have difficulty reaching a conclusion even in such a readily studied area as this, it should be no surprise that a question about whether there is any genetic basis for observed population differences in IQ scores cannot be answered with current knowledge and methods.

Health and Disease
The concepts of adaptation and environmental influences are crucial for explaining the state of health or disease of a person or population. I know of no better illustration of the impact of a changing environment than the profound alteration of disease patterns which has occurred in the United States during the past 100 years. Infectious diseases such as influenza, pneumonia, and tuberculosis were the

major killers a century ago, but their place has now been taken by the so-called "chronic degenerative diseases," notably cardiovascular disease and cancer. No one doubts that the decrease in the infectious diseases has been due to environmental alteration—modern sanitation, better nutrition, vaccines, and (to a much lesser extent) specific medical therapies. The assumption is frequently made that, with the "conquest" of the serious infectious diseases, people are simply living to be old enough to eventually succumb to the chronic degenerative diseases. However, this assumption rests upon another: that these modern killers strike only the aged, and are inexorable concomitants of aging. But such is not the case. The incidence of these diseases has been steadily rising in young age groups as well; moreover, many societies have much lower incidences of the disease within the same age groups. Perhaps the recent research on cigarette smoking and cancer and on diet and heart attacks has been most important in making people (including physicians!) finally recognize that the emergence of these diseases, like that of the microbial diseases before them, depends largely on environmental factors. Just as malnutrition made people far more susceptible to tuberculosis, many factors in our contemporary environment increase our susceptibility to diseases that might better be called "diseases of industrialized societies" rather than "diseases of aging."

Spurred on by a growing number of epidemiological studies, we have come to recognize that it may be more meaningful to speak of "risk factors" rather than of "specific causes" in describing the etiology of disease. Risk factors are variables which influence the rate at which a disease progresses. Thus, atherosclerosis ("hardening of the arteries") does not have a simple "cause"; instead, it results from many interplaying environmental factors (diet, smoking, physical activity, psychosocial stress, etc.), and the degree of influence of each factor will differ to some degree from person to person because of genetic differences.

Another point of great importance is that these environmental factors have many years over which to act, so that extremely small and subtle effects may ultimately be multiplied into gross manifestations. As a result, it may be extremely difficult to find out just which factors are the pathogenic ones—and this difficulty is a major reason why we should be extremely cautious about introducing new environmental influences.

It is not surprising that one's state of health or disease reflects, in large part, the success or failure of one's adaptive responses to the environment. It is surprising, however, that many of the body's responses actually turn out to be maladaptive, that they lead to disease instead of preventing it. To some extent, this maladaption results from the very complexity of the body: the response to one stressor automatically alters many other components of body function, including responses to other, simultaneous stressors. For example, when the body responds adaptively to a foreign chemical by increasing the rate at which the liver breaks down the chemical, the rate at which certain endogenous chemicals are broken down is also changed, with potentially harmful results.

Another type of maladaption occurs when a response which is highly adaptive in the short term turns out to be harmful if continued for long periods. For example, low levels of irritating air pollutants increase mucus production, which is useful in eliminating the irritants from the respiratory tract. However, chronic excessive mucus accumulation can lead to lung infection.

There may well be a common denominator to such maladaptive responses, namely, the fact that most people now live in environments which, because of human intervention, are very different from the environments in which early man evolved. Accordingly, it should not be surprising that physiological activities which were selected by evolution because of their adaptiveness in one environment might prove to be maladaptive when the environment changes. For example, when food supply is quite precarious, it is adaptive to possess a strong "drive" for food; in contrast, when food is relatively plentiful, such a "drive" may cause overeating, obesity, and its attendant consequences.

I

NUTRITIONAL
INFLUENCES

I

NUTRITIONAL INFLUENCES

INTRODUCTION

During the course of human history, the nature of man's food resources has undergone several revolutions. During their early hunting and plant-gathering phase, humans consumed a great variety of different plant and animal foods. This variety assured that all essential nutrients were consumed when food was available, but, because of the precariousness of the hunt, there were often relatively long periods when enough food was not available. Accordingly the body evolved mechanisms for dealing with alternating periods of fasting and feasting.

The agricultural revolution (after 10,000 B.C.) brought about a shift in importance from animal to vegetable sources of food, more specifically, created an almost complete dependency on a single type of cultivated plant, and this pattern remains the norm today for most nonindustrialized countries. With it came a new set of problems. First, the great reduction in the variety of food resulted in chronic deficiencies of several essential nutrients (protein, iron, and vitamin A, for example). Second, the expansion of the total food supply made possible by agriculture ironically ushered in the age of great famines. These famines occur whenever the production of the staple crop decreases (because of weather or social upheavals), and they affect the large populations whose growth was originally made possible by the shift from hunting and gathering to agriculture. (Today, of course, this population growth is further stimulated by the introduction of modern public health and medical practices.) Thus, the paradox: a major result of the agricultural revolution has been widespread chronic malnutrition, interrupted by periods of acute starvation.

The second nutritional revolution has occurred in modern times, in those countries which have industrialized. The diet of most of the inhabitants of these countries differs profoundly from both the diet of hunter-gatherers and that of the inhabitants of nonindustrialized agricultural countries. There has been a large increase in protein intake, an even larger increase in fat intake, and a shift from fat of plant origin to that of animal origin. Total calories have increased, as has the supply of simple sugars, salt, and food additives, whereas fiber (roughage) intake has greatly decreased. Moveover, our food supply is now continuous, so that periods of fasting are no longer necessary. Only recently have we begun to recognize that these changes are almost certainly a mixed blessing, that our "overnutri-tion" may be another form of malnutrition, in that it may be contributing to those diseases which have increased so dramatically in the West (hypertension, atherosclerosis, and diabetes mellitus, for example).

In summary, we have seen that humans evolved in one nutritional setting (hunting-gathering) but have profoundly changed their nutritional patterns, with important consequences for both health and disease. Accordingly, a starting point for delving into the physiology of human adaptation to these different nutritional patterns is to study those basic mechanisms which must have evolved quite early and which permit continued function and survival during alternating periods of "fasting and feasting." These are described by Young and Scrimshaw in "The Physiology of Starvation."

As described in the General Introduction, most adaptations ultimately reflect changes in the enzymes of the body, and the physiological changes which occur during starvation provide a splendid illustration of this principle. Fasting triggers changes in the concentrations or activities of a large number of enzymes in the liver, adipose tissue, muscle, brain, and other cells of the body; these changes, in over-all effect, promote fat breakdown, produce glucose from protein, and conserve this glucose for use by the brain. These integrated enzyme changes are, in large part, mediated by changes in the production of insulin and other hormones, but the precise causal sequences of events are still unclear.

Young and Scrimshaw also describe how fasting elicits another important adaptive response: the total energy used by the body is reduced largely because body mass is reduced (a smaller body uses less energy). Here again is an excellent illustration of a general concept described in the General Introduction, namely, the fact that environmental factors have their most profound effects early in life. If a normal adult fasts for weeks or months, he can readily regain all the lost weight after the fast is ended. In contrast, a malnourished infant will be permanently stunted. In terms of body size, this could be viewed as a useful adaptation, in that the smaller body size requires fewer calories; in contrast, the smaller brain size which also results is certainly not a beneficial adaptation, and the relationship between early malnutrition, brain growth, and intelligence is now being intensively studied.

Many other intriguing questions about the body's response to fasting have barely been looked at. For example, given our evolutionary history, might intermittent fasts of several days be beneficial in some way? Popular opinion and at least one prominent scientist claim they are, but the fact is that there is virtually no evidence on the question. Does the frequency of meals (with total calories a constant) have any long-range physiological implications? Again, data are scanty, but they do suggest that plasma cholesterol and the rate of development of certain diseases might be importantly influenced by meal frequency.

Gillie's article, "Endemic Goiter," shifts our focus from total calories to a specific essential nutrient, iodine. The essential nutrients are a group of substances which are required for normal or optimal body function, but which the body either cannot synthesize or synthesizes in amounts too small to meet its needs; these nutrients must therefore be supplied by the diet. Approximately fifty in number, they are water, eight amino acids, several unsaturated fatty acids, about twenty vitamins, and about twenty inorganic minerals (most of which, including iodine, are known as trace elements). Dietary deficiency of an essential nutrient calls forth an adaptive physiological response, usually one that causes the body to conserve that nutrient (frequently by reducing the rate at which the substance is excreted) or to use the nutrient more efficiently. Thus, as Gillie describes, the enlargement of the thyroid secondary to iodine deficiency is the result of an adaptive neuroendocrine response, and mitigates the iodine deficiency by changing the way this element is used by the thyroid.

Many of the other aspects of iodine metabolism as described in the article can, in general, be paralleled for the other essential nutrients, and for the trace elements in particular. First, the amount of iodine required in the diet is not constant, but is determined by the total environment of the individual. For example, the ingestion of foods which contain goitrogens (compounds which oppose the use of iodine) will increase the need for iodine. Second, more of a good thing is not necessarily better; large amounts of iodine, or of any trace element, exert toxic effects quite different from their normal physiological effects. Third, the effects of iodine deficiency early in life (cretinism) are more profound and may be irreversible. Finally, it is easy to detect the presence of extreme iodine deficiency (or excess), but the more subtle, long-term effects of a mild imbalance are extremely difficult to detect.

All these problems must be taken into account when one is trying to answer the question, "What is the *optimal* intake of each essential nutrient?" The solution to this question is the ultimate goal of nutritional physiology, and current recognition of how far we are from answering it has led to a renaissance of research in this field. We can often recognize, although with difficulty and only after taking individual variation into account, the minimal intake below which the physiological adaptations are unsuccessful and gross disease results, or, at the other end of the spectrum, the huge intake which clearly causes toxicity. But where between these extremes should we be? An enormous amount of research remains to be done on the more subtle effects of the essential nutrients and the interactions between them before this question can be answered.

If starvation and deficiency of essential nutrients exist at one end of the nutritional continuum, then obesity and atherosclerosis occupy prime positions at the other. "Atherosclerosis" by Spain introduces the reader to the disease which claims the lives of close to one million Americans per year, and which has been termed the worst epidemic since the Black Plague. Atherosclerosis (the underlying defect in heart attacks and strokes) is the classic example of a multicausal disease, i.e., one which is not "caused" by any simple factor, but whose development is facilitated by many different risk factors. As Spain points out, one of the dominant risk factors for this disease is the diet of industrialized man, specifically, the large animal-fat content of the diet. In the decade since this article was written, a huge amount of additional evidence has been collected which further substantiates this hypothesis, although the "fourth criterion" described by Spain has not yet been adequately validated. The prevalence of atherosclerosis has stimulated a wealth of studies into fat metabolism and cardiovascular function; moreover, the recognition that excesses of specific dietary constituents could facilitate the development of a disease has stimulated investigations into other possible interactions between specific diseases and nutrient excesses (between salt and hypertension, for example) relevant to the food patterns of industrialized peoples. Indeed, the possibility that our increased protein, fat, and caloric intake may be significantly shortening our life-span will be discussed in Section VI, "Aging."

"Nutrition and the Brain" by Fernstrom and Wurtman introduces the reader to an extremely recent and exciting new field of nutrition, one which might be on its way to validating the old saying, "You are what you eat." It has been known that severe nutritional stresses (such as protein malnutrition during infancy) can alter the concentrations of certain transmitter chemicals in the brain, thereby leading to altered brain function, but this fact is totally different from the phenomenon described by Fernstrom and Wurtman. They provide the first demonstration that a neurotransmitter's concentration can be altered rapidly by varying the contents of a single meal. It may turn out that such changes are highly localized, and merely help regulate food intake; even so, the phenomenon will still be very important, since we do not now fully understand either the short-term or the long-term regulation of total calories, and we know virtually nothing of any possible physiological regulation of the distribution of these calories between fat, carbohydrate, and protein. On the other hand, it is also possible that the specific types of food eaten have a more widespread action on the nervous system, perhaps (as suggested in the article) influencing sleep, sexual activity, and other behaviors.

The final article in this section, "Lactose and Lactase," by Kretchmer, provides examples of many of the important concepts concerning gene-environment interactions. The fact that most adults have no intestinal lactase (and cannot, therefore, digest milk) is a striking illustration of the turning-off of a gene (the gene which coded for lactase synthesis during infancy is still present, but is no longer active). Given this fact, the major question becomes: what turns off the gene, or, to put the matter in a different way, what keeps the gene functioning in people who do synthesize lactase in adult life? Does the difference between these groups represent acclimatization to the presence or absence of milk in adult life, or does it reflect genetic differences? Kretchmer describes the approaches required to answer this question and the great difficulties involved in coming to any definite conclusions in this classic "Nature — Nurture" problem.

1

The Physiology of Starvation

by Vernon R. Young and Nevin S. Scrimshaw
October 1971

*How does the human body adapt to prolonged starvation?
Studies of fasting subjects indicate how best to utilize food
when food is scarce and also how protein and calorie
requirements are related.*

The human body has a remarkable capacity for surviving without food for long periods. There is the well-authenticated case of Terence Mac-Swiney, the Irish revolutionist and mayor of Cork, who in his famous hunger strike in a British prison in 1920 survived for 74 days before dying of starvation. It has been shown many times that a fast for the biblical period of 40 days and 40 nights is well within the capability of a healthy adult. Recent tests of total fasting by obese persons for weight reduction have yielded remarkable results. Some obese individuals have gone without food for as long as eight months and emerged from the ordeal in good condition.

How does the body accommodate it-self to prolonged starvation? Although a knowledge of how to survive though hungry could not do much to ameliorate the chronic hunger and the famines that afflict a large portion of mankind as a result of poverty and droughts and wars, the question does not lack practical importance. From investigation of the body's responses to food deprivation we can learn much about its specific nutritional needs. Studies of the physiological and biochemical adaptations to starvation have also thrown light on a wide range of other questions, from appropriate diets for reducing weight to more effective regimes of food use during a food-shortage emergency. Moreover, they have advanced the understanding of the starvation disease called maras-

mus, which is increasing in many developing countries because mothers are giving up prolonged breast-feeding and their infants are not receiving an adequate substitute diet during a critical time in development.

Experimental studies of the effects of food deprivation over very long periods began around the turn of the century. One classic study was conducted in 1915 by F. G. Benedict of the Carnegie Nutrition Laboratory in Boston; he studied a volunteer subject who fasted for 31 days. In the 1940's Ancel Keys and his collaborators at the University of Minnesota tested a group of volunteer subjects kept on a semistarvation diet (about 1,600 calories per day) for 168 days. These experiments have been followed by trials of abstinence from food for the treatment of obesity, pioneered by Garfield G. Duncan of the University of Pennsylvania School of Medicine and Walter Lyon Bloom of Piedmont Hospital in Atlanta.

A large number of obese patients have now undergone the total-fasting treatment for extensive periods under careful observation at centers in North America and Europe, and in almost all cases there have been no serious complications. The longest reported fasts were by two women treated by T. J. Thompson and his coworkers at the Stobhill General Hospital and Ruchill Hospital in Glasgow. One was a 30-year-old woman who ate no food for 236 days and reduced her weight from 281 pounds to 184; the other patient, a 54-year-old woman, fasted for 249 days and reduced from 282 pounds to 208. Of 13 fasting patients in Thompson's group none showed any significant adverse side effects that could be attributed to lack of food.

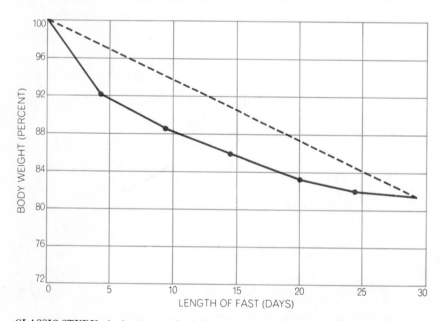

CLASSIC STUDY of a fasting man by L. Luciani in 1890 shows how body weight decreases with fasting time. The decrease is not linear (*broken curve*) but slows with time (*solid curve*). When subject began fast, he weighed 139.5 pounds; after 29 days he weighed 113.

There have been several deaths elsewhere among fasting obese patients, but in all but one case the deaths apparently were due to preexisting medical conditions that had been aggravated by the obesity rather than by the fasting itself. The one exception was a 20-year-old girl who in 30 weeks of total fasting cut her weight from 260 pounds to 132 pounds. On the seventh day after she had resumed eating, her heartbeat became irregular, and she died of ventricular fibrillation on the ninth day. E. S. Garnett and his co-workers at the General Hospital in Southhampton, England, found that this patient not only had lost fatty tissue but also had consumed, during her fast, half of the lean-tissue mass in her body, including part of the fibrous tissue of the heart muscle.

To explain the body's ability to mobilize its inner resources for survival in the absence of food intake we must begin with a review of its chemical needs. The primary need, of course, is fuel to supply energy for the vital functions. Normally the principal fuel is glucose, and its most critical user is the brain, for which glucose is fully as essential as oxygen. A rapid drop of the sugar level in the blood, which must continuously deliver glucose to the brain, brings about behavioral changes, confusion, coma and, if prolonged, structural damage to the brain resulting in death. In the body at rest the brain consumes about two-thirds of the total circulating glucose supply (compared with about 45 percent of the oxygen supply). Most of the remaining third of the glucose supply goes to the skeletal muscles and the red blood cells.

The human brain requires between 100 and 145 grams of glucose (equivalent to about 400 to 600 calories) per day. The body's main reserve of glucose, in the form of glycogen in the liver, amounts to considerably less than 100 grams, and part of this store is not ordinarily available because the liver tends to conserve some glycogen for stressful emergencies the body must be prepared to meet. As a result the liver's store of fuel can supply the brain's need for only a few hours. In fact, the stored glucose is not sufficient for the duration of the overnight fast between dinner and breakfast. Between meals the liver begins to draw on the tissues of the body for materials to synthesize the required glucose. We have found by examination of subjects in our laboratory at the Massachusetts Institute of Technology that after a person has eaten a meal at 10:00

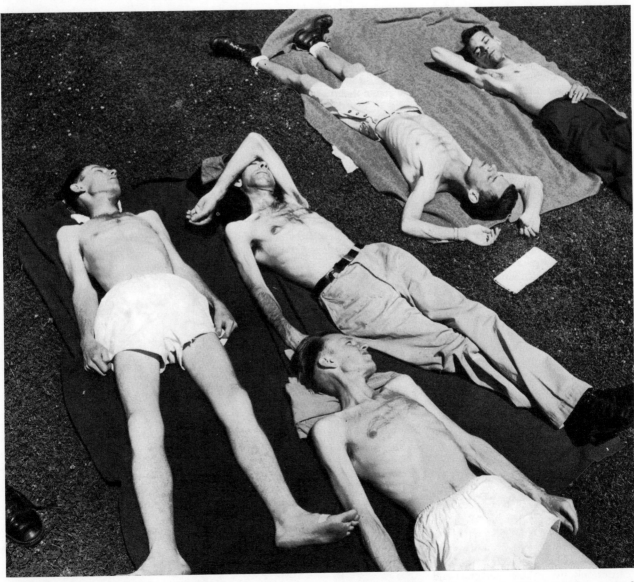

SEMISTARVED VOLUNTEER SUBJECTS rest in sun during an experiment undertaken in 1944 by Ancel Keys and his colleagues at the University of Minnesota. Volunteers were conscientious ob- jectors of World War II. Their fast was only partial; they received a ration of 1,600 calories per day for 168 days. This photograph was made by Wallace Kirkland of *Life* and is copyrighted by Time Inc.

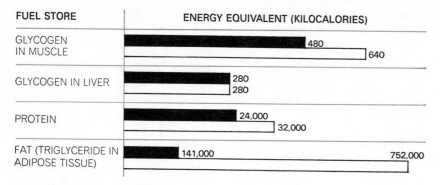

FUEL STORE	ENERGY EQUIVALENT (KILOCALORIES)		
GLYCOGEN IN MUSCLE	480		
	640		
GLYCOGEN IN LIVER	280		
	280		
PROTEIN	24,000		
	32,000		
FAT (TRIGLYCERIDE IN ADIPOSE TISSUE)	141,000		752,000

FUEL STORES in normal adult (*black bars*) and obese adult (*white bars*) are compared. Each pair of bars is on a different scale. The main store is fat, and in obese people this store is five or six times larger than in normal people. The data are from George F. Cahill, Jr.

P.M. certain amino acids that are precursors for the synthesis of glucose begin to accumulate in the blood plasma by 1:00 A.M., and they continue to increase until breakfast. The rise in amino acids is an indication that proteins in the skeletal muscles are being broken down to provide material for the production of glucose by the liver. Analysis of the blood also shows that at the same time the blood contains free fatty acids, which are derived from the breakdown of triglycerides in the fatty tissues and are capable of supplying energy to tissues other than those of the nervous system.

Clearly if the breakdown of protein continued at the initial rate, the skeletal muscles would rapidly waste away and the body could not survive for long. As starvation is prolonged, other sources of energy for the brain come into play, as we shall see. Let us first, however, follow the contribution of protein.

During the early period of starvation the body of an average man (143 pounds) synthesizes about 160 grams of glucose per day. Most of this is produced by the liver, but the kidney cortex also synthesizes an appreciable amount of glucose. The loss of protein involved and substantial losses of body minerals (such as calcium, potassium and magnesium) cause a loss of the water associated with these substances in the body, and this is mainly responsible for the initial loss of weight. As starvation continues, however, a progressively greater proportion of the weight loss is accounted for by the consumption of body fat. Gram for gram, fat is much richer in energy than other nutrients are: fat represents about nine calories per gram of weight in the body, whereas protein in the body carries only two calories per gram and carbohydrate one calorie per gram. Thus each unit of consumption of body fat donates much more energy to the

starved body. This is probably a major factor in slowing the loss of weight as starvation is prolonged [*see illustration on page 12*]. Eventually the fat consumed during continued loss of weight in obese people provides for essentially all the energy needed by the body.

There is an interesting question concerning what the weight loss means in terms of cells. Does the loss take the form of shrinkage of the cells' size or reduction of their number? Animal studies have shown that total or nearly total starvation can reduce the number of cells or fibers in the skeletal muscles. Little direct study of this question has been conducted in man. Radiographs of the chest in persons on a starvation diet have indicated that the heart shrinks in size, but not whether this is due to a reduction of the cells' size or of their number. Jules Hirsch of Rockefeller University obtained somewhat more definite

information. He studied a group of obese adults who had been fed only 600 calories per day and had lost upward of 100 pounds of body weight. Examining cells of their fatty tissues aspirated through a hypodermic needle, he found that the cells had shrunk by about 45 percent in size. The number of cells had not changed appreciably, however, except in a few people who had achieved particularly large losses of body fat.

George F. Cahill, Jr., of the Elliott P. Joslin Research Laboratory of the Diabetes Foundation, the leading investigator of the biochemical aspects of starvation in man, has looked into the changes in metabolism of obese people during fasting. Analyzing the blood's content of metabolites from skeletal muscle, he finds that at the beginning of fasting (not long after a meal has been digested and absorbed) the blood shows an increase of amino acids released from the muscle cells. Of these amino acids, which provide the supply of substrate for the liver's synthesis of glucose, the principal one is alanine. Furthermore, it turns out that alanine given by injection can increase the production of glucose, as shown by a rise of the glucose level in the blood.

The amount of alanine released from the muscle cells is surprising, because alanine makes up only 7 percent of the total content of amino acids in the cell proteins. It appears that most of the alanine discharged from the cells during fasting is not produced directly from protein breakdown but must be synthesized from alanine's immediate pre-

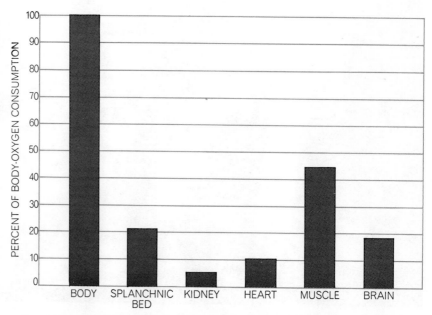

OXYGEN UPTAKE after an overnight fast is apportioned among various organs as shown by bars at right. "Splanchnic bed" refers to viscera, mainly the liver. Pattern of oxygen uptake is quite different from that of glucose uptake (*see illustration on opposite page*).

cursor, pyruvic acid, by the addition of an amino group furnished by other amino acids liberated by the breakdown of protein.

Cahill has proposed a cycle for the conversion of alanine to glucose and reconversion to alanine; it is somewhat analogous to the Cori cycle for lactate [*see illustration on page 18*]. According to Cahill's model, the alanine cycle, like Cori's, merely recycles a fixed supply of glucose. In addition, however, the alanine cycle offers an efficient means of transporting to the liver the nitrogen derived from the amino acids liberated by the breakdown of muscle protein.

As starvation continues, a number of general factors come to the aid of the organism. The basal metabolic rate slows, and the body's need for calories is further reduced by the loss of metabolically active tissue. The starving person engages in less spontaneous activity and becomes more sparing in the expenditure of energy, so that he uses his available energy more efficiently in accomplishing a given work load. His ability to survive will also depend, of course, on individual variables such as his body size and his stores of fat, and on environmental ones such as temperature and humidity.

Paramount, however, is the matter of the expenditure of protein. The starving body soon resorts to strong measures to preserve its integrity. It is confronted with two seemingly irreconcilable demands. The brain still requires a daily supply of energy equivalent to at least 100 grams of glucose; yet the synthesis of glucose at that rate would quickly exhaust the protein on which life depends. The triglycerides of fatty tissue provide a source of synthesis for glucose, but they can furnish only about 16 grams per day. In order to obtain the rest of the daily glucose requirement, some 90 grams, the body would have to break down about 155 grams of muscle protein. This would involve a daily loss of 25 grams of nitrogen. The nitrogen content in the body of an adult amounts to about 1,000 grams, and a loss of more than 50 percent of that amount is lethal. Hence a starving man could not live longer than three weeks if he had to expend his nitrogen at that rate.

Actually the body takes steps to control its loss of protein. The skeletal-muscle cells reduce their release of alanine, and the liver's synthesis of glucose declines. Cahill and an associate, Oliver E. Owen, found that by the fifth or sixth week of an obese adult's fast the liver and kidney were producing only 24 grams of glucose per day, and that essentially all of this glucose was going to the brain.

Where and how did the brain obtain the rest of the energy it required? Cahill discovered that the deficit was made up by a substitute source of energy derived from the fatty tissues. The blood of the starved obese subjects showed an accumulation of ketone bodies: acetoacetic acid and two derivatives from it, acetone and beta-hydroxybutyric acid. These substances yield energy on oxidation, and the brain evidently had adapted to using them as energy substrates in place of glucose.

Ordinarily the metabolism of fatty acids does not create ketones. In response to starvation, however, fatty acids are released from the fat depots and are oxidized in the liver to acetoacetic acid, which is then transported by the blood to other tissues to provide them with energy. The accumulation of ketones in the blood during starvation—and indeed in people on a high-fat diet—has been known for some time as the condition called ketosis. It is now clear that the ketosis of starvation signals a response to depletion of the body's supply of glucose, as Hans A. Krebs of the University of Oxford suggested some years ago. The evidence indicates that the brain promptly adopts the ketone bodies, particularly beta-hydroxybutyrate, as a substitute energy source, possibly within the first week of starvation. The Oxford group has recently shown (in studies conducted with experimental animals) that the brain has the enzymatic machinery to utilize ketone bodies. Their studies suggest that the human brain can probably begin to utilize ketone bodies for meeting its energy needs as soon as these metabolites in the blood supplying the brain reach a high enough level.

The breakdown of body protein is not completely eliminated. Even in prolonged starvation nitrogen in the form of urea and ammonia continues to be excreted in the urine. It reflects the basic turnover of proteins in the body that goes on at all times. In our laboratory

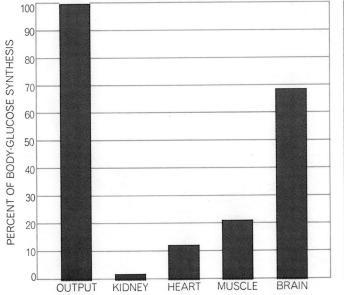
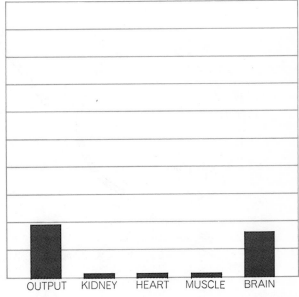
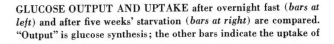

GLUCOSE OUTPUT AND UPTAKE after overnight fast (*bars at left*) and after five weeks' starvation (*bars at right*) are compared. "Output" is glucose synthesis; the other bars indicate the uptake of glucose by various organs. After the overnight fast glucose comes mainly from the liver; after the five weeks' starvation it comes 50 percent from the liver and 50 percent from the kidneys.

we estimated the amount of this basic turnover by measuring the urinary-nitrogen output of subjects who were fed a diet containing no protein but adequate in calories. Comparing their daily nitrogen loss with that reported for starved subjects in the fourth week without food, we find that the starved obese subjects' loss is not markedly higher. This could mean that the starved subjects were producing some five more grams of glucose per day than can be obtained through the basic turnover of body protein. The body cannot do entirely without glucose, because most tissues need it for replen-

ishing the tricarboxylic acid (TCA) cycle, which among other things synthesizes the energy-rich adenosine triphosphate (ATP) on which so much of the body's chemistry depends. Nevertheless, the very small extra loss of protein shown by obese people during prolonged starvation indicates that, thanks to the substitution of ketones for energy, their need for glucose is limited to not much more than is provided by the ordinary turnover of protein in the body.

One of the consequences of the body's conservation of protein during starvation is that urination for the excretion of

nitrogen is reduced. Hence a starving man needs less water intake. If his loss by sweating is minimal, a cup of water a day is sufficient to maintain his body's water balance.

What are the mechanisms that bring about the adaptive changes in metabolism during prolonged starvation? This question remains to be explored. No doubt hormones will be found to play an important part. It is known that the pancreatic hormone insulin is an important regulator of chemical activity in the body's ordinary daily cycle of eating and fasting. During the digestion of a meal

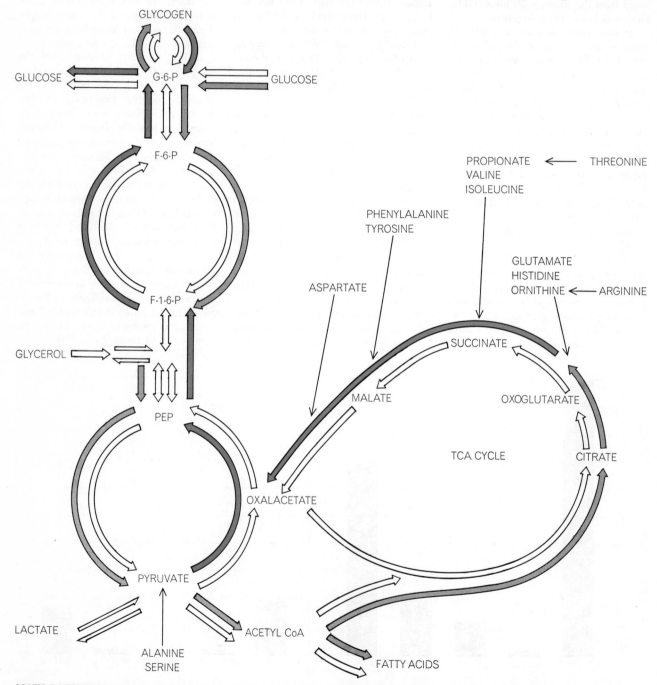

MAIN PATHWAYS in the utilization and production of carbohydrate in the liver are outlined in starvation (*dark-colored arrows*) and nonstarvation (*gray arrows*). "G-6-P" stands for glucose-6-phosphate; "F-6-P," for fructose-6-phosphate; "F-1-6-P," for fructose-1-6-diphosphate; "PEP," for phosphoenolpyruvate; "CoA," for coenzyme A, and "TCA cycle," for tricarboxylic acid cycle.

the absorption of glucose and amino acids from the intestinal tract stimulates the secretion of insulin; the hormone in turn stimulates the synthesis of fat and inhibits its breakdown, promotes the uptake of glucose and amino acids by the muscle cells and inhibits the synthesis of glucose by the liver. After the meal has been absorbed the insulin level in the blood falls, and during prolonged starvation it stands at a level lower than normal. Cahill has found that during prolonged starvation glucagon, the pancreatic hormone whose effect is opposite to that of insulin, is at a higher level in relation to insulin. Glucagon normally acts to stimulate the liver's synthesis of glucose. It is possible, therefore, that the alteration in the balance between the two hormones in the blood during starvation serves to heighten the activity of the liver in forming glucose and in metabolizing fats. The possible participation of other hormones, notably the growth hormone of the anterior pituitary gland and the glucocorticoid hormones of the adrenal gland, is being investigated, but so far it does not appear that these play primary roles in the metabolic adaptation to starvation.

There is a striking change in the role of the cortex of the kidney during prolonged starvation. It is promoted from a relatively minor partner of the liver in the synthesis of glucose to the main producer; by the sixth week of an obese person's fast the kidney cortex is synthesizing more glucose from amino acids than the liver is. This shift is believed to be attributable, at least in part, to the change in the acid-base balance in the blood caused by the increase in the body's production of ketone bodies.

The ability of an adult to survive prolonged starvation is not shared by children, particularly very young children. In a child deprived of food growth stops almost immediately, because of the high requirement of energy necessary to build protein. The child develops the emaciated condition known as marasmus. In cases where a deficiency of protein is more pronounced than a deficiency of calories the child shows the symptoms of the disease called kwashiorkor. A child who has suffered undernourishment very early and for an appreciable length of time will never reach normal size for his age, even though he is later fed well enough to restore a normal rate of growth. This is part of the reason for the small body size of many people in impoverished countries.

Particularly critical is the first year or so of life, the "preweaning" period. Be-

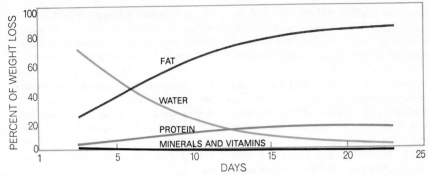

DAILY WEIGHT LOSS in prolonged starvation is analyzed by constituents of the body. Data are from Josef Brožek, Ancel Keys and their co-workers at University of Minnesota.

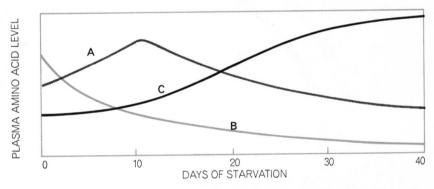

LEVELS OF AMINO ACIDS IN THE BLOOD normally follow these generalized curves. The level of certain amino acids (for example tryptophan, leucine and valine) rises and falls daily (*top curve*). The level of others (for example aspartic acid) remains steady.

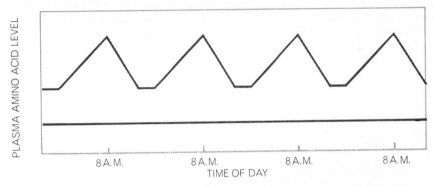

LEVELS OF AMINO ACIDS IN STARVATION follow different curves. Certain of the amino acids (for example valine) rise and then fall (*A*). Other amino acids (for example alanine) fall steadily (*B*). Still others (for example glycine) show a delayed rise (*C*).

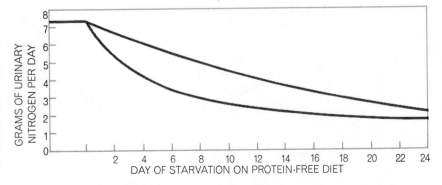

NITROGEN LOST IN URINE by starved subjects (*top curve*) and by subjects fed a protein-free but otherwise adequate diet (*bottom curve*) are compared. The difference between curves reflects starved person's need to synthesize glucose out of his own protein.

cause the brain is still growing and developing during this time, underfeeding is likely to result in permanent physical stunting of the central nervous system. Myron Winick of the Cornell University Medical Center in New York City found in experimental studies of rats, and in analysis of the brains of children who had died of marasmus, that the underfed brain had a subnormal content of DNA. Starvation had interfered with cell division and left the animal or child with a permanent deficit in the number of cells in the brain. Winick's experiments with rats also showed that when the mother was underfed during pregnancy, malnutrition of the offspring after birth had an even more devastating effect on the brain.

In the burgeoning cities of the less de-veloped countries many mothers in low-income families are now abandoning breast-feeding early, either in order to go to work or in imitation of the more affluent classes. As a result infantile ma-rasmus is becoming common in a number of countries. A particularly well-documented report of this trend, and the cause, has been made by Fernando Mönckeberg of the University of Chile, who studied the situation in that coun-try.

What useful conclusions can we draw from the studies conducted so far on the body's adaptations to starvation? First, let us consider the best way of coping with emergency situations in which the food supplies are very limited.

A little food, of course, is better than no food at all. Yet there is a paradox here. The edema of famine is hardly ever seen in cases of total starvation but develops often in semistarvation. More-over, a semistarved person's survival time may actually be shortened if he tries to subsist on a diet consisting main-ly of carbohydrate and deficient in pro-tein. In such circumstances a child may quickly fall victim to kwashiorkor. Why is it that, although a person can be stricken with this disease when he eats a little food, it never shows up in total starvation, when the person gets no pro-tein intake at all?

The typical clinical signs of kwashior-kor are apathy, loss of appetite, edema and changes in the skin and hair. On close examination of the blood and other tissues it is found that there is a marked drop in the concentration and activity of

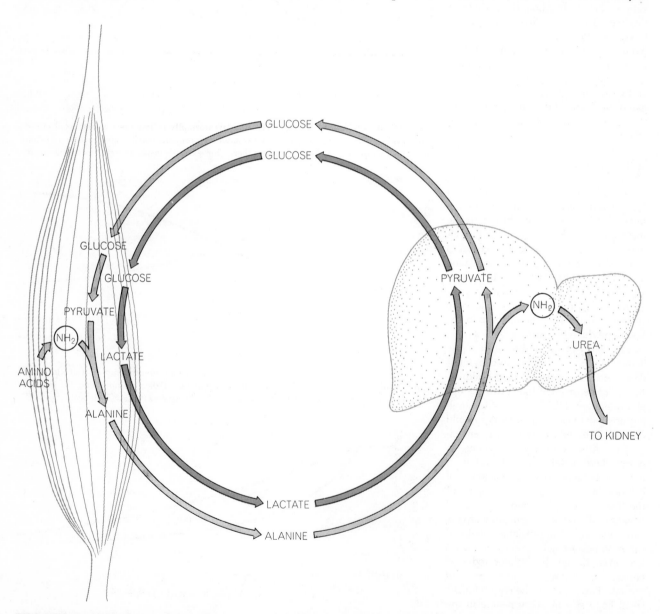

TWO METABOLIC-SUBSTRATE CYCLES operate between mus-cles (*left*) and the liver (*right*) through the blood. Lactate cycle worked out by Carl F. Cori (*color*) results in a net gain for muscle of two molecules of adenosine triphosphate (ATP). Alanine cycle proposed by Cahill would also facilitate the removal of nitrogen from amino acids liberated by the breakdown of muscle protein.

key enzymes. In the light of the known facts about the body's adjustment to a lack or shortage of food we can deduce the reason for the enzyme deficiency. In a semistarved child or adult the brain probably continues to depend mainly on glucose for energy. With some glucose being supplied by way of food, the need to synthesize glucose from body proteins would be reduced. Consequently there is only a modest release of amino acids from skeletal muscle into the bloodstream. If the semistarved individual is receiving little or no protein in his food, the amount of free amino acid in the blood is not sufficient for the body's synthesis of essential enzymes and other tissue proteins. Therefore the body shows the devastating results of protein deficiency. This is precisely what was observed during the recent famine in Biafra. The population was subsisting almost solely on the starchy roots of the cassava plant. Edema and other symptoms of acute protein deficiency were most conspicuous in the children. A high frequency of kwashiorkor is now being found among the East Pakistan refugees in India because many of the young children are not receiving protein foods.

We are seeing in such phenomena an indication of the conditions that gave rise to the evolution of man's present metabolic resources. In the hunting and plant-gathering phase of his early history his hungry periods took the form of general undernourishment, and the body evolved adaptations to improve metabolic efficiency for that contingency. It is only recently that human populations have come to depend heavily on a single cultivated plant staple for food—a situation with which the human body is not prepared to cope.

We do not yet have precise knowledge about the mechanisms that cause the brain to switch from glucose to ketone bodies as its main energy source to induce this switch artificially for preservation of the body's integrity. All that can be suggested is that in a food-shortage emergency it may be best to spread out the consumption of the limited supply of protein and/or carbohydrate over the day, taking nibbles at frequent intervals, so that the periods of fasting and consequent breakdown of body protein for glucose synthesis will be shortened.

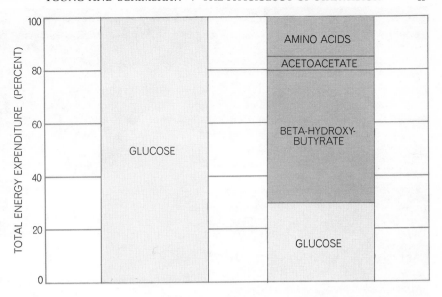

ENERGY NEEDS OF THE BRAIN are met differently in normal circumstances (*bar at left*) and after five to six weeks of starvation (*bar at right*). Glucose normally suffices, but in starvation it can meet only 30 percent of the requirement and other substances fill in.

More information of practical usefulness is available on dieting for weight reduction, since most of the studies of adaptation to starvation have been carried out in obese subjects. It is quite clear that there is no way of achieving a permanent weight reduction without reducing the intake of calories to less than the outgo. The greater the difference between the intake and the expenditure of caloric energy, the faster one will lose weight. What about the various special diets that have become popular?

A high-protein or high-carbohydrate diet in theory should tend to minimize the body's loss of protein. It has also been argued that part of the protein and carbohydrate intake is spent in generating body heat after a meal and therefore does not go into the building of body fat. In practice, however, these considerations are probably too small in effect to be significant in preserving health or reducing weight.

On the whole it must be said that bizarre reducing diets have no scientific basis; any apparent success they may have appears to be due solely to their poor palatability or, as in the case of low-carbohydrate, high-protein diets, a rapid initial weight loss due to loss of body water. The best diet for reducing is still one that is balanced in food ingredients and sufficiently low in total calories to produce weight loss at the desired rate.

From a purely biochemical standpoint the most efficient way to lose weight, as the starvation tests have shown, is complete fasting into the stage where body fat is being consumed as the main source of energy for the brain and other tissues. Total fasting for an extensive length of time can be dangerous, however. It should not be prescribed for high-risk patients, and in all cases one must take care to avoid too much exercise in the initial phases and refrain from continuing the fast too long. Duncan of the University of Pennsylvania School of Medicine, who has perhaps had the most experience with this method of dealing with obesity, has treated fasting patients in a total of more than 1,300 hospital admissions without a fatality. Each fast has been limited to 10 days or two weeks, with patients returning for repeated fasts at varying intervals. Duncan cautions that any total fast for more than two weeks should still be considered in the category of a research enterprise. It must be emphasized that no one should undertake total fasting for weight reduction without prior medical screening, hospitalization and continuous medical supervision.

MODERATE GOITER is evident in this detail from a portrait of Maria de' Medici, wife of Henry IV of France. It was painted in 1625 by Rubens and now hangs in the Prado in Madrid. Moderate goiter was considered an adornment in the late Renaissance.

Endemic Goiter

by R. Bruce Gillie

June 1971

The disorder has a long record because its principal sign is so apparent. It is now a disease of the poor, because an unbalanced diet often cannot correct for a deficiency of iodine in the soil

The "regular and rounded neck" with which Maria de' Medici was endowed by Rubens [*see illustration on opposite page*] is a goiter, or compensatory hypertrophy of the thyroid gland. The thyroid is a pinkish pad of tissue wrapped partly around the trachea and esophagus; it is a ductless gland of vertebrates that secretes into the blood the hormones that regulate the rate of development and metabolism. Goiter is an unusually obvious manifestation of an endocrine disorder, and as such it has drawn attention, sometimes admiring and sometimes fearful, since man's earliest days.

There are many different causes of goiter: disease, developmental defects and environmental conditions. Endemic goiter, so designated because it affects a significant proportion of a given population, is almost always the result of a dietary deficiency of iodine, an essential substrate for the synthesis of the thyroid hormones thyroxine and tri-iodothyronine. Iodine-deficiency goiter is now easily prevented or cured by the ingestion of minute quantities of iodine, but over the centuries it has been one of the most persistent and ubiquitous diseases of mankind. As recently as 1960, 200 million people were still afflicted with it.

The secretion of thyroid hormones is a link in one of the exquisitely balanced feedback systems that regulate the internal environment of vertebrate organisms [see "The Thyroid Gland," by Lawson Wilkins; SCIENTIFIC AMERICAN, March, 1960]. Impulses from the nervous system cause the hypothalamus at the base of the brain to release a neurosecretion, the thyrotropin-releasing factor (TRF), into portal veins leading directly to the pituitary, the pea-sized master gland that regulates the activity of the thyroid and other endocrine glands. The thyrotropin-releasing factor stimu-lates the pituitary to secrete into the blood thyrotropin, or thyroid-stimulating hormone (TSH), which in turn causes the thyroid to synthesize and secrete its hormones. The system is self-regulating: an excess of thyroid hormones in the blood suppresses hypothalamus and pituitary activity and reduces the secretion of the thyroid-stimulating hormone; when the thyroid hormone concentration is too low, the pituitary responds by secreting more thyroid-stimulating hormone to restore the normal thyroid-hormone level [*see illustration on page 24*].

If the thyroid is healthy and there is enough iodide (ionic iodine) in the blood, the thyroid-stimulating hormone steps up the trapping of iodide by the thyroid and in other ways abets the synthesis of thyroxine and tri-iodothyronine within the follicles of the thyroid gland [*see illustration on page 26*]. In the absence of sufficient iodide thyroxine synthesis is inhibited; the flow of thyroid-stimulating hormone is unchecked and its effect is to increase the number and change the shape of the cells that form the follicles; in time the follicles become distended. This compensatory proliferation of cells and distention of the follicles, which constitute goiter, may restore thyroid-hormone production to a satisfactory level for normal life.

A Chinese document from about 3000 B.C. is the earliest known record of goiter. Remarkably, it not only described the symptoms but recommended an effective cure: the ingestion of seaweed and burned sponge, which contain large amounts of iodine. Speculating on the causes of what was apparently a common affliction, Chinese scholars listed poor quality of drinking water, mountainous terrain and emotional vicissitudes, all of which are indeed associated with a higher incidence of goiter. The Chinese even administered desiccated thyroid glands of deer as a treatment for goiter. (Nowadays extracts of beef, sheep or hog thyroid are given for hypothyroidism.)

The Ebers Papyrus of Egypt, dating from about 1500 B.C., described two possible treatments for goiter: surgical removal of the gland (which must have been a high-risk procedure if it was ever attempted) and the ingestion of salt (presumably containing iodine) from a particular site in lower Egypt.

Hippocrates blamed goiter on the drinking water in certain places. Juvenal, Vitruvius and Julius Caesar were impressed by the enlarged neck of residents of some alpine regions; Caesar, in fact, believed that the large neck was a national characteristic of the Gauls. The word "goiter," incidentally, is from the Latin *guttur*, or "throat."

Roman physicians noticed that even in a normal person the size of the thyroid may fluctuate somewhat during times of physiological stress such as puberty, menstruation and pregnancy. They noticed in particular that the physical and emotional circumstances surrounding the initial sexual activity of a bride brought about a slight swelling of her thyroid. The Romans thereupon originated the ritual of measuring the circumference of a bride's neck with a ceremonial ribbon before and after her first week of marriage. If the circumference increased, the marriage was considered consummated.

Because moderate goiter is quite compatible with normal life, causing no pain and often no impairment, it was not necessarily perceived as an affliction; if in some cultures it was considered a divine stigma, in others it was a mark of beauty. In Europe it was often attributed to some serious religious or social transgression—robbing the graves of saints, for example. In Germany during the Middle

Ages it was thought that the condition could be caused by strenuous work, including childbirth. That was the rationale for a now forgotten custom of tying a cord around the neck of a woman in labor. In India inhaling the odor of people dying of malaria was said to cause goiter. At one time or another the condition has been blamed on indolence, drunkenness and debauchery. In 1867 a French student of the matter named J. Saint Leger listed more than 40 different possible causes then being cited—among them a lack of electricity in the atmosphere, incest, alcoholism and coitus interruptus.

Cures were not so easy to find. A procedure that appears to have persisted for many centuries was piercing the thyroid gland with a red-hot needle. That presumably created an inflammation, and the resulting fibrosis may well have reduced the size of the gland. Actual surgery could not have been effective until the end of the 19th century. One reason is that the thyroid is so richly supplied with blood vessels that in the early days of surgery an incision would have resulted in excessive and uncontrollable bleeding. Even after the advent of satisfactory techniques surgical removal was dangerous at best before the discovery of the parathyroid glands. These tiny glands, nesting on the surface of the thyroid lobes, regulate the concentration of calcium in the blood, and their inadvertent removal along with the goitrous thyroid would threaten life.

The first attempt at an epidemiological survey was made at the request of Napoleon I, who was impressed (as Caesar had been) by the many cases he saw in the course of his alpine campaigns. Napoleon was also disturbed by the loss of potential recruits who had to be rejected because the military uniform would not fit their goitrous necks.

The basic mystery surrounding goiter, of course, was the function of the thyroid gland in health. The early anatomists were impressed by the gland's large blood supply and puzzled by the fact that (like the other endocrine glands) it had no duct and therefore, it seemed to them, could have no secretory function. In the Middle Ages some anat-

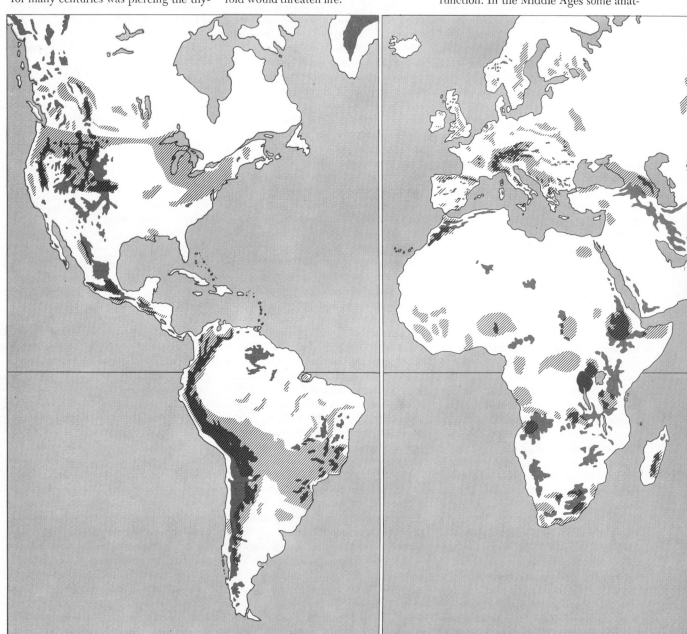

REGIONS OF ENDEMIC GOITER and the mountainous terrain with which it is often associated were mapped by the World Health Organization. Areas where iodine-deficiency goiter is endemic are indicated by the black hatching. Populations near seacoasts are sel-

omists thought of the thyroid as the seat of the soul. Others were more practical. The Italian anatomist Giulio Casserio wrote in 1600: "Kind nature has especially beautified the gentle female sex with many sorts of ornaments. And not the least among them is this one, that the empty spaces which exist around the larynx being filled up, they show to our eyes, to the great joy of our sight, a regular and rounded neck." Paintings by artists of the time, including Dürer and Rembrandt, suggest that Casserio's view was the general one, since madonnas and other female subjects are often depicted with moderately goitrous necks.

In 1656 the British anatomist Thomas Wharton wrote a complete description of the thyroid and also named it after the Greek word for a large oblong shield: *thyreos*. Wharton agreed with Casserio that it served to beautify the neck ("particularly in females to whom for this reason a larger gland has been assigned"), but he suggested that it might also keep the tracheal cartilages warm, since they were "rather of a chilly nature," and lubricate the larynx, rendering the voice more melodious. Other students believed the thyroid shunted blood away from the brain to protect it from sudden changes in blood pressure, or that it was a cushion to support and protect the structures of the larynx.

It was not until 1895, after surgeons had seen the effects of removal of the thyroid gland and after treatment with thyroid extract had been attempted, that Adolf Magnus-Levy of Germany demonstrated that the thyroid regulated the basal metabolic rate: the rate at which the cells of the body consume oxygen, which is to say the rate at which they convert nutrients into the energy of life. In the same year the German biochemist Eugen Baumann learned that the thyroid is particularly rich in iodine. It was a serendipitous discovery. Baumann was trying to analyze the protein content of thyroid tissue, and his usual procedure was to precipitate the protein from an extract with sulfuric acid. One day, reaching for the sulfuric acid on a shelf above his workbench, he picked up a bottle of nitric acid instead, and before he had realized his mistake he had added some of its contents to the extract. To his astonishment the characteristic brownish-purple fumes of iodine gas swirled up from the preparation. Baumann went on to describe the role of iodine in thyroid physiology. In 1914 Edward C. Kendall of the Mayo Foundation first crystallized some thyroid hormone. It was a large and difficult task: the 37 grams of crystallized hormone that Kendall subsequently obtained were derived from three and a quarter tons of pig thyroid! Finally in 1927 Charles Robert Harington of the University College Hospital Medical School in London and George Barger of the University of Edinburgh established the definitive structure of thyroxine, confirming Baumann's observations. Well before that time Baumann's work had led on the one hand to the understanding that endemic goiter was the result of environmental iodine deficiency and on the other to simple and effective iodine therapy.

As investigators looked into the ecology of goitrous populations they first found a correlation between goiter and the accessibility of a population to the sea and thus to a seafood diet rich in iodine. A map compiled by the World Health Organization makes it clear that it is in inland areas, particularly mountainous ones, that goiter may be endemic [*see illustration on these two pages*]. The Alps, the Pyrenees, the Himalayas and the Andes are strikingly goitrous. So are inland plains regions in Italy, in the Congo and in the Great Lakes basin of North America.

The geography of goiter is not simple, however. Many inland and mountainous regions do not support goitrous populations, and there are coastal areas that unpredictably have goitrous populations. A factor that is more closely correlated with the incidence of endemic goiter than mere distance from the ocean is the

dom affected because of the iodine content of seafood. Not all inland areas are equally affected; the geology and remoteness of mountainous regions (*color*) make them most susceptible.

iodine content of the soil. As long as the soil content of iodine is adequate, enough iodine (about 100 to 200 micrograms per person per day) will be ingested in locally grown produce to prevent the onset of goiter. Although the iodine content of soils is generally higher in coastal regions than it is inland, soil content is determined by more complex factors than distance from the ocean alone. The most seriously depleted soils are in areas that were subjected to the most intense glaciation. Such glaciation did two things. By crushing virgin igneous rocks that had never been exposed to atmospheric iodine, it left behind vast amounts of new, iodine-poor topsoil, and

it leached the soluble iodine salts out of the original soil.

Leaching may also make soils along the shores of rivers that periodically overflow their banks deficient in iodine. An interesting example of this process was noted in a study of two villages on opposite banks of the Congo River in an area where heavy rain and periodic flooding had reduced the iodine content of the soil. On one bank the village population was 80 percent goitrous; on the alluvial soil of the opposite bank the population was hardly goitrous at all. Iodine being leached by the heavy rain out of land upstream was being redeposited in the alluvial soil around the second vil-

lage, making the iodine concentration there just sufficient to prevent goiter.

The steady replenishment of iodine in terrestrial soils from atmospheric iodine tends in time to reverse the effects of glaciation. The degree of replenishment is complexly affected by the distance of an area from the ocean, the prevailing wind conditions and the amount of iodine in the precipitation. In addition some areas of the world have natural terrestrial iodine deposits that may also help to determine the iodine content of the local soil. In other words, the ecology of a human society as well as its principal staple diet is a major factor in the etiology of endemic goiter.

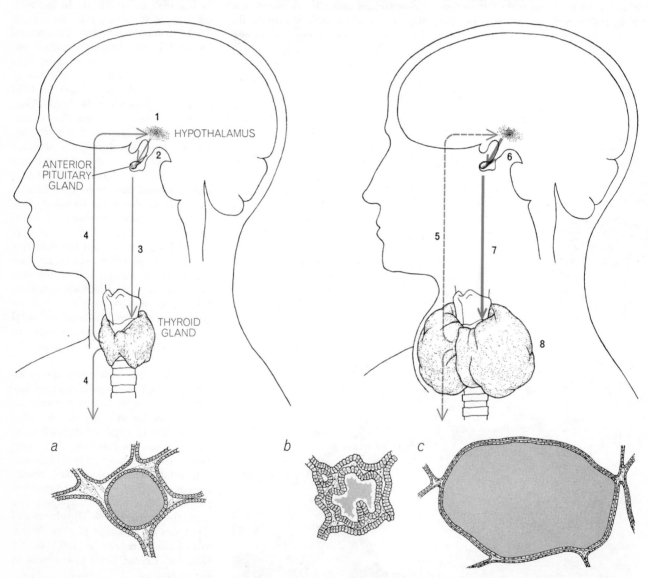

NEGATIVE-FEEDBACK SYSTEM controlling production of thyroid hormones begins with the neurosecretion from the hypothalamus (1) of thyrotropin-releasing factor (TRF), which goes directly to the pituitary (2) and causes it to release thyrotropin, or thyroid-stimulating hormone (TSH), into the bloodstream (3). In the thyroid gland TSH acts to bring about the synthesis and secretion into the circulation of the thyroid hormones thyroxine and tri-iodothyronine (4); the amount of thyroid hormones reaching the hypothalamus in turn controls the secretion of TSH, completing the

negative-feedback loop. In the absence of iodine, an essential substrate for thyroid hormones, not enough hormone is produced (5) to "turn off" the system; excessive TRF (6) and TSH (7) are secreted, stimulating the iodine-depleted thyroid tissue to grow (8). A normal thyroid follicle, in which hormones are synthesized and stored, consists of an envelope of cells containing a colloid, thyroglobulin (a). In the absence of iodine TSH causes the cells to proliferate and become more columnar (b) and then to produce more colloid, so that the follicles become distended (c), forming a goiter.

Beginning in 1907 the extensive investigations of David Marine and O. P. Kimball of the Western Reserve University School of Medicine with laboratory animals provided the first direct experimental evidence that endemic thyroid hypertrophy is caused by iodine deficiency. These workers subsequently carried out the first large-scale program of goiter prophylaxis in Akron, Ohio. The study, completed in 1920, involved 4,500 schoolgirls between the fifth and the 12th grade. Half of them received two grams of iodized salt twice a year and the other half served as untreated controls. At the end of two and a half years 65.4 percent of the treated group showed regression of goiter and only five treated girls evinced thyroid enlargement. Meanwhile only 13.8 percent of the girls in the untreated group showed a regression of goiter and 495 untreated girls had developed thyroid hypertrophy. The study was a dramatic demonstration of the efficacy of iodine in the treatment and the prevention of simple goiter.

When these findings were published, many individuals and groups of health enthusiasts took to consuming iodine to the extent of a fetishism. Some people even hung around their neck little bottles of iodine from which they would occasionally take a swig. Iodine became the magic ingredient in the nostrums of certain charlatans. To everyone's surprise, rather than preventing goiter, iodine sometimes served to stimulate it. This apparently paradoxical effect of iodine on the etiology of goiter was later explained by Jan Wolff and Israel L. Chaikoff of the School of Medicine of the University of California at Berkeley, who found that very high iodine concentrations in the blood plasma inhibit the biosynthesis and secretion of thyroid hormone. This aspect of thyroid physiology, together with increasing reports of severe iodine toxicity and the fear that iodine might lead to toxic hyperthyroidism, elicited strenuous opposition to the iodization of table salt by many medical experts, lay people and politicians. The political and ethical controversy over the incorporation of iodine into table salt was even more intense than the present-day controversy over the fluoridation of water. It was not until the mid-1920's that iodine prophylaxis was generally accepted.

Although the idea that some positive agent in food or drink was responsible for goiter had antedated the discovery of iodine deficiency as a cause, it was not until 1941 that a goitrogenic substance was identified. Curt P. Rich-

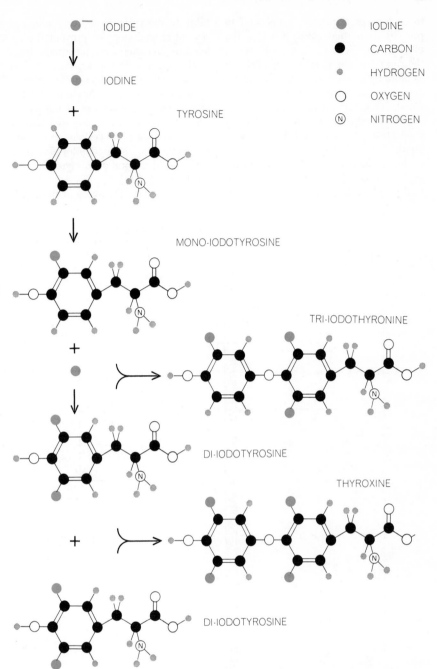

BIOSYNTHESIS of the thyroid hormones depends on the presence of ionic iodine, primarily as sodium iodide. The iodide is oxidized to elemental iodine and combines with the amino acid tyrosine to form mono-iodotyrosine and di-iodotyrosine. Two molecules of di-iodotyrosine may combine to form thyroxine, or mono- and di-iodotyrosine molecules may combine to form tri-iodothyronine. (Only hormone products are shown, not by-products.)

ter and Kathryn H. Clisby of the Johns Hopkins University School of Medicine were investigating the effects of certain rat poisons. When they fed laboratory rats the drug thiourea, they observed to their surprise that the rats survived but their thyroid began growing and soon became goitrous. At about the same time Julia B. and Cosmo G. MacKenzie, in another laboratory at Johns Hopkins, were studying the effect of a new sulfonamide drug on the bacterial flora of the rat intestine. They observed the same

phenomenon: the thyroid of their treated rats became hypertrophic, as if the animals had been maintained for several weeks on an iodine-deficient diet. The drugs were apparently preventing the proper utilization of iodine, which was present in normal concentrations in the animals' food and water. Since that time many additional antithyroid compounds have been discovered.

Theoretically these drugs could act by any of three different mechanisms. First, they could operate at the intestinal level to chelate, or sequester, iodine and so prevent its normal absorption into the bloodstream. Second, they could act at the surface of the thyroid epithelial cell to inhibit the selective absorption of iodine from the blood passing through the gland. (This trapping of iodine ions is an amazingly efficient process: the thyroid concentrates the ions to a level several hundred times higher than their concentration in the plasma.) Third, the goitrogenic compounds could gain admission to the cells of the thyroid and there in-hibit the biosynthetic pathway at any of several crucial steps or prevent the release of thyroxine from its storage form in the follicles. It appears that the last two mechanisms are the significant ones. Thiocyanate and perchlorate inhibit the active transport processes of the iodide trap, thiouracil blocks the oxidation of iodide to iodine by certain peroxidase enzymes, and sulfonamides interfere with the incorporation of tyrosine [*see illustration below*].

Soon after the discovery of these goi-trogenic compounds it was found that goiter endemic to some areas was a result of similar, naturally occurring compounds in local foods. Soybeans and members of the genus *Brassica*, which includes Brussels sprouts, cabbages, turnips and other vegetables, are among the foodstuffs containing significant amounts of goitrogenic compounds. In a nutritionally varied diet such foods do no harm, but they are a more serious matter in societies that survive on less varied diets.

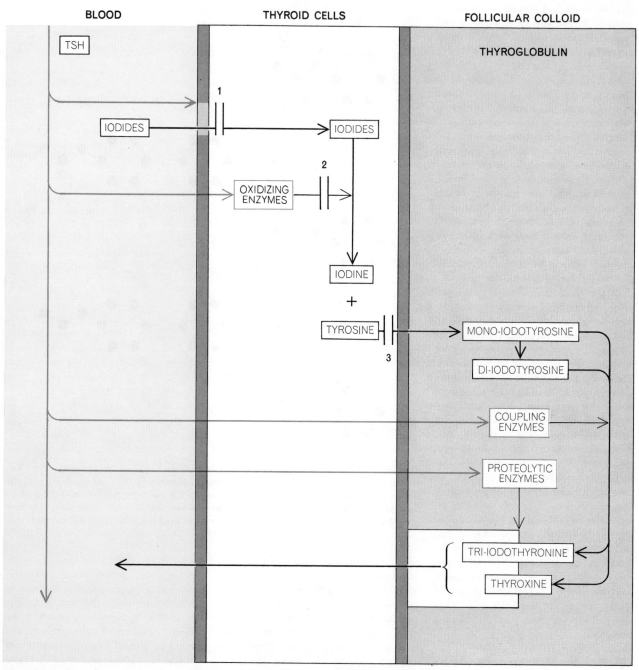

THYROID ACTIVITY is diagrammed schematically. Iodide trapped by thyroid-follicle cells is converted by oxidizing enzymes in the cells into iodine, which combines with tyrosine to form the thyroid hormones. The hormones are stored attached to thyroglob-ulin; on demand they are freed by proteolytic enzymes and pass in-to the blood. TSH stimulates hormone production by acting to abet iodide-trapping and the activity of three sets of enzymes. Goitrogenic substances interfere with hormone production. Thiocyanates and perchlorates block the iodide trap (*1*), thiouracil the oxidizing enzymes (*2*), sulfonamides the combination with tyrosine (*3*).

The effect of the *Brassica* goitrogen was demonstrated not long ago in Tasmania, off the coast of southern Australia. The island was an area of endemic goiter, and so in 1949 a program was instituted to supply iodine-containing tablets to schoolchildren up to 16 years old. Five years later a survey revealed that the incidence of goiter in these children had not decreased; indeed, it had increased. The investigators verified their data and reevaluated their methods, and still they found that goiter had increased. F. W. A. Clements and J. W. Wishart, who had been instrumental in setting up the program, thereupon proposed that something other than iodine deficiency might be promoting goiter in these schoolchildren. As it happened, in 1950 the Australian government had begun a free-milk program in the schools. The increased demand for milk forced local dairies to keep their cows at pasture during seasons when grass was not available. As a consequence the cows were eating marrow-stem kale, which is more frost-resistant than grass and grows well all year. Marrow-stem kale is a member of the genus *Brassica* and contains a large amount of the goitrogenic compound. Further study revealed that this compound, present unaltered in the milk, was blocking utilization of the iodine being supplied in the tablets.

Clearly there are dangers inherent in administering to patients drug preparations that contain significant amounts of potentially goitrogenic compounds or elemental iodine. Although it is not common in this country, iatrogenic goiter—goiter caused by medical treatment—is becoming a more significant factor. Sulfonamides prescribed for urinary-tract infections, thiouracil drugs given routinely for the relief of hyperthyroidism and many iodine-containing compounds administered as expectorants in the treatment of asthma are potentially goitrogenic. The unborn infants of pregnant mothers who take these drugs have in some instances been killed *in utero* by goiters that develop when the drugs diffuse across the placenta and enter the fetal circulation. Because these drugs are concentrated in the lactating breast they may also induce goiter in a nursing infant. The most serious effect of these drugs in pregnancy is that they decrease the availability of maternal thyroxine to the early fetus, and thyroxine is of fundamental importance in the physical and mental development of the baby.

Several epidemiological surveys have indicated that goiter can arise spontaneously in a society, persist for a short time and then regress, all without any

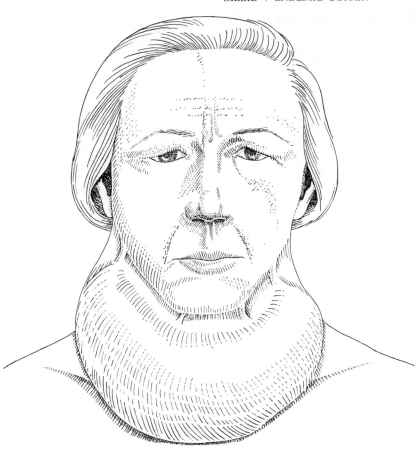

LARGE GOITER is seen frequently in regions where iodine-deficiency goiter is endemic. The drawing is based on a photograph made in the Alps near Innsbruck in Austria. Goiters have been reported that weighed four or five pounds, sometimes hanging below the chest.

apparent change in the food or living habits of the people. The possibility of an infectious origin of goiter has been proposed to explain such epidemics, but no instance of this has been proved. It does seem possible that a strain of iodine-trapping bacteria could become resident among the normal flora of the intestine and decrease the availability of dietary iodine for absorption.

There is at least one documented case of goiter related to bacteria, although in a different way. While studying goitrous populations in the Himalayas in 1906, Robert McCarrison visited several neighboring villages in the valley of the Gilgit River. He was immediately impressed by the fact that whereas the village that was farthest upstream showed a low incidence of goiter (12 percent), as he moved downstream the incidence increased in each village, until in the lowest village more than 45 percent of the population had goiter. An isolated village that was near the river but whose residents did not drink river water did not have goiter. McCarrison undertook a controlled experiment with 30 volunteers divided into two groups. One group drank the muddy river water

after boiling it and the other (including McCarrison himself) drank unboiled river water. Within a month most of the people in McCarrison's group began to develop goiter; those in the other group did not. He concluded from the experiment that bacteria were to blame for the goiter. Poor sanitation meant that the waste material from the villages went into the river, which became more contaminated as it flowed past each village, increasing the dose of bacteria in the drinking water of villagers in proportion to their distance downriver. It has since been shown that some strains of *Escherichia coli,* a bacterium normally found in fecal material, can produce thiouracil.

There is a vicious circle aspect to endemic goiter. Poor societies with an unvaried diet are likely to be the most susceptible to goiter and the most vulnerable to its biological, social and economic consequences. Where iodine-deficiency goiter is endemic in a human population domestic animals will probably be hypothyroid too. Goitrous sheep often produce less wool; goiter in cattle causes sterility, poor milk production and sickly calves; horses do less work;

hens with decreased thyroid activity produce eggs with insufficient calcium in the shell, leading to egg breakage and higher chick mortality. A poor society can scarcely afford to have these serious handicaps afflict the animals on which its survival may depend.

Long-standing endemic goiter has had particularly serious consequences in some remote communities, such as alpine valleys, where inbred populations have persisted for many generations in an iodine-deficient environment. Familial goitrous hypothyroidism can lead to a high incidence of individuals with the severe developmental defects of cretinism. Cretins manifest varying degrees of idiocy and are also physically dwarfed and often malformed. The mental retardation is believed to result from a deficiency of thyroxine during the first three months of pregnancy, when it must be supplied by the mother; the physical anomalies are probably due to a deficiency of the baby's own thyroxine during maturation.

The role of mountainous topography and isolation in cretinism is evident. There are many goitrous regions that do not show a high incidence of cretinism; the "goiter belt" in the Great Lakes region is an example. Presumably population mobility through this channel of westward migration supplied enough biological and social diversity so that cretinism did not develop.

The cretin is only the most extreme example of the consequences of decreased availability of thyroxine during the developmental stages of life. Because all the residents of a community affected by endemic goiter are potentially exposed to a suboptimal supply of thyroxine during their development, there may be serious but subtle effects on the quality of the society at large. Motivation, spontaneity, creativity and native intelligence may be diminished, and the resulting social stagnation may lead to further inbreeding.

Medical science and public health will eventually eliminate iodine-deficiency goiter as an endemic affliction. One must hope that this age-old and benign disorder will not be replaced by a different, nuclear-era thyroid dysfunction resulting from the ingestion of large amounts of radioactive iodine isotopes from nuclear fallout. The iodine is concentrated in the thyroid gland, where the radioactivity may damage cells irreversibly. The study of endemic goiter demonstrates the seriousness of this potential hazard and the effects it might have on the course of human evolution.

Atherosclerosis

by David M. Spain

August 1966

The artery disease that is responsible for most heart attacks is on the rise in the U.S. Its prevalence seems to be related to diet, so there is hope that the epidemic can be controlled

The incidence of heart attacks among adult white males in relatively affluent occupations in the U.S. has reached epidemic proportions. From such attacks (coronary artery occlusions) the overall U.S. death rate is now 500,000 a year, and 200,000 more die from strokes. At least 5 percent of the adult males in the nation show signs of some form of heart disorder. The basic disease responsible for most of these disorders and deaths is atherosclerosis. There is every indication that the prevalence of atherosclerosis in the U.S. is steadily increasing.

It used to be thought that atherosclerosis was a disease of old age and that its rising incidence might be due simply to the lengthening of the average life-span by the control of infectious diseases. This idea has now been refuted by a number of studies. My colleagues and I have made a comparative examination in Westchester County, New York, of two samples of the population taken 20 years apart. The samples were comparable in that both groups covered the same age range (from 20 to 60), had records of good health before a fatal episode, had died of sudden causes not connected with heart disease and had been autopsied after death, so that the extent of atherosclerosis in their coronary arteries and aortas was known. The first sample consisted of people who had died of acute infections in the period between 1931 and 1935; the second was made up of people who had been killed in automobile or industrial accidents between 1951 and 1955. We found that the second group, representing a period 20 years later than the first, had a significantly greater amount of atherosclerosis. This was true for every age level: the young people in the second group had more atherosclero-

sis than the young ones in the first. A similar autopsy study in Sweden yielded the same finding; the degree of coronary atherosclerosis in a population sample in 1958 was greater than in a sample from 1934.

Laboratory studies of experimental animals and postmortem examinations of human infants have established that the development of atherosclerosis often begins shortly after birth. Fatty streaks signaling the beginning of atheromas have been found in many human aortas

as early as the age of three. In a group of U.S. soldiers killed in the Korean war whose average age was only 23, examination showed that most had extensive formations of atherosclerotic "plaques" in their arteries. In our study of accident victims in Westchester County we found that many 35-year-old males who had shown no indication of heart disease nevertheless had their coronary arteries so thickened by atherosclerosis that the channels were narrowed by 50 percent. It has become

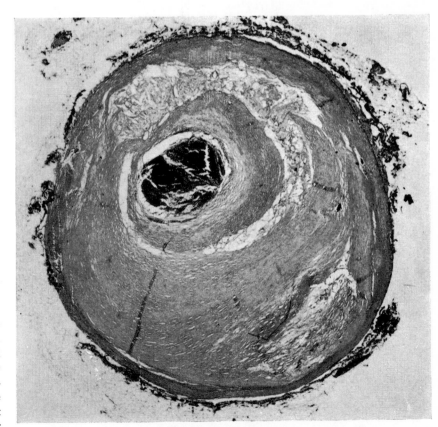

CRITICAL STAGE in the process under way in illustration at bottom of page 36 occurs when the atherosclerosis has advanced and blood clot almost fills the constricted channel.

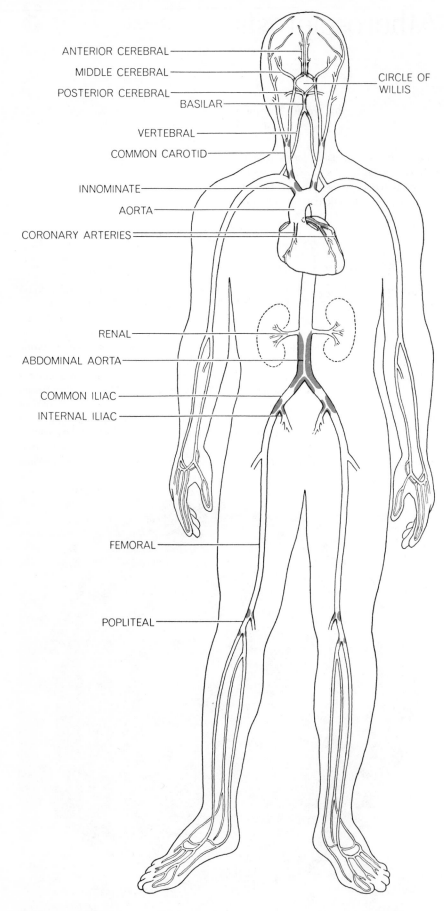

ANTERIOR CEREBRAL

MIDDLE CEREBRAL

POSTERIOR CEREBRAL

BASILAR

VERTEBRAL

COMMON CAROTID

INNOMINATE

AORTA

CORONARY ARTERIES

RENAL

ABDOMINAL AORTA

COMMON ILIAC

INTERNAL ILIAC

FEMORAL

POPLITEAL

CIRCLE OF WILLIS

FAVORED SITES of atherosclerosis are indicated in color in this diagram of the major arteries. There is a tendency for the disease to begin at points where an artery branches.

quite clear that atherosclerosis is a widespread disease of the young as well as the old.

The Nature of the Disease

Atherosclerosis appears to be at least as old as the civilization of mankind. The aortas of some Egyptian mummies that were entombed more than 3,500 years ago show the typical lesions of atherosclerosis. The modern name of the disease is derived from two Greek words: *athere,* meaning "porridge" or "mush," and *skleros,* meaning "hard." This apparently contradictory combination describes the fact that the lesion begins as a soft deposit and hardens as it ages. Materials that have been deposited from the bloodstream in the inner lining of the major arteries penetrate the arterial wall; they form plaques that gradually grow and thicken the wall, thus narrowing the blood channel. Eventually the thickening may close the channel entirely, or pieces of the plaques may break off and travel with the bloodstream until they are stopped in a smaller artery and thereby plug it. When the blockage occurs in the coronary artery, it produces a heart attack by cutting off the blood supply to the heart muscles; in the brain it produces a cerebral stroke; in the lower extremities it can lead to gangrene.

The circulatory system of the human body is a pipeline through which blood is pumped at a rate amounting to 4,300 gallons a day through 60,000 miles of pipe reaching every cell of the body. We might liken the atherosclerotic deposits in an artery to the rustlike encrustations that may form on the inner wall of a pipe. The living system, however, is vastly more complex than any ordinary pipeline. The fluid coursing through the arterial pipes contains living cells and a mixture of liquids that are continually changing in chemistry and physical characteristics. The flow is pulsatile, varying from moment to moment in velocity, volume and pressure. The living walls of the arteries themselves partake of the same changeability. They undergo a continual metabolism, conduct exchanges with the blood and the fluids bathing them externally and are subjected to various kinds of stress. In this dynamic system, subject to so many internal and external influences, unraveling the process that is responsible for atherosclerosis is akin to trying to solve a many-body problem in astronomy without knowing how many bodies are involved.

The atherosclerotic lesion is a complicated affair. When fully developed, it is composed of a considerable variety of structures and substances: blood and blood products, fibrous scar tissue, calcium deposits, complex carbohydrates, cholesterol (a fatlike, waxy substance normally present in the blood and body tissues), fatty acids and lipoproteins. Apparently the fatty acids and cholesterol are the crucial substances responsible for the development of the lesion, because they provoke inflammation and scarring of the arterial-wall tissue.

How does the process leading to atherosclerosis begin? Examination with electron and light microscopes shows that the first visible event is the invasion of the inner lining of the artery by fatty substances. These substances appear mainly in smooth muscle cells and foam cells found within the lining. In the spaces between the cells small amounts of cholesterol can be detected. Very fine fibers of a material that behaves like fibrin (a natural product of blood coagulation) also show up, both within the lining and on its surface. At this early stage the forerunner of the atherosclerotic lesion can be recognized in the form of fatty streaks, which when stained with a suitable dye are visible to the unaided eye as red streaks or spots on the lining surface.

Fats in the Arterial Wall

To solve the mystery of the origin of atherosclerosis one of the first questions we must answer is: How are the fatty materials deposited in the arterial wall? There are several current hypotheses. The one most widely accepted is that the fatty substances are transported into the wall by plasma, the blood fluid, and are trapped within the wall. It is believed that the plasma itself, under the force of the blood pressure, can leak all the way through the wall of the artery in small amounts, which then return to the bloodstream by way of the lymph-circulating system. The large lipoprotein molecules or complexes, on the other hand, cannot filter through the wall so easily; consequently they may tend to pile up within the wall, particularly if the plasma carries an excessive quantity of them.

The known structure of the walls of the major arteries gives support to this view. The wall of an artery consists essentially of three layers: outer, middle and inner. The outer layer and part of the middle one are nourished by a system of fine blood vessels (called vasa

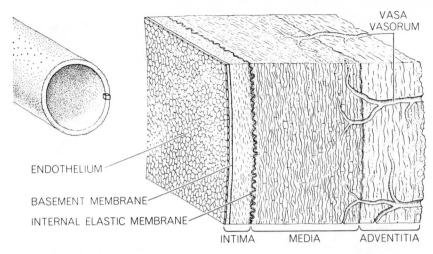

STRUCTURE of the wall of an "elastic" artery, the type usually involved in atherosclerosis, is shown in a somewhat schematic diagram. There is an inner lining of endothelial cells. The wall itself has three ill-defined layers of elastic tissue and muscle: the intima, media and adventitia. The vasa vasorum are small blood vessels that supply the artery wall.

vasorum) that come from the outer coat and go inward only as far as the middle layer. The nourishment of the inner portion of the arterial wall is taken care of by nutrients filtering into it from the bloodstream in the channel of the artery. Between the inner and the middle layer of the wall there is a curtain of elastic tissue. This tends to impede the flow of fluids through the wall. Hence it may act to trap substances that follow the gradient of flow, which in an artery is from inside to outside. Significantly, atherosclerosis rarely develops in veins, where the flow gradient is from outside to inside.

That lipids and other large molecules from the bloodstream can penetrate the arterial walls has been demonstrated conclusively by experiments with radioactively labeled cholesterol and other materials. These labeled materials usually turn up in highest concentration in areas in the walls that are already atherosclerotic. It has also been learned that in the early lesions of atherosclerosis the pattern of lipids present in the lesion is strikingly similar to the pattern in the blood.

Are there special conditions that favor the infiltration of lipids into the arterial wall? At least one interesting finding points in that direction. It was found that the administration of high doses of vitamin C to experimental animals inhibited the accumulation of lipids in the walls of their arteries. Conversely, when animals were fed a diet on which they developed a vitamin C deficiency, the inner lining of their arteries became more permeable: its cells were more widely spaced.

Abel Lazzarini-Robertson, Jr., of the

Cleveland Clinic, who has studied the behavior of arterial-lining cells in tissue cultures, suggests that there may be a self-feeding process that generates expansion of an atherosclerotic lesion. Some of these cells, he says, respond to an excess of lipids by requiring more oxygen. When they fail to obtain enough oxygen to satisfy this increased need, the cell membranes become more permeable to lipids. Thus a vicious circle is set up: the more lipid enters the cells, the easier it is for additional lipid to invade them.

As I have mentioned, hypotheses other than the lipid-infiltration theory have been proposed to account for the

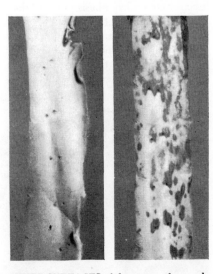

INNER SURFACES of the aortas of two rabbits were stained to visualize fatty material. Both animals had been fed cholesterol but one had also received estrogen, a female hormone. The aorta of that rabbit (*left*) was found to be almost free of atherosclerosis, whereas the other one was heavily affected.

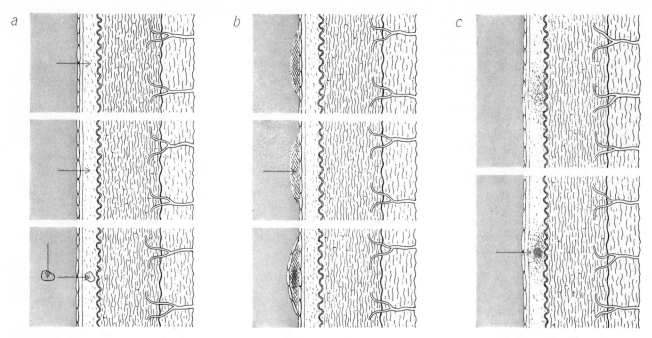

a   *b*   *c*

ENTRY OF FATTY MATERIAL into the arterial wall is the first
sign of atherosclerosis. One theory holds that fat molecules infil-
trate from the bloodstream (*a*) either by moving between endothelial cells or through them or by being carried between them by
large cells called foam cells. Another theory is that the first event
is the formation of a blood clot that is subsequently invaded by fats
(*b*). A third idea is that an injury alters the wall's cementing substance so that fats can invade from the blood or be synthesized (*c*).

origin of atherosclerosis. One of these
suggests that a disturbance of the blood-
clotting mechanism, in combination with
an injury to the inner wall of the artery,
may result in the formation of a fibrin
clot on the surface of the wall. Fatty
substances from the bloodstream may
then accumulate in the clot, particularly if there is a considerable amount of
such substances in the blood, and this
focus may generate the atherosclerotic
lesion. Another hypothesis proposes that
some alteration of the cementing substances (mucopolysaccharides) in the
artery wall that occurs after an injury to
the wall may open the way to local in-

vasions or synthesis of lipids. It may be,
indeed, that there is some truth in all
the hypotheses. There is reason to believe that atherosclerosis can originate
in a number of different ways.

Once the process has started it develops a kind of life of its own. Enzymes within the artery wall break
down the fatty complexes, liberating
cholesterol and fatty acids. These act
as noxious foreign agents and excite inflammation of the wall tissues. Scar
tissue develops. Fragile capillaries growing into this tissue tend to rupture and
thus lead to more inflammation. The
artery lining may ulcerate, and blood

from the bloodstream clots around these
breaks. Gradually the atherosclerotic
lesion expands in size, and as the scar
tissue and calcium deposits accumulate, the lesion stiffens and renders the
arterial wall brittle and weak.

Atherosclerosis may occur simultaneously in many of the body's major arteries. There are, however, certain favored sites. Much depends on the shape
and position of the vessel. For example,
the coronary arteries, which receive the
full impact of the pulsatile blood pressure against their walls during systole
(the heart's pumping cycle), have a high
tendency to develop atherosclerotic le-

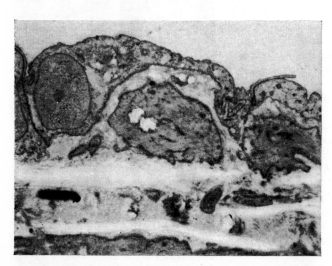

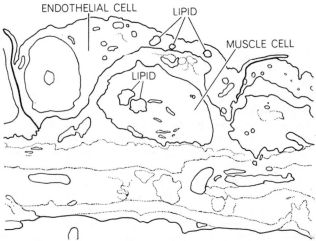

ENDOTHELIAL CELL   LIPID
LIPID
MUSCLE CELL

SECTION OF AORTA of a rabbit that had been fed cholesterol for
two weeks is enlarged 5,000 diameters in an electron micrograph
(*left*) made by Frank Parker and George F. Odland of the University of Washington School of Medicine. As the drawing (*second
from left*) shows, an endothelial cell of the artery lining contains
vacuoles filled with lipid, or fatty material. A single smooth-muscle

sions; on the other hand, the renal arteries, which branch from the aorta at right angles and have a low resistance to blood flow and therefore do not feel the pulsatile impact nearly as much, are relatively free of the disease. The vessels that are most often, and most critically, attacked are the coronary arteries, the aorta, the arteries in the neck and brain and the iliac and femoral arteries supplying blood to the lower extremities.

When thrombosis (formation of a blood clot) occurs in a narrowed coronary artery, the resulting partial or complete shutoff of blood supply to a portion of the heart muscle may have various effects: angina pectoris (pain in the chest), a myocardial infarct (destruction of part of the heart wall), irregularity or weakening of the heartbeat or sudden death due to complete failure of the heart. If the arteries to the brain become clogged, the result is a massive stroke causing paralysis or death. "Small" strokes, causing only slight or temporary paralysis of particular functions, may arise from fragments that break off from the atherosclerotic lesion and flow on to clog small vessels in the brain, thereby killing small areas of brain tissue. When atherosclerosis and clots clog a major artery to the legs, the result may be severe pain in these extremities and sometimes so much destruction of tissue that the gangrenous limb must be amputated.

In the aorta the lower section, passing through the abdomen, is particularly subject to atherosclerosis. The disease may so weaken the arterial wall that a portion of it balloons out, forming an aneurysm. Aneurysms of the aorta may press on the important organs in this area, interfering with their functions and causing pain. The rupture of one of these aneurysms usually produces massive hemorrhage and death.

One of the peculiarities of atherosclerosis is that even among the susceptible arteries it often selects particular ones for attack. For example, an individual may have severe atherosclerosis in his coronary arteries but very little of the disease in the cerebral arteries, or extensive lesions in the aorta with very little involvement of the coronary arteries. This form of selectivity is reflected in the disease rates of certain peoples. In Japan, for instance, strokes are common but heart attacks are relatively rare.

The Role of Diet

It is natural to suspect that diet has a great deal to do with atherosclerosis, and for more than a century the primary suspicion has focused on cholesterol. As early as 1847 a German anatomist, J. Vogel, reported that atherosclerotic arteries invariably contained cholesterol. In 1909 a Russian army medical officer, A. Ignatowski, observing that the army officers, who were of the meat-eating class, had many more heart attacks than the vegetarian peasants, undertook an experiment. He fed rabbits animal products and found that their aortas did indeed develop atherosclerotic lesions. A few years later a pair of Russian investigators, N. Anitschkow and S. Chalatow, followed up with a series of careful studies that became classic references in this field. When they fed rabbits fat and cholesterol, they observed that the cholesterol level in the animals' blood rose and atherosclerotic plaques appeared in their arteries. After cholesterol feedings were discontinued, lipids gradually disappeared from these plaques.

Cholesterol is indeed an inevitable suspect, because the formation of the atherosclerotic lesion is essentially an inflammatory response to this substance. The involvement of cholesterol in the disease has been demonstrated in many different ways. Experimenters have produced the disease by cholesterol feeding in many animals, including rabbits, rats, guinea pigs, chickens, dogs and monkeys. In almost every case in which the disease is induced experimentally the animals' serum shows a rise in cholesterol as a prelude to the atherosclerosis. In primates the disease exhibits all the features that occur in human beings. At the human level a cooperative study in Britain and the U.S. found that peptic ulcer patients who were treated with the Sippy diet (rich in milk and cream) had elevated levels of cholesterol in their serum and suffered twice as high a rate of heart attacks from coronary atherosclerosis as ulcer patients who did not use this diet. Conversely, patients with multiple myeloma, a malignant disease that tends to lower the serum-cholesterol level as one of its effects, have an unusually low rate of heart attacks. People who have died of so-called wasting diseases (essentially malnutrition) show a low lipid content in their arteries, which suggests that the loss of fat may have reduced their atherosclerotic lesions. On the other hand, people with diseases or conditions that are usually accompanied by a high cholesterol level (diabetes, nephrosis, hereditary elevation of lipids in the body) tend to develop atherosclerosis at

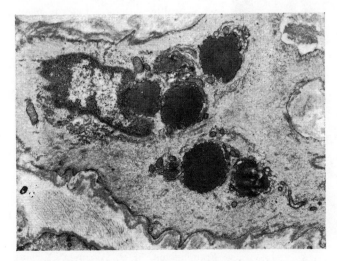

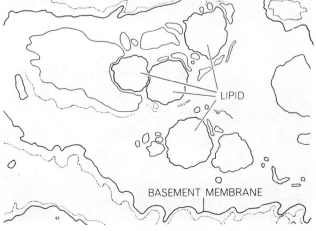

cell under the lining has two lipid vacuoles. In the other electron micrograph (*third from left*), made by Jack C. Geer and Marion A. Guidry of the Louisiana State University School of Medicine, a portion of a fatty streak from a human aorta has been enlarged 8,000 diameters. As indicated in the drawing (*right*), a smooth-muscle cell in the intima contains four lipid inclusions.

an earlier age and more extensively than usual.

Yet it has become increasingly clear that atherosclerosis cannot be explained simply in terms of cholesterol, or even a fatty diet. Certain species of pigeons spontaneously develop atherosclerotic lesions closely resembling those of human beings although these birds eat no animal fat. (Spontaneous atherosclerosis is also found in dogs, baboons, ostriches, pigs and whales.) Laboratory experiments have shown that exposure to cold, elevation of the blood pressure, antithyroid substances, high doses of vitamin D, lack of oxygen and other factors can contribute to the development of atherosclerosis. On the other hand, the disease process can be inhibited in animals by undernourishment, thyroid hormones, heparin (the anticlotting agent), fat-eliminating agents, unsaturated fats and sitosterol (a precursor of steroid hormones).

Epidemiology

Even the evidence of epidemiology is not entirely clear. It is true that populations whose diet contains a relatively small amount of saturated animal fats and cholesterol tend in general to have a low blood-cholesterol level and a low incidence of heart attacks. To illustrate with some often cited statistics: The South African Bantu, among whom death from coronary atherosclerosis is exceedingly rare, have a diet very low in fats (average: 17 percent of the total caloric intake) and a mean serum-cholesterol level of only 166. In Europe, where death from this disease is common, the average fat intake amounts to 35 percent of the diet and the serum-cholesterol level is 234. In the U.S., where the coronary death rate is very high, the average fat intake is between 40 and 45 percent and the serum-cholesterol level is about 250. Moreover, there is some evidence that people who migrate from a country with a low heart-disease death rate to one with a high death rate, and adopt the diet and cultural pattern of the latter country, tend to acquire a rise in the cholesterol level and an increase in the rate of heart attacks. This, at least, has been found to be true of Yemenite Jews and Japanese who have migrated to the U.S.

Nonetheless, it is not easy to determine exactly what factors separate the immune populations from the vulnerable ones, or the sheep from the goats. Among the peoples distinguished by exceptionally low rates of heart attack are the farmers of Guatemala, the Yemenite Jews, the South African Bantu, the Chinese, the Japanese and the Apache Indians living on reservations. Very high heart attack rates, on the other hand, are found among the adult white male inhabitants of New York, New Orleans, England, Sweden and parts of Finland. What do the latter populations have in common that differentiates them from the first group? This is one of the principal problems that today engages the attention of many investigators of the causes of atherosclerosis.

To narrow down the search for the significant environmental, biological or dietary factors, it would be very helpful if we could identify the individuals in each population who have atherosclerosis. Unfortunately this is difficult to do in a live population. Atherosclerosis has aptly been called an "iceberg" disease, because only five to 10 percent of those whose arteries are affected show any clinical sign of illness. Recently a method has been developed for examining the arteries in the body by X ray. In this method, called angiography, a radiopaque dye is injected into the bloodstream and the artery is then X-rayed to show whether or not the flow is normal. A narrowing or other abnormality of the channel is taken to indicate atherosclerosis. The technique is not, however, sufficiently simple, accurate or safe to be used as a screening procedure for the general population.

The cholesterol level in the blood is not itself a reliable index of the disease. In any population the level varies as a continuous spectrum, and one cannot find a dividing line that separates the atherosclerotic individuals from those with healthy arteries. Indeed, many people with low serum-cholesterol levels have heart attacks whereas many with high levels do not.

The investigation of atherosclerosis must therefore rely mainly on postmortem examinations and studies of people who clearly show signs of the disorder by their coronary disease or heart attacks. As everyone knows, there is now a very large accumulation of epidemiological studies that have sought to shed light on the factors associated with heart disease. These include the worldwide studies of Ancel Keys of the University of Minnesota and his associates, the famous mass studies in Framingham, Mass., Albany, N.Y., and Chicago and our own recently completed study of 10,000 males in Westchester County. All these studies have

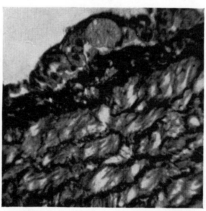
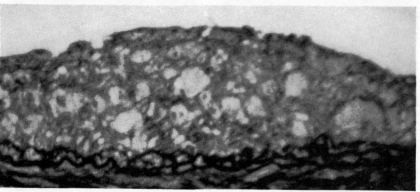

DEVELOPMENT of an atherosclerotic "plaque" in an arterial wall of a cholesterol-fed rabbit is traced in a sequence of photomicrographs made by the author. A normal section is shown at the upper left. A few foam cells penetrate the endothelium (*upper right.*) The bottom micrograph shows a larger accumulation of lipid—an early atherosclerotic plaque.

arrived at remarkably similar conclusions about the high-risk factors associated with coronary atherosclerosis. To sum them up in one profile, the most vulnerable person would be an adult male who has a high lipid content in his blood and high blood pressure, who engages in little physical activity and is markedly obese and who is a heavy smoker of cigarettes.

The difference between men and women in the rate of heart attacks from coronary atherosclerosis is striking. Our observations indicate that, in the age levels up to 55, deaths from such attacks are at least 10 times more common among men than among women. It seems that the factor protecting women is the female hormone estrogen. Women who have had their ovaries removed (thus reducing the estrogen output) tend to have more atherosclerosis than those with their ovaries intact. Injections of estrogens have been found to be capable of lowering the cholesterol level in the blood and of altering lipoproteins from the type associated with the development of atherosclerosis. At the Michael Reese Cardiovascular Research Institute in Chicago, Jeremiah Stamler and his colleagues demonstrated that the development of atherosclerosis in the coronary arteries of young male chickens that had been fed cholesterol could be stopped by injecting estradiol benzoate, a variant of the female hormone. On the strength of all the experimental evidence, estrogen injection has been tried as a treatment to inhibit atherosclerosis in men, but it is not promising for widespread use because of its feminizing effects.

High blood pressure is a serious contributor to atherosclerosis only when it is combined with a high cholesterol level in the blood, in which case the pressure forces cholesterol into the artery walls. The Apache Indians commonly have high blood pressure but seldom suffer heart attacks, probably because their blood content of cholesterol is low. Our studies of New York men indicated that the combination of high blood pressure and high cholesterol carries a high risk. In the age group between 36 and 50 men with this combination had a rate of atherosclerotic heart disease more than four times higher than that of men with normal blood pressure and lower serum cholesterol; the respective disease rates were 7.6 percent and 1.8 percent.

In the cases of the other risk factors revealed by the epidemiological studies —lack of physical activity, obesity, cigarette smoking—no direct tie to the athero-

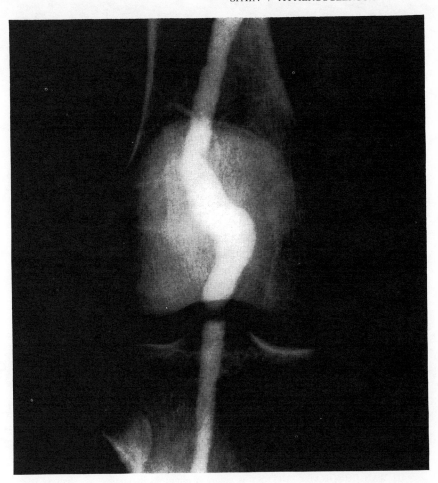

ANGIOGRAM, an X-ray photograph of a blood vessel injected with a radiopaque dye, can sometimes locate atherosclerotic damage. This one shows an aneurysm, or abnormally dilated segment, in a popliteal artery caused by atherosclerotic weakening of the arterial wall.

sclerotic process has been found. Just how these conditions contribute to heart disease remains to be determined.

Nondietary Factors

Many other elements that are suspected of contributing to atherosclerosis have been investigated. Undoubtedly heredity is an important factor. Atherosclerosis is frequently associated with diabetes and hypertension, diseases that are known to stem from genetic causes. Moreover, it seems likely that an individual's relative ability to metabolize and otherwise handle lipids plays a large part in his susceptibility to atherosclerosis. Studies have shown that identical twins tend to have about the same blood-cholesterol level, whereas twins who are not identical are much more likely to differ from each other in this respect. There have also been dramatic cases in which identical twins have had heart attacks at the same time in the prime of life.

Another factor that has had much attention is emotional stress, arising either from the individual's mode of life or his constitutional disposition. Unfortunately most of the studies of this factor have been so poorly conceived or executed that the conclusions are uncertain or questionable. There is no firm information so far to prove or disprove the hypothesis that emotional stress contributes to heart attacks.

We come back finally to the diet, which today holds the center of research attention as the factor most likely to be primarily responsible for the epidemic of atherosclerotic heart disease. There is no gainsaying the fact that this disease is a dominant feature of industrialized, affluent societies. If we look at metropolitan New York, where the disease has increased strikingly in the past 30 years, we can see that in the same period there has been a marked change toward a more luxurious and more passive manner of life, characterized by great increases in the use of the automobile, in automation of occupations and domestic tasks and in the animal-fat content of the average diet. The insurance companies have been compelled at frequent intervals (about 10 times in the past 30 years) to revise

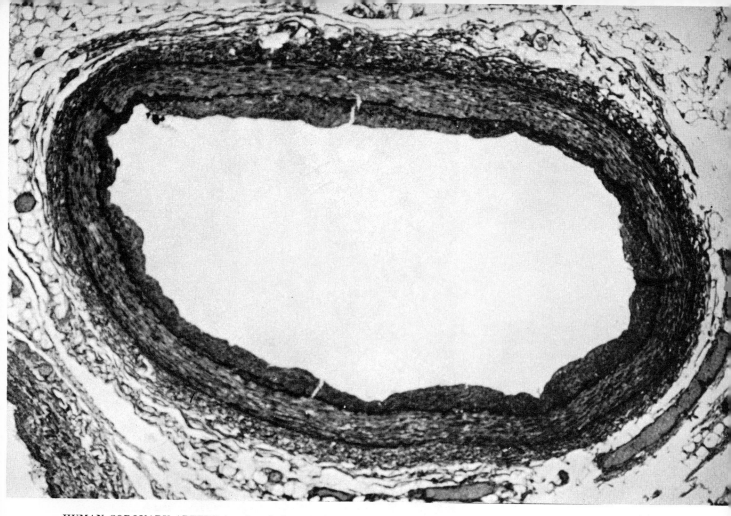

HUMAN CORONARY ARTERY is enlarged about 38 diameters in these photomicrographs made by the author. A normal artery is seen in cross section (*above*). In a diseased artery (*below*) the channel is partially occluded by atherosclerosis. Fibrous scar tissue, with fatty deposits (*clear areas*) in it, and other materials have thickened the arterial wall, reducing the blood-carrying capacity.

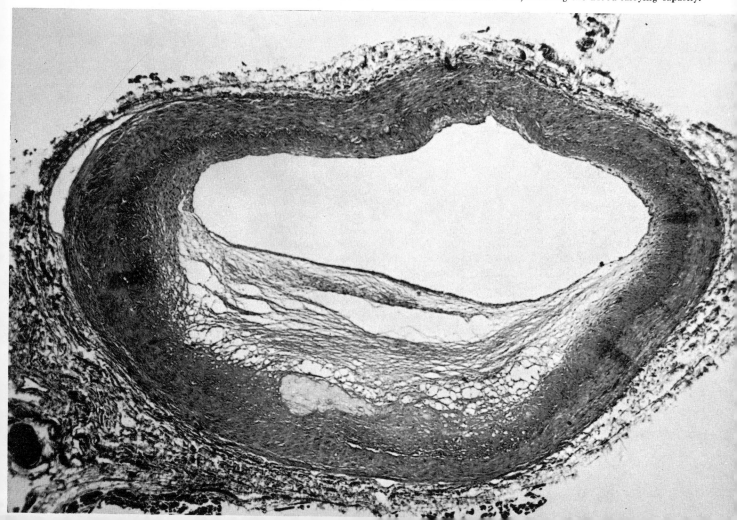

upward their statistical tables of average weights.

In a report to the White House Conference on Children and Youth, Stanley Marion Garn of the Fels Research Institute noted a disquieting trend in the eating habits of the younger generation in the U.S. today.

"If 35 percent of his calories comes from fats, is Junior being prepared, starting in nursery school, for a coronary occlusion?" asked Garn. "Reviewing the dietaries of some of our teen-agers, I am struck by the resemblance to the diet that Olaf Mickelsen uses to create obesity in rats. Frappes, fat-meat hamburgers, bacon-and-mayonnaise sandwiches, followed by ice cream, may be good for the farmer, good for the undertaker and bad for the population.... Through the stimulation of advertising, tap water is being replaced by sugared juices, milk and carbonated drinks. Snacks have become a ritualized part of the movies and are inseparably associated with television viewing."

To what extent can the animal-fat diet be specifically incriminated on the basis of the research done so far? The epidemiologist Ernest L. Wynder has suggested four criteria to determine whether or not a given factor can be regarded as a cause of a disease: (1) the incidence of the disease in a population must be proportional to the population's exposure to the factor; (2) the distribution of the disease—in geography, time, by sex and among various population groups—should be consistent with the distribution of the suspected factor; (3) the factor should produce the same disease, or one corresponding to it, in experimental animals in the laboratory, and (4) the removal of the factor or the reduction of exposure to it by the human population should reduce the incidence of the disease in the population.

The animal-fat diet has fulfilled the first three of these criteria for atherosclerosis in many tests. The fourth piece of incriminating evidence—reduction of the disease in man by reduction of the exposure—has not yet been established. It is currently being tested, however, in a massive dietary study, expected to involve ultimately 100,000 men, that is being conducted at five major centers under the auspices of the National Institutes of Health. If this and similar studies demonstrate that the fatty diet is indeed a major cause of atherosclerosis, there may be hope that the epidemic increase of the disease can be halted and reversed.

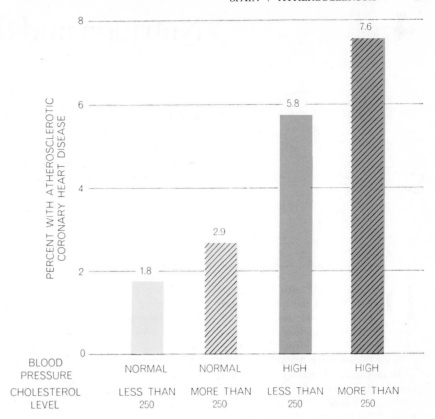

STUDY by the author of some 6,000 men showed that blood pressure and cholesterol level are correlated with the incidence of atherosclerosis. A diastolic blood pressure of 90 or less was called "normal"; that figure and the cholesterol count of 250 are necessarily arbitrary.

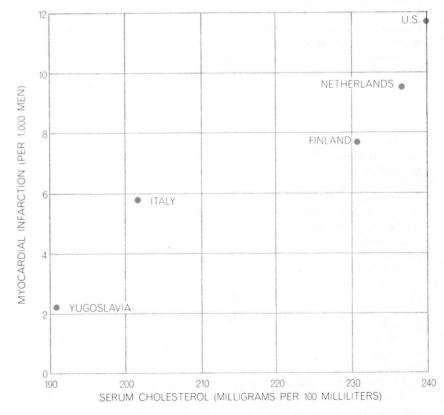

STATISTICS collected by Ancel Keys's group at the University of Minnesota indicate a correlation between the average cholesterol level in male populations and the amount of myocardial infarction, or heart-muscle damage. Cholesterol count may in turn be related to diet.

4 Nutrition and the Brain

by John D. Fernstrom and Richard J. Wurtman
February 1974

*When a meal rich in carbohydrate is eaten,
the brain makes more of the nerve-impulse transmitter
serotonin. The mechanism may be part of a
closed circle in which diet influences food consumption*

Does eating influence brain function? To put the question more specifically, do the changes in blood chemistry that follow the intake of nutrients produce corresponding changes in the tissues of the brain? If so, could such diet-induced changes affect the functional activity of the brain? Finally, might the "hungry" brain feed itself, so to speak, by "ordering" the body to eat?

In the past the usual answer to these questions has been no. The brain extracts oxygen, glucose and other nutrients, such as amino acids, from the bloodstream at a rate that seems unrelated to the concentration of those substances in the blood. For this reason transient fluctuations in the composition of the blood following the ingestion of food or during a period of fasting have not been thought to influence the brain. There is, however, an exception to this general rule and we and our colleagues in the Department of Nutrition and Food Science at the Massachusetts Institute of Technology have been investigating it for several years.

The exception is as follows. After a meal rich in carbohydrate there is an increase in the rate at which the brain synthesizes the neurotransmitter serotonin. Neurotransmitters are the substances present in the neurons of mammals that, when released, transmit signals across synapses to other neurons within the brain or to muscle cells or secretory cells outside the brain. Serotonin is one of six compounds that are reasonably well established as being neurotransmitters in the brains of mammals. The others are acetylcholine, gamma-amino butyric acid and the three catecholamines: epinephrine (or adrenalin), norepinephrine and dopamine.

In laboratory rats the brain's response to the ingestion of carbohydrate begins within an hour after eating. It is preceded by a change in the normal concentration of amino acids in the blood. The concentration of most of the amino acids decreases, but the concentration of the amino acid tryptophan increases. As a result there is a proportional increase in the brain's uptake of tryptophan. Now, tryptophan is a precursor in the synthesis of serotonin. Therefore after the ingestion of carbohydrate the brain neurons that convert tryptophan to serotonin are able to do so at a higher rate, and the level of serotonin in the brain rises. Since an increased amount of neurotransmitter will probably be released whenever the neurons that store serotonin are excited, it appears likely that brain function will be modified.

That the brain should be sensitive to diet-induced changes in the blood seems at first not only surprising but also disconcerting. What physiological advantage can there be to the individual when an organ as important as the brain can be affected by the vagaries of food intake? If in fact there is no advantage, how has such an open-ended mechanism for modifying brain function survived the tests of natural selection? Whatever the answer may be, there is no escaping the conclusion that nutrient intake does alter the amount of serotonin present in the serotoninergic neurons of the brain.

Like the three catecholamines, serotonin is a monoamine; specifically it is 5-hydroxytryptamine. It was first discovered in 1948 by three workers at the Cleveland Clinic: Maurice M. Rapport, Arda Green and Irvine H. Page. They isolated the substance from the blood of mammals, primarily the blood platelets. Soon afterward it was learned that relatively large quantities of serotonin are present in the mammalian brain, and within a decade the Swedish neuroanatomists Annica Dahlström and Kjell Fuxe,

working with the fluorescence microscope, had located the neurons where serotonin is stored. They found that virtually all the serotonin present in the brain is confined to a distinct group of neurons known as raphe nuclei. The cell bodies of the raphe-nuclei neurons are situated in the brain stem; their fibers ascend into the rest of the brain and descend through the spinal cord to form widely distributed synapses.

Tryptophan is one of the "essential" amino acids, that is, it cannot be synthesized by mammalian cells. Whatever store of tryptophan an animal has can be obtained only by the ingestion of protein that contains the substance. This is not the case for many of the other amino acids, for example tyrosine, the precursor of both norepinephrine and dopamine. Tyrosine can be synthesized in the mammalian liver, with the amino acid phenylalanine as a precursor. Tryptophan not only has to be obtained by ingestion; it is also the least abundant amino acid in dietary protein. As a result the mammalian body's supply of tryptophan is usually quite small.

The transformation of tryptophan and tyrosine into neurotransmitters begins in the same way: by the addition of a hydroxyl group (OH) through the catalytic action of an enzyme. The two enzymes involved, however, differ significantly. Tryptophan hydroxylase is a low-affinity enzyme. This is to say that a molecule of tryptophan and a molecule of the enzyme have no strong inclination to combine. Therefore it is only when the concentration of tryptophan is much higher than normal that the enzyme can function at the maximum rate. In contrast, tyrosine hydroxylase is a high-affinity enzyme; it operates at maximum efficiency even when the concentration of tyrosine is no greater than normal.

We first became interested in the relation between nutrition and serotonin synthesis in 1968, when a graduate student in our laboratory, William Shoemaker, found that the brains of weanling rats given a diet deficient in protein contained abnormally low levels of dopamine and norepinephrine. In view of the fact that tryptophan is present only in small amounts even in a normal diet and, unlike the precursor of the two catecholamines, is available only from proteins, it seemed to us that a tryptophan-poor diet should depress the level of brain serotonin even more strongly.

We proceeded to feed rats for several weeks a diet that had only one source of protein: corn, a food known to be particularly deficient in tryptophan. When we measured the levels of tryptophan and serotonin in the rats' brains, the levels proved to be far below normal. With the long-term dependence of brain-serotonin levels on the levels of brain tryptophan thus established, we wondered whether the same dependence could be demonstrated when the change was a rapid, short-term one. It seemed possible that such acute changes might occur spontaneously during the course of a day; we had earlier observed that in man the level of blood tryptophan increases and decreases with a characteristic daily rhythm.

Working with rats, we found that animals fed a normal diet and assayed at four-hour intervals day and night also exhibited rhythmic changes in the level of blood tryptophan. Moreover, a brain-tryptophan rhythm existed and paralleled the blood-tryptophan rhythm.

While the study was in progress we learned of a new chemical assay for serotonin that was far more sensitive than the methods that had been available. Using the new assay, we compared brain-serotonin and brain-tryptophan concentrations in groups of rats sacrificed as before. Because the level of serotonin in the brain would depend as much on its rate of utilization as on its rate of synthesis, we did not anticipate that the rhythms would be perfectly in phase. Indeed, they were not, but they were nevertheless quite close. A slight afternoon increase in the brain-tryptophan concentration was not matched by a corresponding change in the brain-serotonin concentration; the brain serotonin actually decreased. Thereafter, however, the two rhythms were generally in phase. (Being nocturnal feeders, rats have a daily rhythm that is opposite in phase to the rhythm in diurnal feeders such as man.)

We completed these rhythm studies in 1970 and were satisfied that the nat-ural variations in rats' brain-tryptophan concentrations influenced the brain's synthesis of serotonin. Shortly thereafter we undertook a search of the scientific literature on factors affecting the synthesis of other neurotransmitters. Having worked with daily changes, we were somewhat surprised to find that, at least insofar as studies of the effect of precursor availability were concerned, most other experiments had dealt only with long-term changes. It was our view that, if availability of precursors was a significant physiological factor affecting neurotransmitter synthesis, then the rapid short-term changes in the level of precursor concentrations would be the most important ones. We decided to see if a rapid artificial elevation of the brain-tryptophan level in rats, produced by injection of the amino acid, would increase the brain-serotonin concentration.

Our first injections were large enough to immediately supply the rat with about half the amount of tryptophan in a day's food ration. Within an hour the tryptophan concentration in the rat's brain reached nine times its normal level and the serotonin concentration nearly doubled. Even when we reduced the amount of tryptophan injected to a tenth of the original dose, the levels of brain tryptophan and serotonin still rose.

Having demonstrated that even small increases in brain tryptophan cause a rise in the concentration of brain serotonin, we decided to see if a decrease in the level of brain tryptophan would have the effect of diminishing the brain's rate of serotonin synthesis. Injections of insulin are known to lower the concentrations of amino acids in the bloodstream of both mammals and reptiles, and so we set about testing the possibility of lowering the brain-serotonin level in rats by injecting them with insulin. Far from lowering the concentration of tryptophan in the bloodstream, the injection increased it. Two hours after the injection the concentration of blood tryptophan reached a maximum and the levels of brain tryptophan and brain serotonin were similarly elevated. Not only was the result of our experiment paradoxical; it also was the first experimental evidence that a hormone (in this instance insulin) can affect the synthesis of a neurotransmitter in the brain.

But was our analysis of the experiment correct? Perhaps the increase in brain serotonin was unrelated to the increase

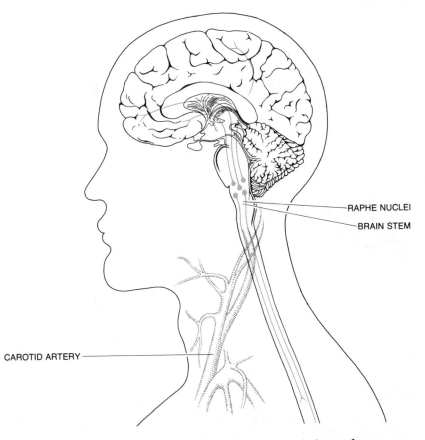

BRAIN NEURONS where most of the neurotransmitter serotonin is stored are a group known as the raphe nuclei. The neuron cell bodies are situated in the brain stem; the nerve fibers ascend into the rest of the brain and descend through the spinal column (*color*).

in brain tryptophan. It might be due to reflex changes in neuronal activity caused by the precipitous fall in the blood-sugar level induced by the insulin injection. To test this possibility we decided to let the rats introduce the insulin into their bloodstream themselves. We would allow the rats to eat a carbohydrate-rich diet, thereby stimulating the pancreas to secrete insulin naturally.

We fasted groups of rats overnight and then gave them a ration containing carbohydrate and fat but no protein. The natural secretion of insulin that followed produced the same effects that injections of insulin had: tryptophan levels in the blood rose significantly within an hour and reached a peak in two hours. During the first hour the level of brain tryptophan rose more than 20 percent and after two hours it reached a peak of 65 percent above normal. The concentration of brain serotonin increased in the first hour and was nearly 20 percent above normal after two hours [see illustration on page 43].

Reviewing our results, we reached two conclusions and ventured a prediction. First, the natural release of a hormone (insulin) in response to the consumption of carbohydrates elevated the level of a neurotransmitter (serotonin) in the brain of the rat; evidently it was the increase in the concentration of brain tryptophan that accelerated the brain's synthesis of serotonin. Further,

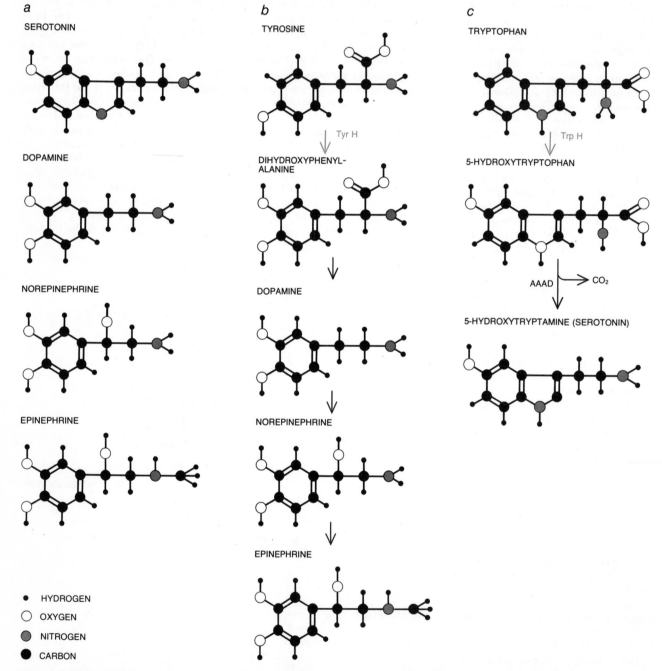

FOUR NEUROTRANSMITTERS found in the brain of mammals (*a, left*) are serotonin, dopamine, norepinephrine and epinephrine. The precursor of serotonin is the amino acid tryptophan (*c, right*); the first step in the synthesis is catalyzed (*colored arrow*) by the enzyme tryptophan hydroxylase. An "essential" amino acid, tryptophan is available only in dietary protein. The other neurotransmitters (*b, center*) are formed in successive steps from the same precursor amino acid, tyrosine; the initial reaction (*colored arrow*) is catalyzed by the enzyme tyrosine hydroxylase. Not an essential amino acid, tyrosine can be synthesized in the liver of mammals.

not only this "natural" insulin stimulus but also the artificial provision of insulin or of tryptophan by injection gave rise to identical effects: an increase in the level of tryptophan in the blood, paralleled by an increase in the levels of tryptophan and serotonin in the brain. On the basis of these conclusions we predicted that any increase in the level of blood tryptophan would be followed by an increase in the concentration of brain serotonin. For example, if we fed the rats a diet that included protein (the natural source of tryptophan) as well as carbohydrate and fat, we would expect the levels of both brain tryptophan and brain serotonin to increase in proportion to the increase in blood tryptophan.

We tested our prediction by feeding special diets to two groups of rats. For one group we fortified a ration of carbohydrate and fat with natural protein by adding 18 percent of the milk protein casein. For the other group in place of the casein we substituted a complete mixture of amino acids. Only one pair of feedings was needed to show that our prediction was wrong. On either of the fortified diets the level of tryptophan in the rats' bloodstream increased but neither brain tryptophan nor brain serotonin was elevated at all!

We soon learned what was wrong with our prediction. As two groups of independent investigators had already established, amino acids are conveyed from the bloodstream to the brain by a carrier system that embraces several separate subsystems. For example, one carrier transfers only amino acids that are not ionized at normal blood acidity (*p*H 7). A second carrier transfers only those that are positively ionized and a third transfers only those that are negatively ionized. As it happens, tryptophan is not ionized at normal blood acidity. Once present in the bloodstream following the ingestion of protein, it competes with five similarly neutral amino acids (leucine, isoleucine, valine, phenylalanine and tyrosine) for transport. The levels of the five competing amino acids in the bloodstream of our rats rose sharply when they received a diet including protein. The rise in the level of tryptophan, however, was much smaller because tryptophan is far less abundant in protein than the other amino acids are. Evidently when the concentration of the competing amino acids became high enough, it impeded the passage of tryptophan into the brain by swamping the transport system.

This sequence of events could also account for the action of injected and se-

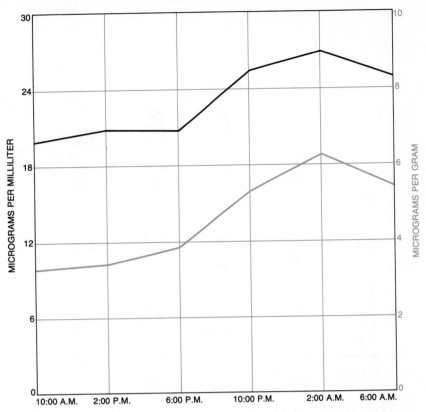

DAILY FLUCTUATION was noted in the level of tryptophan in the bloodstream of laboratory rats (*black curve*). The rhythm was correlated with the rats' time of food intake. The level of brain tryptophan also fluctuated (*colored curve*); the rhythms were in phase.

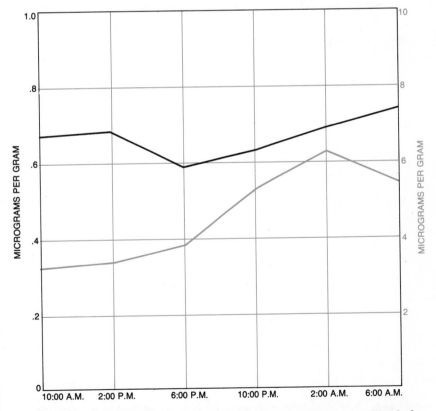

SEROTONIN CONCENTRATION in rats' brains also fluctuates (*black curve*). The rhythm reflects two interacting processes: accumulation of the neurotransmitter by tryptophan synthesis and its expenditure in neuron discharge. The rhythm is not quite in phase with the brain-tryptophan rhythm (*colored curve*), but a correlation between the two is clear.

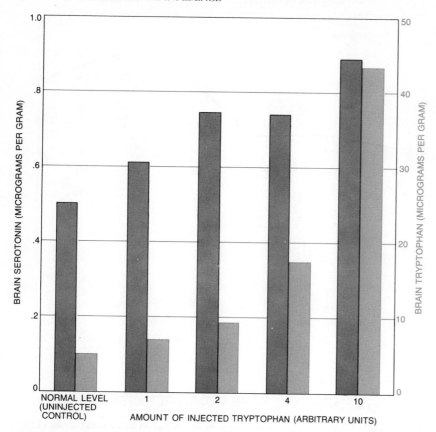

BODY INJECTIONS of different amounts of tryptophan rapidly raised the level of brain tryptophan (*colored bars*) and brain serotonin (*gray bars*) in rats. In one hour a dose equivalent to half a day's normal intake in food (*far right*) raised the tryptophan level ninefold and serotonin by 40 percent compared with an uninjected rat (*far left*). As little as one-tenth that dose still raised both precursor and neurotransmitter levels significantly.

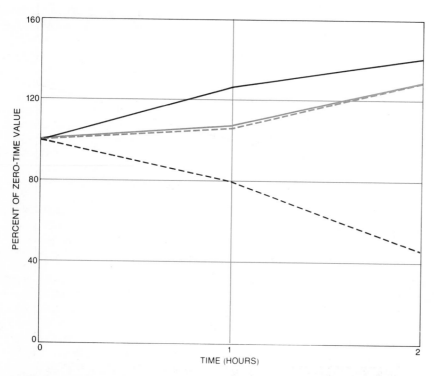

UNFORESEEN RESULT followed an attempt to lower the levels of brain tryptophan and brain serotonin in rats. In general the injection of insulin decreases the amount of amino acids normally found in the bloodstream, as it does the amount of blood sugar (*broken black line*). The injection instead increased the tryptophan level in both blood and brain (*solid black and solid colored line*) and also the brain-serotonin level (*broken colored line*).

creted insulin on brain tryptophan. By lowering the bloodstream concentration of other neutral amino acids that compete with tryptophan for transport into the brain the insulin would facilitate a rise in the level of brain tryptophan. The simultaneous insulin-induced rise in the level of blood tryptophan would contribute further to the increase in brain tryptophan.

To test this interpretation we prepared two new diets to be fed to further groups of rats. One ration contained carbohydrate, fat and all the amino acids that would normally be present in an 18 percent casein diet. The second ration was identical except that the five neutral amino acid competitors of tryptophan were omitted. We found that the level of tryptophan in the bloodstream increased among the groups of rats fed either ration. Only among rats fed the ration without competing amino acids, however, did brain-tryptophan and brain-serotonin levels increase substantially [*see top illustration on page 44*].

As a further test we fed groups of rats a diet omitting two other amino acids, aspartate and glutamine, whose charge is such that they do not compete with tryptophan for transport. The blood-tryptophan level was elevated but neither brain tryptophan nor brain serotonin increased. It was clear that when it is relatively unimpeded by competitors, the transport of tryptophan from bloodstream to brain is accelerated.

We summarized our new conclusions as follows. The concentration of tryptophan in the brain (and the resulting synthesis of serotonin) reflects the ratio of tryptophan to other neutral amino acids in the bloodstream more accurately than it does the quantity of blood tryptophan alone. A carbohydrate-rich diet, by stimulating insulin secretion, will tend to raise that ratio; a protein-rich diet, by elevating the level of competitive amino acids in the bloodstream, will tend to lower it. We recently tested this conclusion further by feeding a group of rats a diet containing a very high concentration of protein: 40 percent casein, about the proportion of protein found in beefsteak. The rate of serotonin synthesis in the brain of the rats actually decreased.

Nonetheless, the paradox with respect to insulin was not entirely dispelled. Our measurements showed that in rats insulin acted in the bloodstream not only to elevate the ratio of tryptophan to the other amino acids but also to increase the amount of tryptophan present. How could this be? The search for an answer to the question proved to involve a di-

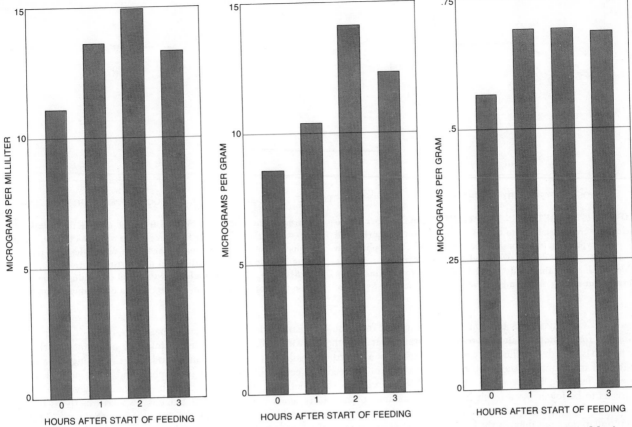

INSULIN PARADOX was investigated further by feeding a high-carbohydrate diet to hungry rats so that they would secrete their own insulin. From normal levels (*zero bar in each group*) the blood tryptophan (*left*), brain tryptophan (*center*) and brain serotonin (*right*) increased substantially in the first two hours after the rats were given access to food that induced insulin secretion.

etary component we had more or less ignored: fat.

In common with a number of other substances that circulate in the bloodstream (bilirubin, certain hormones and free fatty acids), tryptophan has a tendency to bind to albumin, a protein that is normally present in the blood serum. Roughly 80 percent of all the tryptophan molecules in the bloodstream are bound in this way; the phenomenon was first demonstrated in 1958 by Rapier H. McMenamy and John L. Oncley at the Harvard Medical School. Ray W. Fuller of the Eli Lilly & Company Research Laboratories, who has followed our work and made many helpful suggestions, pointed out in 1971 that some of our observations might be the result of changes in the ratios of albumin-bound and free tryptophan in the subjects' bloodstream.

We discussed this possibility with our colleague Hamish Munro and spent the next year working with Bertha Madras and Deborah Lipsett in search of the most efficient way to separate free tryptophan in the blood serum from bound tryptophan. We then began an experiment with human volunteers aimed at testing Fuller's suggestion. After an overnight fast each volunteer drank 75 grams of glucose; blood samples were then taken at regular intervals. As we expected, the ingestion of carbohydrate caused the volunteers to secrete insulin, and the levels of all the amino acids in their blood except tryptophan fell markedly. (Unlike the response in rats, the human response to insulin does not include an actual increase in the amount of tryptophan in the bloodstream.) Although the total amount of the volunteers' blood tryptophan remained the same, the amount of free tryptophan in the blood samples fell by as much as 40 percent. The decrease was matched by a proportional, albeit smaller, increase in the amount of bound blood tryptophan (which of course had originally made up the greater part of the total). What was causing the change?

Like the concentration of amino acids in the blood, the concentration of free fatty acids is known to be lowered by the presence of insulin in the bloodstream. It seemed possible that, if the insulin forced some molecules of free fatty acid to give up their bonds with molecules of albumin and leave the bloodstream, the increased availability of unbound albumin might allow more molecules of tryptophan to shift from the free state to the albumin-bound one.

To test this conjecture we measured the concentration of free fatty acids in the blood serum of our volunteers. The concentration changed in a pattern that was strikingly like the change in the concentration of free tryptophan. After the ingestion of glucose both concentrations fell within a similar length of time and in similar proportions. When we repeated the experiment with rats instead of human volunteers, the result was the same.

This led us to the conclusion that the free fatty acids in the blood serum are intermediaries in the effect of insulin on tryptophan circulating in the bloodstream. On this hypothesis tryptophan does not decrease in the bloodstream in the presence of insulin as the other neutral amino acids do because the action of insulin simultaneously frees albumin from existing fatty-acid bonds. The freeing of albumin, in turn, allows additional molecules of free tryptophan to become albumin-bound. As the concentration of free tryptophan in the blood falls, it is partially replenished by additional molecules of tryptophan that diffuse out of other tissues, such as skeletal

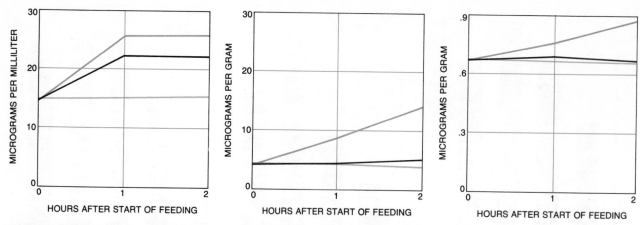

COMPETITION BETWEEN TRYPTOPHAN and five other amino acids for "transportation" to the brain accounts for the unchanged level of tryptophan in the brain after ingestion of a tryptophan-rich diet. Two groups of rats were fed contrasting rations; in one (black) all the amino acids found in dietary protein (including tryptophan and its five competitors) were present. The five competing amino acids were omitted from the other ration (dark color) so that the level of the five competitors would be lowered in the blood of rats fed it. Eating either ration raised the rats' level of blood tryptophan above the level of a fasting control group (left, light color). Brain tryptophan (center) and serotonin (right) levels rose far more, however, in the rats fed no competing amino acids.

muscle. Evidently in rats the affinity of albumin for tryptophan in the presence of insulin is so high that the total concentration of tryptophan in the bloodstream rises. In humans, however, the affinity appears to be lower. As a result the increase in the amount of albumin-bound tryptophan does not depress the level of free tryptophan in the blood to the point where additional tryptophan molecules diffuse from the tissue into the blood. Thus in humans insulin does not give rise to a net increase in the blood-tryptophan level.

The existence of such an albumin-bound and insulin-immune reservoir of tryptophan in the bloodstream gives this amino acid a competitive advantage over other neutral amino acids with respect to

transport from the bloodstream into the brain. Our hypothesis awaits testing at least in one regard. It would be a useful experiment to feed subjects a high-fat diet in order to see whether an overabundance of free fatty acids locks up enough extra albumin in the bloodstream to limit or suppress the binding of tryptophan to albumin and thus shrink the size of the insulin-immune reservoir of bound tryptophan.

We are satisfied that the consumption of foods low in protein produces an increase in the synthesis of serotonin in rats' brains, although the reason has proved to be a quite complex one. The same assay methods cannot be used with human volunteers, so that we have no tangible proof that the same is true of

man. Nonetheless, by feeding our volunteers protein-free and protein-rich substances and then measuring the amount of tryptophan and other neutral amino acids in their blood, we have established that humans respond exactly as rats do to the ingestion of the same diets. It seems logical to assume that the changes in the level of human brain tryptophan are similarly reflected in the level of brain serotonin.

We have not yet found direct evidence that the increased concentration of this particular neurotransmitter in the neurons of the raphe nuclei necessarily induces changes in the functional activity of the neurons. Intuition, however, tempts us to this conclusion

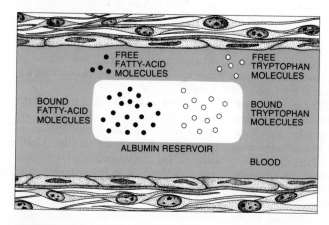

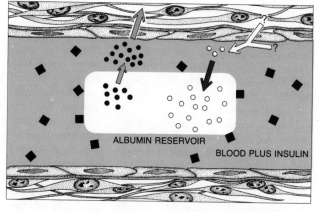

FURTHER EXPLANATION of the insulin paradox involves competition between tryptophan molecules and molecules of free fatty acid in the bloodstream for binding to available molecules of albumin. Some 80 percent of all the tryptophan in the blood is linked to albumin in this way; in the illustration this is schematically represented by the areas labeled "Albumin reservoir." When insulin enters the bloodstream (right, black rectangles), the bound fatty

acids are reduced in number and many of them are driven from the blood (gray arrows). This allows more free tryptophan molecules to bind to the released albumin molecules (black arrow); the shift into the insulin-immune reservoir helps to keep the total blood-tryptophan level in rough equilibrium. In the rat still other tryptophan molecules migrate from red blood cells or adjacent tissue (double arrow at right), thus raising the tryptophan level.

and leads us to suspect that the susceptibility of brain-serotonin concentrations to diet-induced changes probably provides mammals, man included, with a useful transduction mechanism.

It seems reasonable to speculate that diet-induced changes in the concentration of serotonin in the raphe-nuclei neurons would be paralleled by changes in the amount of serotonin released into the synapses whenever the neurons elect to transmit signals or are forced to do so through chemical intervention. If this is the case, then the serotoninergic neurons would function as sensors, converting a signal that is circulating in the bloodstream into a statement in brain language. A regular protein-rich diet would result in a minimum release of serotonin by the raphe-nuclei neurons, whereas even a brief period of a carbohydrate-rich diet would provide for a maximum release of the neurotransmitter.

It does not seem unreasonable that the brain could make good use of such responses to the protein-to-carbohydrate proportions of the diet, and perhaps to other nutritional and hormonal inputs that were sensed in the same way. The serotoninergic neurons may convey to the rest of the brain important data not only about the metabolic consequences of short-term and long-term changes in nutrition but also about growth and development processes and about stress.

The serotoninergic neurons of some mammals are known to be associated with a number of neuroendocrine mechanisms. For example, cats deprived of brain serotonin become insomniac, and the reduction of the brain-serotonin level in male and female rats gives rise to exaggerated sexual activity. Conversely, an increase in the brain-serotonin level decreases a rat's sensitivity to pain. The secretion of a number of pituitary hormones is also correlated with the level of brain serotonin. Recent work we have done with our colleague Loy Lytle suggests that changes in the brain-serotonin level of rats can influence both motor activity and food consumption. The latter finding suggests a kind of closed circle, with food consumption affecting brain biochemistry and brain biochemistry in turn affecting food consumption.

We may eventually find that this particular closed circle is a spiral viewed end on, or even a curved line that leads nowhere. Yet it remains a fact that, at least in our mammalian relative the laboratory rat, diet does control the synthesis of a significant neurotransmitter. We are loath to dismiss such a finding as mere coincidence.

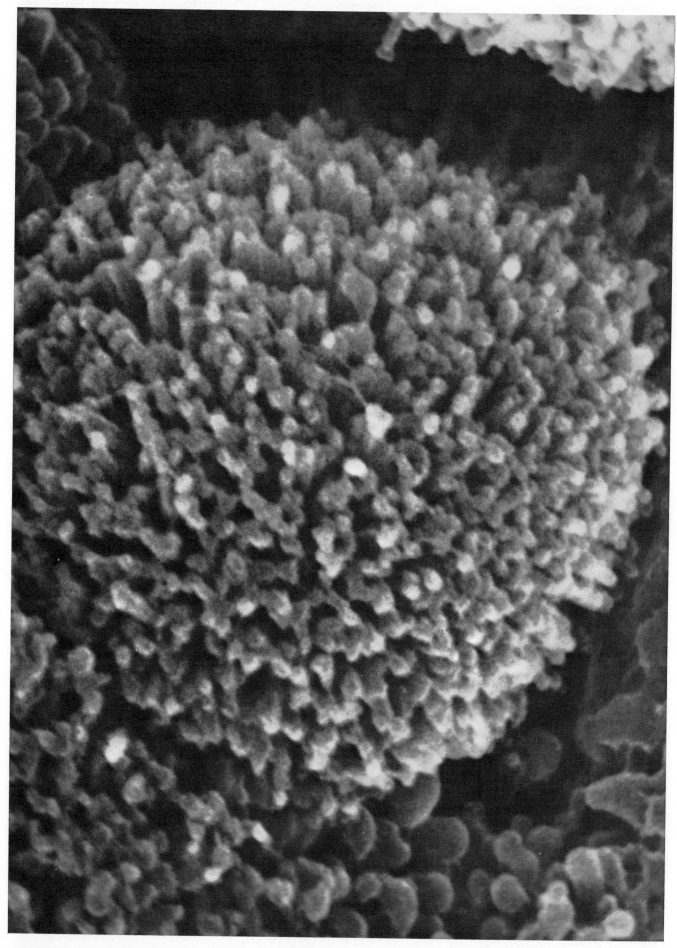

LACTOSE IS DIGESTED BY LACTASE in the intestine, a single epithelial cell of which is enlarged 37,500 diameters in this scanning electron micrograph made by Jeanne M. Riddle of the Wayne State University School of Medicine. The cell, on the surface of one of the finger-like villi that stud the lining of the intestine, is in turn covered by innumerable fine processes called microvilli.

Lactose and Lactase

by Norman Kretchmer
October 1972

Lactose is milk sugar; the enzyme lactase breaks it down. For want of lactase most adults cannot digest milk. In populations that drink milk the adults have more lactase, perhaps through natural selection

Milk is the universal food of newborn mammals, but some human infants cannot digest it because they lack sufficient quantities of lactase, the enzyme that breaks down lactose, or milk sugar. Adults of all animal species other than man also lack the enzyme—and so, it is now clear, do most human beings after between two and four years of age. That this general adult deficiency in lactase has come as a surprise to physiologists and nutritionists can perhaps be attributed to a kind of ethnic chauvinism, since the few human populations in which tolerance of lactose has been found to exceed intolerance include most northern European and white American ethnic groups.

Milk is a nearly complete human food, and in powdered form it can be conveniently stored and shipped long distances. Hence it is a popular source of protein and other nutrients in many programs of aid to nutritionally impoverished children, including American blacks. The discovery that many of these children are physiologically intolerant to lactose is therefore a matter of concern and its implications are currently being examined by such agencies as the U.S. Office of Child Development and the Protein Advisory Group of the United Nations System.

Lactose is one of the three major solid components of milk and its only carbohydrate; the other components are fats and proteins. Lactose is a disaccharide composed of the monosaccharides glucose and galactose. It is synthesized only by the cells of the lactating mammary gland, through the reaction of glucose with the compound uridine diphosphate galactose [*see illustrations on next page*]. One of the proteins found in milk, alpha-lactalbumin, is required for the synthesis of lactose. This protein apparently does not actually enter into the

reaction; what it does is "specify" the action of the enzyme galactosyl transferase, modifying the enzyme so that in the presence of alpha-lactalbumin and glucose it catalyzes the synthesis of lactose.

In the nonlactating mammary gland, where alpha-lactalbumin is not present, the enzyme synthesizes instead of lactose a more complicated carbohydrate, N-acetyl lactosamine. Test-tube studies have shown that alpha-lactalbumin is manufactured only in the presence of certain hormones: insulin, cortisone, estrogen and prolactin; its synthesis is inhibited by the hormone progesterone. It is when progesterone levels decrease late in pregnancy that the manufacture of alpha-lactalbumin, and thus of lactose, is initiated [see "Milk," by Stuart Patton; SCIENTIFIC AMERICAN Offprint 1147].

The concentration of lactose in milk from different sources varies considerably. Human milk is the sweetest, with 7.5 grams of lactose per 100 milliliters of milk. Cow's milk has 4.5 grams per 100 milliliters. The only mammals that do not have any lactose—or any other carbohydrate—in their milk are certain of the Pinnipedia: the seals, sea lions and walruses of the Pacific basin. If these animals are given lactose in any form, they become sick. (In 1933 there was a report of a baby walrus that was fed cow's milk while being shipped from Alaska to California. The animal suffered from severe diarrhea throughout the voyage and was very sick by the time it arrived in San Diego.) Of these pinnipeds the California sea lion has been the most intensively studied. No alpha-lactalbumin is synthesized by its mammary gland. When alpha-lactalbumin from either rat's milk or cow's milk is added to a preparation of sea lion mammary gland in a test tube, however,

the glandular tissue does manufacture lactose.

In general, low concentrations of lactose are associated with high concentrations of milk fat (which is particularly useful to marine mammals). The Pacific pinnipeds have more than 35 grams of fat per 100 milliliters of milk, compared with less than four grams in the cow. In the whale and the bear (an ancient ancestor of which may also be an ancestor of the Pacific pinnipeds) the lactose in milk is low and the fat content is high.

Lactase, the enzyme that breaks down lactose ingested in milk or a milk product, is a specific intestinal beta-galactosidase that acts only on lactose, primarily in the jejunum, the second of the small intestine's three main segments. The functional units of the wall of the small intestine are the villus (composed of metabolically active, differentiated, nondividing cells) and the crypt (a set of dividing cells from which those of the villus are derived). Lactase is not present in the dividing cells. It appears in the differentiated cells, specifically within the brush border of the cells at the surface of the villus [*see illustrations on page 50*]. Lactase splits the disaccharide lactose into its two component monosaccharides, glucose and galactose. Some of the released glucose can be utilized directly by the cells of the villus; the remainder, along with the galactose, enters the bloodstream, and both sugars are metabolized by the liver. Neither Gary Gray of the Stanford University School of Medicine nor other investigators have been able to distinguish any qualitative biochemical or physical difference among the lactases isolated from the intestine of infants, tolerant adults and intolerant adults. The difference appears to be

LACTOSE, a disaccharide composed of the monosaccharides glucose and galactose, is the carbohydrate of milk, the other major components of which are fats, proteins and water.

merely quantitative; there is simply very little lactase in the intestine of a lactose-intolerant person. In the intestine of Pacific pinnipeds, Philip Sunshine of the Stanford School of Medicine found, there is no lactase at all, even in infancy.

Lactase is not present in the intestine of the embryo or the fetus until the middle of the last stage of gestation. Its activity attains a maximum immediately after birth. Thereafter it decreases, reaching a low level, for example, immediately after weaning in the rat and after one and a half to three years in most children. The exact mechanism involved in the appearance and disappearance of the lactase is not known, but such a pattern of waxing and waning activity is common in the course of development; in general terms, one can say that it results from differential action of the gene or genes concerned.

Soon after the turn of the century the distinguished American pediatrician Abraham Jacobi pointed out that diarrhea in babies could be associated with the ingestion of carbohydrates. In 1921 another pediatrician, John Howland, said that "there is with many patients an abnormal response on the part of the intestinal tract to carbohydrates, which expresses itself in the form of diarrhea and excessive fermentation." He suggested as the cause a deficiency in the hydrolysis, or enzymatic breakdown, of lactose.

The physiology is now well established. If the amount of lactose presented to the intestinal cells exceeds the hydrolytic capacity of the available lactase (whether because the lactase level is low or because an unusually large amount of lactose is ingested), a portion of the lactose remains undigested. Some of it passes into the blood and is eventually excreted in the urine. The remainder moves on into the large intestine, where two processes ensue. One is physical: the lactose molecules increase the particle content of the intestinal fluid compared with the fluid in cells outside the intestine and therefore by osmotic action draw water out of the tissues into the intestine. The other is biochemical: the glucose is fermented by the bacteria in the colon. Organic acids and carbon dioxide are generated and the symptoms can be those of any fermentative diarrhea, including a bloated feeling, flatulence, belching, cramps and a watery, explosive diarrhea.

At the end of the 1950's Paolo Durand of the University of Genoa and Aaron Holzel and his colleagues at the University of Manchester reported detailed studies of infants who were unable to digest lactose and who reacted to milk sugar with severe diarrhea, malnutrition and even death. This work stimulated a revival of interest in lactose and lactase, and there followed a period of active investigation of lactose intolerance. Many cases were reported, including some in which lactase inactivity could be demonstrated in tissue taken from the patient's intestine by biopsy. It became clear that intolerance in infants could be a congenital condition (as in Holzel's two patients, who were siblings) or, more frequently, could be secondary to various diseases and other stresses: cystic fibrosis, celiac disease, malnutrition, the ingestion of certain drugs, surgery and even non-

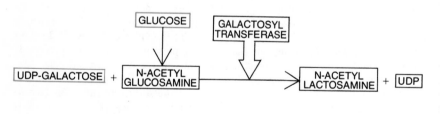

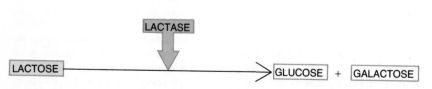

SYNTHESIS OF LACTOSE in the mammary gland begins late in pregnancy when specific hormones and the protein alpha-lactalbumin are present. The latter modifies the enzyme galactosyl transferase, "specifying" it so that it catalyzes the synthesis of lactose from glucose and galactose (top). In the nonlactating gland the glucose takes part in a different reaction (middle). In intestine lactase breaks down lactose to glucose and galactose (bottom).

specific diarrhea. During this period of investigation, it should be noted, intolerance to lactose was generally assumed to be the unusual condition and the condition worthy of study.

In 1965 Pedro Cuatrecasas and his colleagues and Theodore M. Bayless and Norton S. Rosensweig, all of whom were then at the Johns Hopkins School of Medicine, administered lactose to American blacks and whites, none of whom had had gastrointestinal complaints, and reported some startling findings. Whereas only from 6 to 15 percent of the whites showed clinical symptoms of intolerance, about 70 percent of the blacks were intolerant. This immediately suggested that many human adults might be unable to digest lactose and, more specifically, that there might be significant differences among ethnic groups. The possibility was soon confirmed: G. C. Cook and S. Kajubi of Makerere University College examined two different tribes in Uganda. They found that only 20 percent of the adults of the cattle-herding Tussi tribe were intolerant to lactose but that 80 percent of the nonpastoral Ganda were intolerant. Soon one paper after another reported a general intolerance to lactose among many ethnic groups, including Japanese, other Orientals, Jews in Israel, Eskimos and South American Indians.

In these studies various measures of intolerance were applied. One was the appearance of clinical symptoms—flatulence and diarrhea—after the ingestion of a dose of lactose, which was generally standardized at two grams of lactose per kilogram (2.2 pounds) of body weight, up to a maximum of either 50 or 100 grams. Another measure was a finding of low lactase activity (less than two units per gram of wet weight of tissue) determined through an intestinal biopsy after ingestion of the same dose of lactose. A third was an elevation of blood glucose of less than 20 milligrams per 100 milliliters of blood after ingestion of the lactose. Since clinical symptoms are variable and the biopsy method is inconvenient for the subject being tested, the blood glucose method is preferable. It is a direct measure of lactose breakdown, and false-negative results are rare if the glucose is measured 15 minutes after lactose is administered.

By 1970 enough data had been accumulated to indicate that many more groups all over the world are intolerant to lactose than are tolerant. As a matter of fact, real adult tolerance to lactose has so far been observed only in northern Europeans, approximately 90 percent of whom tolerate lactose, and in the

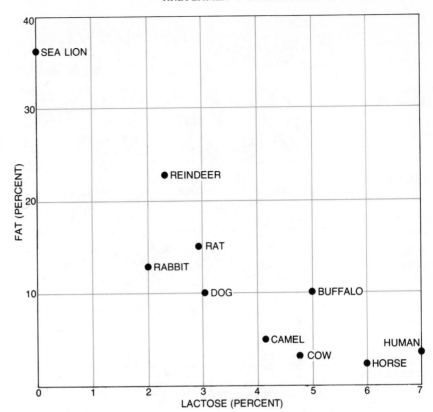

CONCENTRATION OF LACTOSE varies with the source of the milk. In general the less lactose, the more fat, which can also be utilized by the newborn animal as an energy source.

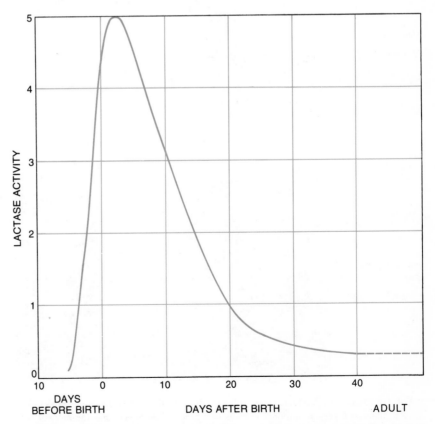

LACTASE is present in mammals other than man, and in most humans, in the fetus before birth and in infancy. The general shape of the curve of enzyme activity, shown here for the rat, is about the same in all species. Enzyme activity, given here in relative units, is determined by measuring glucose release from intestinal tissue in the presence of lactose.

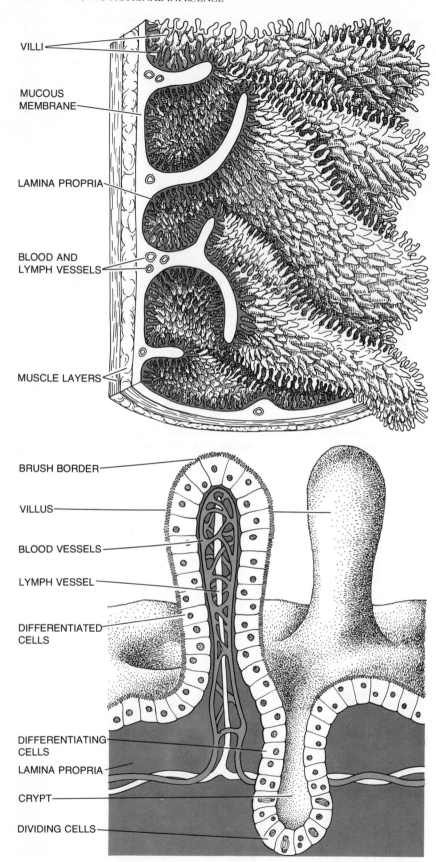

VILLI

MUCOUS
MEMBRANE

LAMINA PROPRIA

BLOOD AND
LYMPH VESSELS

MUSCLE LAYERS

BRUSH BORDER

VILLUS

BLOOD VESSELS

LYMPH VESSEL

DIFFERENTIATED
CELLS

DIFFERENTIATING
CELLS

LAMINA PROPRIA

CRYPT

DIVIDING CELLS

WALL OF SMALL INTESTINE, seen in longitudinal section (*top*), has outer muscle layers, a submucosa layer and an inner mucous membrane. The mucous membrane (*bottom*) has a connective-tissue layer (lamina propria), which contains blood and lymph capillaries, and an inner surface of epithelial cells. The cells multiply and differentiate in the crypts and migrate to the villi. At what stage the lactase is manufactured is not known; it is found primarily in the microvilli, which constitute the brush border of the differentiated cells.

members of two nomadic pastoral tribes in Africa, of whom about 80 percent are tolerant. Although many other generally tolerant groups will be found, they will always belong to a minority of the human species. In this situation it is clearly more interesting and potentially more fruitful to focus the investigation on tolerant people in an effort to explain adult tolerance, a characteristic in which man differs from all other mammals.

There are two kinds of explanation of adult tolerance to lactose. The first, and perhaps the most immediately apparent, originates with the fact that most people who tolerate lactose have a history of drinking milk. Maybe the mere presence of milk in the diet suffices to stimulate lactase activity in the individual, perhaps by "turning on" genes that encode the synthesis of the enzyme. Individual enzymatic adaptation to an environmental stimulus is well known, but it is not transferable genetically. The other explanation of tolerance is based on the concept of evolution through natural selection. If in particular populations it became biologically advantageous to be able to digest milk, then the survival of individuals with a genetic mutation that led to higher intestinal lactase activity in adulthood would have been favored. An individual who derived his ability to digest lactose from this classical form of Darwinian adaptation would be expected to be able to transfer the trait genetically.

These two points of view have become the subject of considerable controversy. I suspect that each of the explanations is valid for some of the adult tolerance being observed, and I should like to examine both of them.

The possibility of individual adaptation to lactose has been considered since the beginning of the century, usually through attempts to relate lactase activity to the concentration of milk in the diet of animals. Almost without exception the studies showed that although there was a slight increase in lactase activity when a constant diet of milk or milk products was consumed, there was no significant change in the characteristic curve reflecting the developmental rise and fall of enzymatic activity. Recently there have been reports pointing toward adaptation, however. Some studies, with human subjects as well as rats, indicated that continued intensive feeding of milk or lactose not only made it possible for the individual to tolerate the sugar but also resulted in a measurable increase in lactase activity. The discrepancy among the findings could be partly attributable to improvement in methods for assaying the enzyme activity.

On balance it would appear that individual adaptation may be able to explain at least some cases of adult tolerance. I shall cite two recent studies. John Godell, working in Lagos, selected six Nigerian medical students who were absolutely intolerant to lactose and who showed no physiological evidence of lactose hydrolysis. He fed them increasing amounts of the sugar for six months. Godell found that although the students did develop tolerance for the lactose, there was nevertheless no evidence of an increase of glucose in the blood—and thus of enzymatic adaptation—following test doses of the sugar. The conjecture is that the diet brought about a change in the bacterial flora in the intestine, and that the ingested lactose was being metabolized by the new bacteria.

In our laboratory at the Stanford School of Medicine Emanuel Lebenthal and Sunshine found that in rats given lactose the usual pattern of a developmental decrease in lactase activity is maintained but the activity level is somewhat higher at the end of the experiment. The rise in activity does not appear to be the result of an actual increase in lactase synthesis, however. We treated the rats with actinomycin, which prevents the synthesis of new protein from newly activated genes. The actinomycin had no effect on the slight increase in lactase activity, indicating that the mechanism leading to the increase was not gene activation. It appears, rather, that the presence of additional amounts of the enzyme's substrate, lactose, somehow "protects" the lactase from degradation. Such a process has been noted in many other enzyme-substrate systems. The additional lactase activity that results from this protection is sufficient to improve the rat's tolerance of lactose, but that additional activity is dependent on the continued presence of the lactose.

Testing the second hypothesis—that adult lactose tolerance is primarily the result of a long-term process of genetic selection—is more complicated. It involves data and reasoning from such disparate areas as history, anthropology, nutrition, genetics and sociology as well as biochemistry.

As I have noted, the work of Cuatrecasas, of Bayless and Rosensweig and of Cook and Kajubi in the mid-1960's pointed to the likelihood of significant differences in adult lactose tolerance among ethnic groups. It also suggested that one ought to study in particular black Americans and their ancestral populations in Africa. The west coast of Africa was the primary source of slaves for the New World. With the objective of studying lactose tolerance in Nigeria, we developed a joint project with a group from the University of Lagos Teaching Hospital headed by Olikoye Ransome-Kuti.

The four largest ethnic groups in Nigeria are the Yoruba in western Nigeria, the Ibo in the east and the Fulani and Hausa in the north. These groups have different origins and primary occupations. The Yoruba and the Ibo differ somewhat anthropometrically, but both are Negro ethnic groups that probably came originally from the Congo Basin; they were hunters and gatherers who became farmers. They eventually settled south of the Niger and Benue rivers in an area infested with the tsetse fly, so that they never acquired cattle (or any other beast of burden). Hence it was not until recent times that milk appeared in their diet beyond the age of weaning. After the colonization of their part of Nigeria by the British late in the 19th century, a number of Yoruba and Ibo, motivated by their intense desire for education, migrated to England and northern Europe; they acquired Western dietary habits and in some cases Western spouses, and many eventually returned to Nigeria.

The Fulani are Hamites who have been pastoral people for thousands of years, originally perhaps in western Asia and more recently in northwestern Africa. Wherever they went, they took their cattle with them, and many of the Fulani are still nomads who herd their cattle from one grazing ground to another. About 300 years ago the Fulani appeared in what is now Nigeria and waged war on the Hausa. (The Fulani also tried to invade Yorubaland but were defeated by the tsetse fly.) After the invasion of the Hausa region some of the Fulani moved into villages and towns.

As a result of intermarriage between the Fulani and the Hausa there appeared a new group known as the town-Fulani or the Hausa-Fulani, whose members no longer raise cattle and whose ingestion of lactose is quite different from that of the pastoral Fulani. The pastoral Fulani do their milking in the early morning and drink some fresh milk. The milk reaches the market in the villages and towns only in a fermented form, however, as a kind of yogurt called *nono*. As the *nono* stands in the morning sun it becomes a completely fermented, watery preparation, which is then thickened with millet or some other cereal. The final product is almost completely

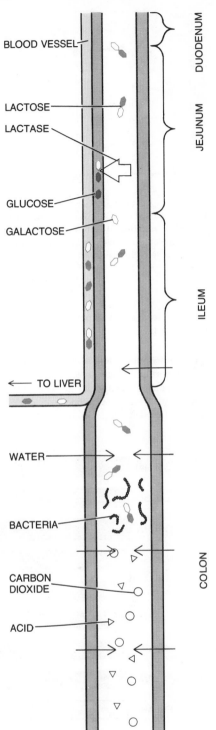

DIGESTION OF LACTOSE is accomplished primarily in the jejunum, where lactase splits it into glucose and galactose. Some glucose is utilized locally; the rest enters the bloodstream with the galactose and both are utilized in the liver. In the absence of enough lactase some undigested lactose enters the bloodstream; most goes on into the ileum and the colon, where it draws water from the tissues into the intestine by osmotic action. The undigested lactose is also fermented by bacteria in the colon, giving rise to various acids and carbon dioxide gas.

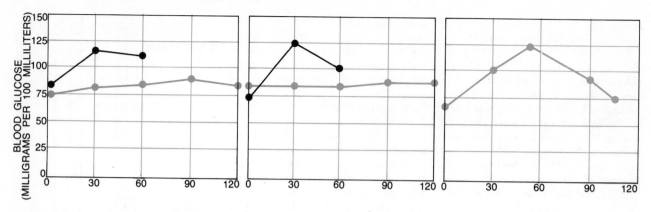

LACTOSE INTOLERANCE is determined by measuring blood glucose after ingestion of lactose. The absence of a significant rise in blood glucose after lactose ingestion (*color*) as contrasted with a rise in blood glucose after ingestion of sucrose, another sugar (*black*), indicates that a Yoruba male (*left*) and an American Jewish male (*middle*) are lactose-intolerant. On the other hand, the definite rise in blood glucose after ingestion of lactose in a Fulani male (*right*) shows that the Fulani is tolerant to lactose.

free of lactose and can be ingested without trouble even by a person who cannot digest lactose.

We tested members of each of these Nigerian populations. Of all the Yorubas above the age of four who were tested, we found only one person in whom the blood glucose rose to more than 20 milligrams per 100 milliliters following administration of the test dose of lactose. She was a nurse who had spent six years in the United Kingdom and had grown accustomed to a British diet that included milk. At first, she said, the milk disagreed with her, but later she could tolerate it with no adverse side effects. None of the Ibos who were studied showed an elevation of glucose in blood greater than 20 milligrams per 100 milliliters. (The major problem in all these studies is determining ethnic purity. All the Yorubas and Ibos who participated in this portion of the study indicated that there had been no intermarriages in their families.) Most of the Hausa and Hausa-Fulani

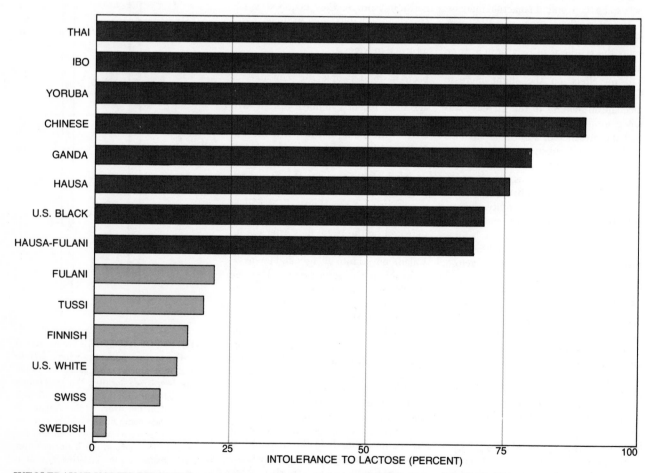

INTOLERANCE VARIES WIDELY among populations. The bars are based on tests conducted by a number of investigators by different methods; they may not be strictly comparable or accurately reflect the situation in entire populations. Among the groups studied to date lactose intolerance is prevalent except among northern Europeans (and their descendants) and herders in Africa.

(70 to 80 percent) were intolerant to lactose. In contrast most of the nomadic Fulani (78 percent) were tolerant to it. In their ability to hydrolyze lactose they resembled the pastoral Tussi of Uganda and northern Europeans more than they resembled their nearest neighbors.

Once the distribution of lactose intolerance and tolerance was determined in the major Nigerian populations, we went on to study the genetics of the situation by determining the results of mixed marriages. One of the common marriages in western Nigeria is between a Yoruba male and a British or other northern European female; the reverse situation is less common. Our tests showed that when a tolerant northern European marries a lactose-intolerant Yoruba, the offspring are most likely to be lactose-tolerant. If a tolerant child resulting from such a marriage marries a pure Yoruba, then the children are also predominantly tolerant. There is no sex linkage of the genes involved: in the few cases in which a Yoruba female had married a northern European male, the children were predominantly tolerant.

On the basis of these findings one can say that lactose tolerance is transmitted genetically and is dominant, that is, genes for tolerance from one of the parents are sufficient to make the child tolerant. On the other hand, the children of two pure Yorubas are always intolerant to lactose, as are the children of a lactose-intolerant European female and a Yoruba male. In other words, intolerance is also transmitted genetically and is probably a recessive trait, that is, both parents must be lactose-intolerant to produce an intolerant child. When the town-dwelling royal line of the Fulani was investigated, its members were all found to be unable to digest lactose—except for the children of one wife, a pastoral Fulani, who were tolerant.

Among the children of Yoruba-European marriages the genetic cross occurred one generation ago or at the most two generations. Among the Hausa-Fulani it may have been as much as 15 generations ago. This should explain the general intolerance of the Hausa-Fulani. Presumably the initial offspring of the lactose-tolerant Fulani and the lactose-intolerant Hausa were predominantly tolerant. As the generations passed, however, intolerance again became more prevalent. The genes for lactase can therefore be considered incompletely dominant.

The blacks brought to America were primarily Yoruba or Ibo or similar West African peoples who were originally

GEOGRAPHICAL EXTENT of dairying coincides roughly with areas of general lactose tolerance. According to Frederick J. Simoons of the University of California at Davis, there is a broad belt (*color*) across Africa in which dairying is not traditional. Migrations affect the tolerance pattern, however. For example, the Ganda, a lactose-intolerant group living in Uganda, came to that milk-drinking region from the nonmilking central Congo.

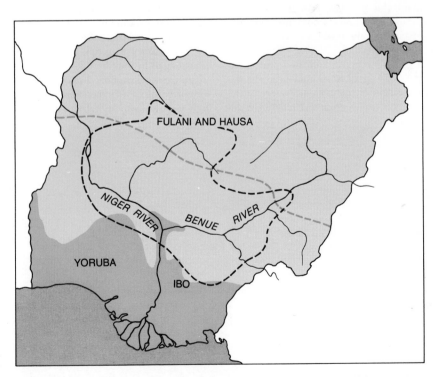

LARGEST ETHNIC GROUPS in Nigeria are the Ibo in the east, the Yoruba in the west and the Hausa and Fulani in the north. Map shows regions of mangrove swamp or forest (*dark color*) and grassland or desert (*light color*). Southern livestock limit (*broken colored line*) is set by climate, vegetation and tsetse fly infestation (*broken black line*).

FULANI WOMAN offers *nono*, a yogurt-like milk drink, for sale in the marketplace of a town in northern Nigeria. The pastoral Fulani drink fresh milk. The partially fermented *nono*, with reduced lactose content, is tolerated by villagers who could not digest milk.

intolerant to lactose. American blacks have been in this country for between 10 and 15 generations, in the course of which a certain complement of white northern European genes has entered the black population. Presumably as a result lactose intolerance among American blacks has been reduced to approximately 70 percent. One can speculate that if this gene flow eventually stopped, lactose intolerance would approach 100 percent among American blacks.

What events in human cultural history might have influenced the development of tolerance to lactose in the adults of some groups? Frederick J. Simoons of the University of California at Davis has proposed a hypothesis based on the development of dairying. It would appear that the milking of cattle, sheep, goats or reindeer did not begin until about 10,000 years ago, some 100 million years after the origin of mammals and therefore long after the mammalian developmental pattern of lactase activity had been well established. Man presumably shared that pattern, and so adults were intolerant to lactose. When some small groups of humans began to milk animals, a selective advantage was conferred on individuals who, because of a chance mutation, had high enough lactase activity to digest lactose. A person who could not digest lactose might have difficulty in a society that ingested nonfermented milk or milk products, but the lactose-tolerant individual was more adaptable: he could survive perfectly well in either a milk-drinking or a non-milk-drinking society.

The genetic mutation resulting in the capability to digest lactose probably occurred at least 10,000 years ago. People with the mutation for adult lactase activity could be members of a dairying culture, utilize their own product for food (as the Fulani do today) and then sell it in the form of a yogurt (as the Fulani do) or cheese to the general, lactose-intolerant population. These statements are presumptions, not facts, but they are based soundly on the idea that tolerance to lactose is a mutation that endowed the individual with a nutritional genetic advantage and on the basic assumption, which is supported by fact, that lactose intolerance is the normal genetic state of adult man and that lactose tolerance is in a sense abnormal.

What are the implications of all of this for nutrition policy? It should be pointed out that many people who are intolerant to lactose are nevertheless able to drink some milk or eat some milk products; the relation of clinical symptoms to lactose ingestion is quantitative. For most people, even after the age of four, drinking moderate amounts of milk has no adverse effects and is actually nutritionally beneficial. It may well be, however, that programs of indiscriminate, large-scale distribution of milk powder to intolerant populations should be modified, or that current moves toward supplying lactose-free milk powder should be encouraged.

II

RESPONSES TO ENVIRONMENTAL CHEMICALS

RESPONSES TO ENVIRONMENTAL CHEMICALS

INTRODUCTION

The body is subject to influence not only by a wide variety of nutrient chemicals, as described in Section I, but also by a huge number of other environmental chemicals, including inorganic nonnutrient elements, naturally occurring fungal and plant toxins, and synthetic chemicals. The closely related questions, "What do all these environmental chemicals do to the body?" and "How does the body handle them?" are of enormous practical significance. How successful are the body's adaptive responses in preventing toxic effects?

There are now more than 10,000 foreign chemicals being commercially synthesized; these are "foreign" in the sense that they are not normally found in nature. Over one million have been synthesized at one time or another. These foreign chemicals inevitably find their way into the body, either because they are purposely administered, as drugs (medical or "recreational"), or simply because they are in the air, water, and food we use.

A central focus for research into the body's handling of any foreign chemical is those factors which determine the effective concentration of the chemical at its sites of action. First, the chemical must gain entry to the body through the gastrointestinal tract, lungs, or skin. Accordingly, its ability to move across these barriers will have an important influence on its blood concentration. In practice, most organic molecules move through the lining of some portion of the gastrointestinal tract fairly readily, either by simple diffusion or by carrier-mediated transport. This should not be surprising, since the gastrointestinal tract evolved to favor absorption of the wide variety of nutrient molecules in the environment; the nonnutrient synthetic organic chemicals are the beneficiaries of these relatively nondiscriminating transport mechanisms.

The blood concentration of the chemical depends not only on its rate of entry but on its rate of removal as well. It may be eliminated from the blood by being moved into storage depots (for example, DDT into fat tissue), by being excreted from the body in the urine or feces (having been deposited in the latter by biliary secretion), or by being transformed into another molecular species. The processes of renal excretion and biotransformation are quite interrelated, since the former frequently depends on the latter, as the following description shows. Many foreign chemicals are handled by the kidneys by means of a combination of glomerular filtration and tubular reabsorption; accordingly, the rate of excretion of the chemical will depend on what fraction of the filtered molecules is reabsorbed. The problem is that many of these foreign chemicals are highly lipid-soluble; so, as the filtered fluid moves along the renal tubules, these molecules passively diffuse along with reabsorbed water through the tubular epithelium and back into the blood (highly lipid-soluble molecules diffuse through biological membranes quite easily). The net result is that little is excreted in the urine, and the chemical is retained in the body. If these chemicals could be transformed into more polar (and, therefore, less lipid-soluble) molecules, their passive reabsorption from the tubule would be retarded, and they would be excreted more readily. This type of transformation is precisely what occurs in the liver (and in other organs to a lesser extent).

For these reasons, the biotransformation mechanisms described by Kappas and Alvares, in "How the Liver Metabolizes Foreign Substances," are perhaps the single most important component of the body's defenses against foreign organic chemicals, since a potentially harmful molecule not only may be made inactive by the transformation, but is usually rendered more excretable as well. Moreover, the enzyme systems which mediate the biotransformation reactions are highly inducible; i.e., their activity can be greatly increased by exposure to a chemical which acts as a substrate for the system. This is another excellent example of acclimatization, in that exposure to a stressor (the chemical) has brought about an adaptive increase in the ability to cope with the stressor. Furthermore, a large number of substrates are acted on by these enzymes, and if an enzyme system is induced by one of these chemicals, it also becomes more responsive to all the other substrates that it can act on.

However, as should be apparent from the article, all is not really so rosy, for the hepatic biotransformation mechanisms vividly demonstrate how an adaptive response may, under some circumstances, turn out to be maladaptive. These enzymes all too frequently "toxify" rather than "detoxify" a drug or pollutant; in fact many foreign chemicals are quite nontoxic until the liver enzymes biotransform them. Of particular importance is the likelihood that many, if not all, chemicals which cause cancer do so only after biotransformation. These enzymes can also cause problems in another way, because they evolved primarily not to defend against foreign chemicals (which were much less prevalent during our evolution) but rather to metabolize endogenous substrates, particularly steroids and other fat-soluble molecules. Therefore, their induction by a drug or pollutant increases metabolism not only of that drug or pollutant, but of the endogenous substrates as well. The result is a decreased concentration in the body of that normal substrate.

An example of such a result is provided by Peakall's "Pesticides and the Reproduction of Birds," which describes how DDT and other pesticides reduce estrogen levels in birds by inducing the hepatic biotransformation enzymes. Although similar phenomena have been reported recently for at least one mammalian species (the sea lion), extrapolation of these data to human populations is, of course, very difficult. The primary difficulty is that we have no information about dose response — although there is little doubt that DDT would, at some dose level, alter human hormone concentrations (other hormones likely to be influenced are testosterone, progesterone, and the adrenocortical hormones). Taking the normal control of estrogen secretion into account, one would predict that the body should respond adaptively to any decrease in estrogen concentration that results from increased catabolism by increasing production and release of the hormone, thereby restoring the blood concentration to normal. The mechanism by which this should occur is as follows (note the flow diagram in the article): estrogen secretion is normally stimulated by two hormones, the gonadotropins, from the anterior pituitary; tonic gonadotropin secretion is, in turn, partly inhibited by estrogen, so that, in effect, estrogen exerts a negative feedback control over its own secretion; therefore a decrease in estrogen concentration in the blood would increase secretion of the gonadotropins, which would, in turn, stimulate estrogen secretion. Of course, there must be a limit to this compensation, and the toxic agent might also interfere with the functioning of the feedback system. Moreover, what are the long-term consequences of chronically altered pituitary function? In short, theory alone will not suffice; studies must be performed to discover exactly what the effects of DDT or any other agent actually are. Finally, of great importance is the finding that biotransformation systems similar to those of the liver exist in the placenta and are quite inducible; what are the effects of increased placental-enzyme activity on maternal-fetal hormonal interactions?

Thus far our discussion of environment chemicals has dealt mainly with organic molecules; "Lead Poisoning," by Chisolm, moves us into the realm of inorganic chemicals, and also illustrates that many potentially dangerous substances are not synthetic chemicals but are normally occurring agents that are present in excess because of human activity. As is true for organic drugs and pollutants, the concentrations of inorganic elements at their sites of action depend on rates of entry, excretion, and storage. However, inorganic molecules differ from organic molecules in two major ways: biotransformation is not important for them (except insofar as certain elements, including lead, inhibit the liver enzymes, thereby altering the toxicity of organic drugs and pollutants); and the gastrointestinal tract offers an important first line of defense against them, since absorption of them is usually quite limited, although the lungs do not, since airborne inorganic elements gain entry to the blood more readily through them. Little is known of the mechanisms by which the kidneys handle lead or other potentially harmful trace elements; specifically, it is not known whether adaptive increases in excretion are induced by exposure to the element.

"Lead Poisoning" vividly illustrates how difficult it is to answer the central question of modern toxicology: what are the effects of long-term exposure to small amounts of a chemical? It is relatively easy to identify the symptoms of acute lead (or DDT) poisoning, but extremely difficult to be certain whether chronic exposure to much smaller amounts of the agent produce subtle but important effects, i.e., act as perhaps one factor among many in the production of disease. Chisolm describes how lead can inhibit the function of critical enzymes in virtually every organ of the body, but also emphasizes that, because of the body's enzyme reserves and ability to make adaptive changes in enzyme synthesis, partial inhibition of an enzyme may not result in any significant malfunction. Conversely, interference with enzyme activity may be so subtle that it is not detectable in experiments, yet produces serious consequences over many years. Thus the question of whether every potential toxic agent has a "threshold" below which zero damage occurs, or whether some effect, no matter how small, always occurs, is raised over and over again, not only for chemicals but also for radiation, as we shall see in the next section. For most environmental chemicals, no answer is available.

6

How the Liver Metabolizes Foreign Substances

by Attallah Kappas and Alvito P. Alvares
June 1975

Among the most significant of the liver's chemical transformations are the inactivation of drugs, the detoxification of environmental pollutants and the activation of chemicals that can cause cancer

The intensity and duration of the action of most drugs is determined in large part by their rate of metabolism. If nothing else happened to a drug after it entered the body and reached its target organ, for example, it might continue to act indefinitely. Something does happen, however: most drugs are transformed into inactive substances and then excreted. The biotransformation can occur in any of several tissues and organs. Some drugs are transformed chemically in the intestine, some in the lung, the kidney or the skin. By far the greatest number of these chemical reactions are carried out in the liver, which metabolizes not only drugs but also most of the other foreign chemicals to which the body is exposed. Biotransformation in the liver is therefore a critical factor not only in drug therapy but also in defending the body against the toxic effects of a wide variety of environmental chemicals such as insecticides, herbicides, dyes, food preservatives and a number of substances that are suspected of inducing cancer. The central step in the metabolism of most of these agents involves an oxidation reaction mediated by a complex of enzymes that has come under intensive study in recent years in our laboratory at the Rockefeller University Hospital and in other laboratories.

The liver is the largest organ in the body (it weighs about three pounds in an adult) and has diverse functions. It serves, first of all, as the primary receiving depot, chemical-processing plant and distribution center for almost everything that enters the body through the walls of the alimentary canal. All the blood that has absorbed digested food and other substances from the intestines en-ters the liver through the large portal vein, which ramifies into fine channels through which the blood perfuses slowly among the liver cells. Here nutrients and other foreign substances are removed, metabolized, in some cases stored and then released into the general circulation. Amino acids, for example, are made into proteins and other nitrogenous compounds; glucose is converted into glycogen and stored, to be converted back into glucose and released as required. And drugs and other toxic substances are detoxified. Not everything is metabolized on the first passage of blood through the liver, of course; drugs, for example, are given in doses such that a sufficient amount of the drug moves through the liver to its site of action and is transformed later, on return visits to the liver [*see illustration on page 63*]. The liver also produces bile, which is a secretion that aids in the digestion of fats when it is released into the small intestine and is also a vehicle for the excretion of transformed substances and other waste products of metabolism.

The biotransformation of drugs and other foreign compounds in the liver is accomplished by several remarkable enzyme systems that can metabolize a wide variety of structurally unrelated drugs, toxic agents and environmental pollutants, which enter the body primarily through ingestion but also through the lungs and the skin. The enzyme systems are built into the membranes of the endoplasmic reticulum of the liver cells, a network of interconnected channels that is present in the cytoplasm of most animal cells. There are two kinds of endoplasmic reticulum, rough and smooth, and they differ in both form and function. The surfaces of the rough mem-branes are studded with ribosomes, small granules that translate the genetic code into the sequences of amino acids that constitute proteins. The smooth membranes have no ribosomes. In the liver a major function of both kinds of membrane is to assemble the enzymatic complexes that transform foreign substances and then to serve as the site of those transformations. Unlike some other cellular subsystems, the endoplasmic reticulum cannot be separated from cells as an intact structure. If liver cells are homogenized and then centrifuged, the tubular reticulum breaks up and bits of the membranes are sealed off to form the tiny vesicles, or sacs, called microsomes. The microsomal fraction thus obtained from liver cells is a convenient natural source of enzymes for laboratory studies of liver-cell metabolism.

Drugs and other foreign compounds are metabolized in the liver by a rather small number of reactions: oxidations, reductions, hydrolyses and conjugations. Their essential effect is to convert lipophilic, or fat-soluble, compounds into hydrophilic, or water-soluble, ones. The hydrophilic compounds are the more readily removed from the blood by the kidneys and excreted.

Oxidation accounts for most of the transformations, largely because there are so many different ways in which a compound can be oxidized [*see illustration on pages 66 and 67*]. The alkyl side chains of barbiturates and some other drugs, for example, are oxidized to form alcohols. In the case of compounds incorporating aromatic rings, including polycyclic hydrocarbons (such as those in cigarette smoke) and many drugs, a hydroxyl group is inserted into the ring.

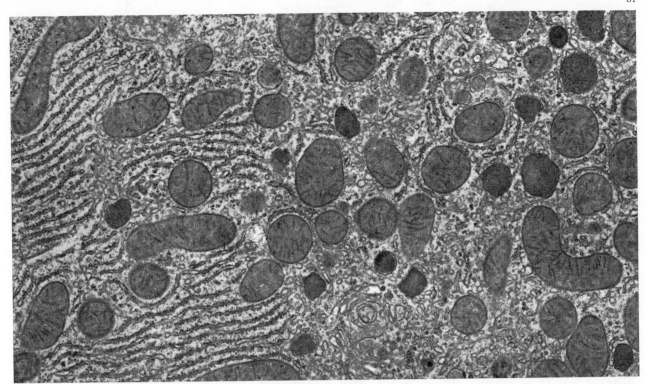

ENDOPLASMIC RETICULUM of a rat liver cell is enlarged 18,000 diameters in this electron micrograph made by Edward S. Reynolds of the Harvard Medical School. Most of the biotransformations of foreign substances take place in this system of membranous tubules. The "rough" endoplasmic reticulum is covered with ribosomes, the structures in which proteins are synthesized; it appears here as fairly linear double membranes studded with black dots. The "smooth" endoplasmic reticulum lacks ribosomes and forms a more branching, tubular network; there are patches of it between the mitochondria, the fine-grained gray objects, at right.

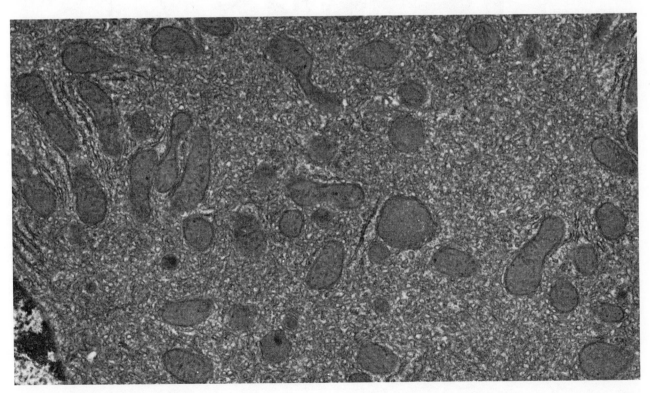

SMOOTH ENDOPLASMIC RETICULUM proliferates when the synthesis of the enzyme systems it contains is stimulated by drugs. This micrograph, also made by Reynolds and reproduced at the same scale as the one at the top of the page, is of a cell from the liver of a rat that was treated with phenobarbital. The drug stimulates the synthesis of cytochrome P-450, an enzyme located in the endoplasmic reticulum membranes. The cytoplasm of the cell has expanded because of a striking increase in the smooth endoplasmic reticulum membranes, which now fill most of the space between the more dispersed mitochondria. Enzyme induction by one drug can significantly affect the metabolism, and thus the activity, of other drugs that are metabolized by the same enzymes.

In other cases alkyl groups are removed from either nitrogen or oxygen atoms, amino groups are removed or sulfoxides are formed. Reduction and hydrolysis are also catalyzed by liver enzymes, but these reactions are less common than oxidation.

Conjugation of a chemical is combination with some natural constituent of the body such as the glucose derivative glucuronic acid, the amino acid glycine or the tripeptide glutathione. In the presence of the appropriate enzyme these natural agents can combine readily with compounds that have carboxyl (COOH), sulfhydryl (SH), amino (NH_2) or hydroxyl (OH) groups. Some drugs have these groups when they are in their active form and are handled in the liver by conjugation; for most drugs, however, conjugation is a second step that comes after metabolism by oxidation, reduction or hydrolysis. Almost without exception the conjugated compound is devoid of pharmacological (or any biological) activity.

Clearly the enzyme systems that catalyze these reactions were not invented by the mammalian organism in order to cope specifically with drugs or novel pollutants. Their basic physiological role has presumably been to metabolize endogenous substrates: substances normally present in the body. For example, liver microsomal enzymes oxidize steroid hormones, cholesterol and fatty acids. The products of these oxidations may then be conjugated (with glucuronic acid, for instance) and excreted. Bili-

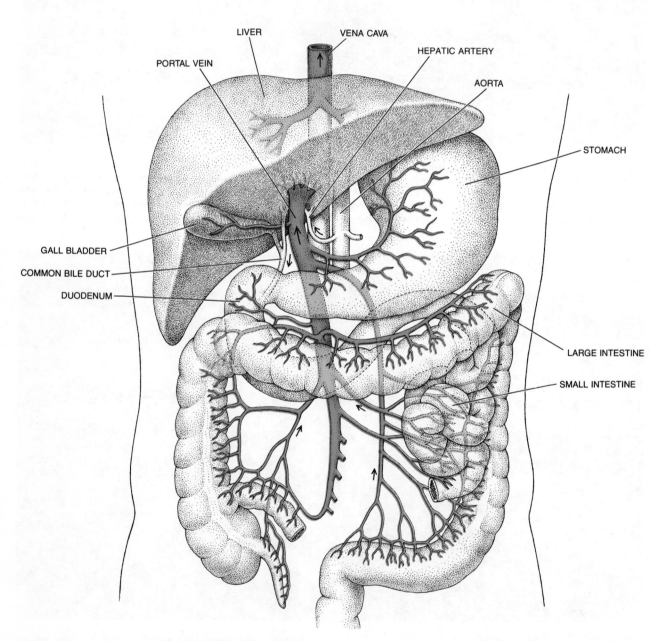

LIVER is a primary site of metabolism of substances entering the body through the alimentary tract. All the fine blood vessels that absorb nutrients and other substances through the wall of the intestines come together and enter the liver through the portal vein. The liver's supply of oxygenated blood from the heart enters through the hepatic artery. After passing through the sinusoids of the liver (see illustration on page 62) the blood is collected by the hepatic veins, which feed into the vena cava. The liver secretes bile, which is collected by the bile duct, stored in the gall bladder and emptied into the duodenum, the first segment of the small intestine.

rubin, a product of the oxidation of heme, the red pigment of hemoglobin, is an example. It is normally prepared for excretion by glucuronide conjugation. The rate of formation of glucuronides is generally low in newborn infants, however, because of a deficiency of the enzyme glucuronyl transferase. As a result bilirubin may not be conjugated and excreted at an adequate rate. Excessive amounts of it can then accumulate and cause grave damage to the brain, a condition called kernicterus.

It has been apparent for some time that the human fetus and the newborn infant are far more sensitive than adults to many drugs. A number of drugs can pass across the placenta, so that obstetricians need to exert care in administering them to an expectant mother. Barbiturates or morphine given to a woman during childbirth can be stored in the infant's tissues and cause respiratory depression and occasionally death. The explanation for the sensitivity of infants to drugs has emerged from a number of reports in recent years on the maturation of the capacity to metabolize and conjugate drugs. These studies make it clear that the capacity to oxidize and conjugate is negligible in the mammalian fetus and newborn animal and increases after birth at a rate that varies with the species, the type of reaction and the drug. Impaired drug metabolism increases the intensity and duration of drug action. For example, newborn mice treated with a dose of the hypnotic drug hexobarbital equivalent to 10 milligrams per kilogram of body weight sleep more than six hours, whereas adult mice given 10 times as large a dose sleep less than an hour.

The adverse effect of drugs on newborn infants as a result of inefficient metabolic conversion is illustrated by the striking "gray-baby syndrome" in infants that may come a few days after treatment with the antibiotic chloramphenicol: abdominal distention, respiratory difficulty, cyanosis (blue skin color as a result of insufficient oxygenation of the blood) and shock. The condition is apparently the result of deficient metabolism of the antibiotic and deficient conjugation with glucuronic acid. In adults about 90 percent of the antibiotic is excreted in the urine in the form of conjugated metabolites within 24 hours after it has been orally administered; only a small amount is excreted unchanged. In comparison a 10-day-old infant in one study excreted less than 50 percent of the drug in 24 hours.

Rates of drug metabolism are very dif-

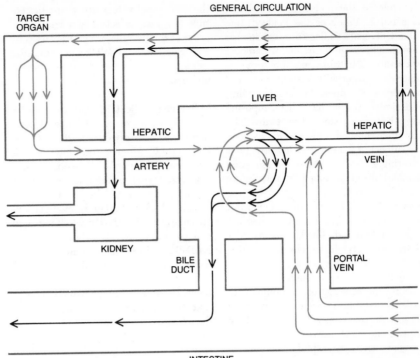

PATHWAY of a drug that is transformed in the liver is shown schematically. The drug (*colored arrows*) enters the liver through the large portal vein, passes into the general circulation, has its effect on the target organ and eventually returns to the liver. On each passage through the liver a fraction of the drug is converted, usually into inactive metabolites (*black arrows*). The metabolites may be carried by the bile into the intestines for excretion or may pass through the circulation to the kidneys, to be excreted into the urine.

ferent in different species, and the effective dosage varies accordingly. The antiinflammatory agent phenylbutazone is metabolized slowly in man; its half-life in the plasma averages about three days. In the horse, the dog, the rabbit, the rat and the guinea pig, however, the drug is metabolized much more quickly and the half-life ranges from three to six hours. (Knowledge of the rapid rate of metabolism of phenylbutazone in the horse is of practical importance because the drug is administered to treat arthritic conditions in racehorses.) A dose of hexobarbital (adjusted for the body weight of the animal) that makes mice sleep for an average of 12 minutes puts rabbits to sleep for 49 minutes, rats for 90 minutes and dogs for 315 minutes. When the enzymatic oxidation of hexobarbital by microsomes from the liver of these animals was measured, the fastest oxidation was carried out, as expected, by microsomes from the mouse; the rates of oxidation by microsomes from the rabbit, the rat and the dog were proportionately lower. Differences in the rates and patterns of drug metabolism in the various animals seem to explain most of the species differences in effect, but there may also be dif-

ferences in the distribution of the drug, the response of the target tissues and excretion.

Even among human patients there are marked individual variations in the metabolism of drugs that are handled primarily by microsomal enzymes. The variability causes some patients to metabolize a drug so quickly that therapeutically effective blood and tissue levels are difficult to achieve and others to metabolize the drug so slowly that they suffer toxic effects. It can therefore be difficult for the physician to predict just what dosage of some drugs will provide a safe and therapeutic effect in an individual patient. Very large individual differences have been noted in the metabolism of the coumarin-derivative anticoagulant drug Dicumarol; the half-life can vary from seven to 74 hours. That makes it hard to predict how much of the drug will provide the desired anticoagulant effect in a patient. Marked variations have also been observed in the metabolism in man of phenylbutazone and of diphenylhydantoin, a drug that is administered to control epileptic seizures.

Genetic factors may play an important role in the metabolism of drugs, as Elliot S. Vesell of the Pennsylvania State University College of Medicine has shown. The large individual differences in Dicumarol half-life persist to some extent when fraternal twins are compared but almost disappear in identical twins, and similar results have been reported for other drugs studied in twins. The antituberculosis drug isoniazid has been the subject of a number of investigations covering different populations. Isoniazid is metabolized in man primarily by an acetylation reaction catalyzed by the enzyme N-acetyl transferase. Soon after the drug was introduced it became apparent that individuals vary in their ability to acetylate it. The distribution of the acetylation rates for a group of subjects tends to be bimodal, that is, plotting the rate against the number of patients exhibiting each rate yields a curve with two peaks. Apparently there are two classes of individuals: those who acetylate isoniazid rapidly and excrete the drug primarily as acetylisoniazid and those who acetylate and excrete the drug more slowly and excrete more of it in an unchanged form. The frequency of rapid inactivators is about 90 percent in a population of Eskimos or of Japanese, whereas among both whites and blacks in North America there are about as many slow inactivators as there are rapid ones.

A clear relation has now been established between this variation in acetylation rate and the incidence of isoniazid toxicity. Several studies have shown that the slow inactivators are more susceptible to a toxic effect of isoniazid: a disorder of the peripheral nerves caused by a specific vitamin B_6 deficiency resulting from an interaction of the drug and the vitamin.

The first indication that microsomal enzymes were special kinds of complexes was developed some 20 years ago by Gerald C. Mueller and James A. Miller at the University of Wisconsin. They discovered that compounds known as aminoazo dyes could be oxidized (specifically, N-demethylated) by liver microsomes and that the oxidation required molecular oxygen (O_2) and the coenzyme NADPH (reduced nicotinamide-adenine dinucleotide phosphate). Soon afterward Bernard B. Brodie's group at the National Heart and Lung Institute showed that there was a similar requirement for the oxidative metabolism of a number of drugs. In 1957 Howard S. Mason of the University of Oregon Medical School proposed that such oxidations are catalyzed by a class of "mixed-function oxidases": enzyme complexes that require oxygen and NADPH, are nonspecific (they catalyze the oxidation of different kinds of compounds) and catalyze the consumption of a molecule of oxygen for each molecule of the drug or other substrate, with one atom of oxygen appearing in the metabolized substrate and the other atom usually combining with two hydrogen atoms to form water.

The key enzyme of these oxidases is cytochrome P-450. A cytochrome is a complex of protein and heme, the iron-containing ring structure that is the oxygen-binding component of hemoglobin. Like hemoglobin, the various cytochromes serve to bind oxygen, which they deliver to their substrates in such processes as cell respiration. Cytochrome P-450 gets its designation from the fact that in the reduced form it binds carbon monoxide and then absorbs light most intensely at a wavelength of 450 nanometers. The amplitude of the absorbance peak is the basis of quantitative studies of the enzyme. In the mixed-function oxidases cytochrome P-450 serves as the terminal oxidase: it accepts electrons passed along by several intermediates, binds oxygen and then delivers the oxygen to oxidize its substrate and (usually) produce water [see illustration on page 68]. In addition to NADPH the system includes the enzyme cytochrome P-450 reductase. Another heme protein, cytochrome b_5, is present in liver microsomes and may participate in drug oxidations, but its precise function is not clear.

Nor is it known just how the various components of the mixed-function oxidase complex are arrayed in the tubular membranes of the endoplasmic reticulum. It appears that the reductase and cytochrome b_5 are on the outside of the membranes, with cytochrome P-450 in the deeper layers. The enzyme glucuronyl transferase, which catalyzes the most important conjugation reaction, is also present within the membranes. The drug or other fat-soluble compound to be transformed is presumably bound and

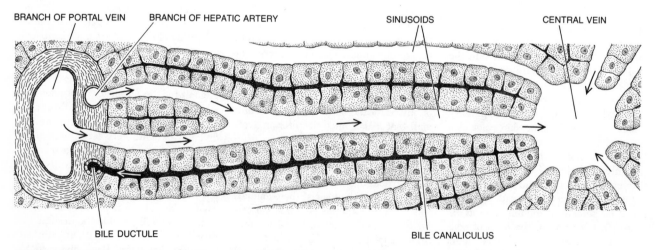

BRANCH OF PORTAL VEIN   BRANCH OF HEPATIC ARTERY   SINUSOIDS   CENTRAL VEIN

BILE DUCTULE   BILE CANALICULUS

LIVER LOBULE, the functional unit of liver tissue, is defined by branches of the portal vein, the hepatic artery, the bile duct and lymphatic vessels, which run together through the tissue, outlining the lobules. (The lymphatics are not shown here.) The venous blood, with its nutrients and other ingested substances, and the oxygenated arterial blood enter the fine sinusoids and perfuse the liver cells, which carry out the metabolic functions discussed in the text. The blood drains into a central vein in each lobule and thence to sublobular and hepatic veins. The liver cells also secrete bile, which is collected by bile canaliculi that feed into bile ductules and eventually into the main bile duct. This two-dimensional view of part of a lobule is highly simplified and diagrammatic.

metabolized by the mixed-function oxidase system that contains cytochrome P-450 and then is often converted into a highly water-soluble compound by conjugation with glucuronic acid; the transformed product passes into the lumen, or central channel, of the tubular membranes and is excreted from the cell into the bile or the bloodstream.

The cytochrome P-450 mixed-function oxidase system has come to be recognized as having a central role in the body's defense against chemical agents, whether they are normal body constituents or are introduced from the environment. It is now clear that the system is responsible for the detoxification of many of the potentially harmful environmental pollutants. The system is highly inducible, that is, its activity can be greatly increased by exposure to a wide variety of environmental agents and drugs that act as substrates for the system. Such chemicals stimulate the synthesis of cytochrome P-450 and other components of the complex. More than 200 steroid hormones, drugs, insecticides, carcinogens and other foreign chemicals are now known to stimulate drug metabolism in experimental animals, and many of them have been shown to do the same thing in man. On the other hand, some substances (such as lead and other heavy metals) inhibit the mixed-function oxidase system, although they are fewer in number and less diverse than the inducers. The P-450 system is also the site of much competitive interaction among drugs and other chemicals that are undergoing transformation.

The consequences of this inducibility, inhibition and competition have important implications in drug therapy. Patients are often given several drugs at the same time. Certain combinations can have unpredictable and often undesirable effects if one drug inhibits or stimulates the metabolism of another or competes with it. For example, phenylbutazone, the coumarin anticoagulants and chloramphenicol compete with the metabolic inactivation of tolbutamide, a drug given to reduce the blood-sugar level in diabetics; the competition can lead to excessive tolbutamide activity and thus to serious hypoglycemia, or low blood sugar.

Phenobarbital is a prime example of a drug that has a different effect: enzyme induction. When rats are treated with phenobarbital, there can be a three- to fourfold increase in the microsomal content of cytochrome P-450 and a two-

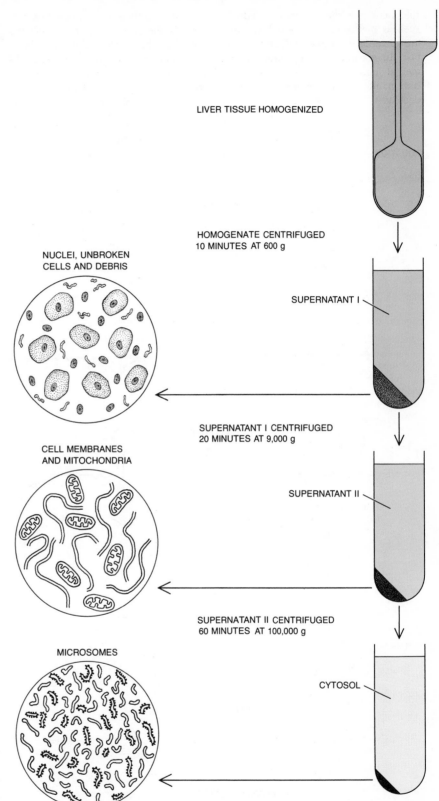

LIVER MICROSOMES, the membrane structures that contain most of the liver enzymes engaged in detoxification, are isolated by spinning homogenized liver tissue at successively higher speeds, which produce successively higher gravity (g) forces, in a centrifuge. Homogenized liver tissue is centrifuged for 10 minutes at 600 g; a pellet of dense material, primarily whole cells, cell debris and cell nuclei, collects at the bottom of the tube. The supernatant, or liquid portion, is centrifuged more strongly, isolating less dense structures such as mitochondria and pieces of membrane. When supernatant II is spun at very high g forces in an ultracentrifuge, the microsomes are separated from the cytosol, or cell fluid.

fold increase in the reductase, and drugs such as methadone and ethylmorphine are metabolized three or four times as fast. In accordance with these experimental observations chronic administration of phenobarbital to patients decreases the effects of many drugs by hastening their inactivation. Sedative doses of the drug reduce the concentration in the blood plasma of phenylbutazone, the analgesic antipyrine and the coumarin anticoagulants; it also decreases their pharmacological actions. Anticoagulant therapy is particularly sensitive to dosage. A patient who is satisfactorily maintained on a given coumarin-drug dosage while he is being given phenobarbital as a sedative may hemorrhage when the phenobarbital is withdrawn because the *P*-450 system is no longer being induced and the anticoagu-

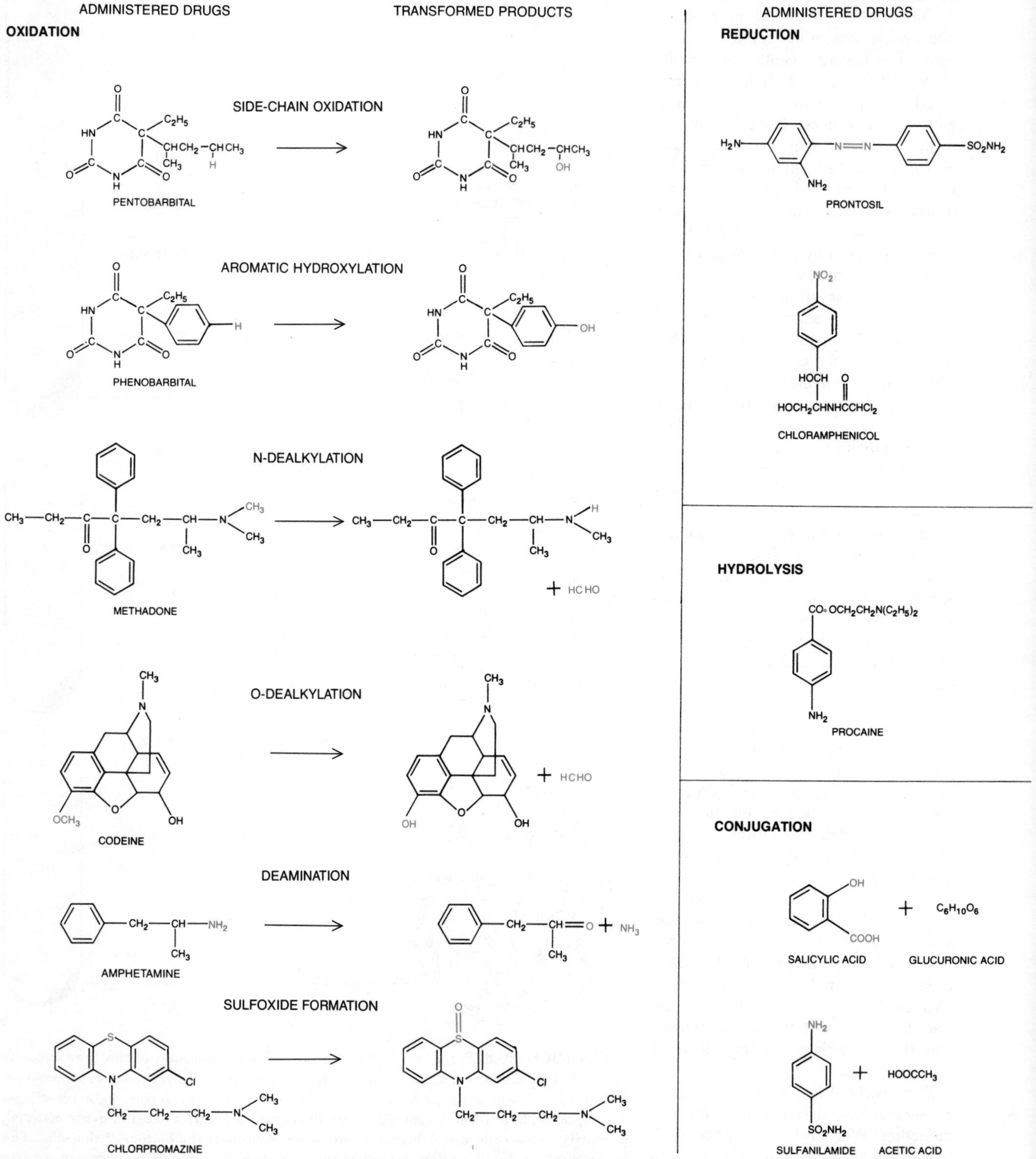

ADMINISTERED DRUGS

OXIDATION

TRANSFORMED PRODUCTS

SIDE-CHAIN OXIDATION

PENTOBARBITAL

AROMATIC HYDROXYLATION

PHENOBARBITAL

N-DEALKYLATION

METHADONE

O-DEALKYLATION

CODEINE

DEAMINATION

AMPHETAMINE

SULFOXIDE FORMATION

CHLORPROMAZINE

ADMINISTERED DRUGS

REDUCTION

PRONTOSIL

CHLORAMPHENICOL

HYDROLYSIS

PROCAINE

CONJUGATION

SALICYLIC ACID   GLUCURONIC ACID

SULFANILAMIDE   ACETIC ACID

BIOTRANSFORMATIONS carried out in the liver include oxidation, reduction, hydrolysis and conjugation, examples of which are given here. The molecular sites of each reaction are indicated in color. In a few cases transformation converts an inactive form of a drug (such as Prontosil) into an active form (sulfanilamide), but most of the reactions lead to inactivation. Conjugation with a nat-

lant is therefore being inactivated more slowly.

Alcohol is converted into acetaldehyde largely in the liver, perhaps to some extent by the mixed-function oxidase system: heavy drinkers are found to have an increased concentration of

TRANSFORMED PRODUCTS

AZO
REDUCTION

SULFANILAMIDE

NITRO
REDUCTION

COOH

NHCOCH₃

ural substance such as glucuronic acid often follows metabolism of drugs and facilitates excretion. Not all the by-products are shown.

cytochrome P-450. The habitual consumption of alcohol therefore stimulates the metabolism of a wide variety of drugs. This helps to explain why heavy drinkers are less affected than other people by barbiturates and other sedatives—when they are sober. A single very large dose of alcohol taken together with another drug, on the other hand, inhibits the drug's metabolism, presumably by competing with the drug for the appropriate enzymes. This effect, in addition to the depressant effect of alcohol on the central nervous system, helps to explain the enhanced sensitivity to barbiturates and other sedatives of a person who has been drinking heavily. The synergistic actions of alcohol and sedatives in the brain can cause death.

The inducibility of microsomal enzymes by drugs suggests a form of therapy for certain conditions in which normal body constituents ordinarily metabolized by such enzymes are present in excessive amounts. Long-term administration of phenobarbital, for example, can lower the concentration of bilirubin, the pigment that produces jaundice, in the blood of patients with chronic obstruction of bile flow in the liver. The excessive bilirubin levels that are normally observed in infants after birth can also be markedly reduced if the mother is given a small dose of phenobarbital for a number of days before delivery. Presumably the drug crosses the placenta and stimulates the conjugating enzyme system that is ordinarily slow to develop in the fetus and the newborn infant.

Halogenated hydrocarbon insecticides such as DDT are potent stimulators of drug and steroid sex-hormone metabolism in mammals and in birds; the breakdown of sex hormones explains in part the devastating effects of DDT on reproduction in some bird populations. The minimum exposure to DDT that will stimulate the metabolism of pentobarbital and decrease its hypnotic action in experimental animals is one that results in concentrations of from 10 to 15 micrograms of DDT per gram of fat; that is a level commonly found in human fat tissues. Among the other insecticides that induce microsomal enzymes in experimental animals are chlordane, aldrin and dieldrin. It is interesting that piperonyl butoxide, a synergist that was added to insecticides to inhibit the enzymatic defenses insects had developed against DDT and its chemical relatives, also inhibits the activity of the microsomal enzymes in the mammalian liver.

The polychlorinated biphenyls (PCB's)

constitute another class of environmental pollutants that have been shown to induce microsomal enzymes. The PCB's are lubricants, heat-exchange fluids, insulators, plasticizers for paints and plastic compounds and a major component of the lens-immersion oil used in microscopy. Whereas some of the consumer-product applications have recently been curtailed, the immersion oils are still handled daily by many laboratory workers. PCB's have been found in the tissues of numerous bird and fish species and in human fat and milk, although the route of entry into the human body has not been accurately determined. In recent experiments we have been able to show that the application of pure PCB's or of microscope immersion oil to the skin of experimental animals in very small amounts (one microliter) causes a marked increase in mixed-function oxidase activity and reduces the pharmacological effect of zoxazolamine, a muscle relaxant, and of hexobarbital in the live animal. These findings suggest that trivial skin exposure to chemicals can have significant and perhaps harmful biological effects in man.

A number of chemicals to which human beings are regularly exposed have been identified as chemical carcinogens, that is, they cause cancers when they are applied to the skin of experimental animals or otherwise administered to them. Clearly a factor that inhibits or stimulates the metabolism of such compounds may affect the development of human cancers. Benzpyrene, benzanthracene and similar polycyclic aromatic hydrocarbons are among the most ubiquitous carcinogens: they are present in tobacco smoke, in polluted city air and in charcoal-broiled and smoked foods. The polycyclic hydrocarbons are metabolized by a mixed-function oxidase enzyme that in this case is called aryl hydrocarbon hydroxylase because of the particular oxidation it catalyzes. Like other microsomal oxidations, this one requires NADPH and molecular oxygen, but the terminal oxidase of the system induced by the polycyclic hydrocarbons is somewhat different from the oxidase induced by drugs. The catalytic properties of the cytochrome are changed, as are its spectral properties: the absorbance maximum of the complex of carbon monoxide with the reduced cytochrome is at 448 nanometers rather than at 450, and so the enzyme is designated cytochrome P-448. Apart from the polycyclic hydrocarbons, only one other class of compounds has so far been noted to induce the formation of cytochrome P-448: we have found that the PCB's induce some of the newly identified cytochrome along with cytochrome *P-450*. This suggests the possibility, for which there is developing evidence in animals, that the PCB's too may have carcinogenic properties.

The aryl hydrocarbon hydroxylase system has been the subject of intensive investigation as a possible link in the causation of some cancers. The reason is that rather than detoxifying its polycyclic-hydrocarbon substrates it seems to make some of them more toxic: intermediates of polycyclic-hydrocarbon metabolism such as epoxides are more active in the malignant transformation of tissue-cultured cells than the parent products are. Moreover, the enzyme system for carcinogen metabolism is available at many sites in the body that are exposed to polycyclic hydrocarbons and is induced not only in the liver but also in the gastrointestinal tract, the kidneys, the skin and the lungs. Cigarette smoking markedly induces aryl hydrocarbon hydroxylase activity even in the human

placenta, as Allan H. Conney, Richard M. Welch and their colleagues showed at the Wellcome Research Laboratories. Little or no such enzyme activity was found in placentas from nonsmokers.

We have investigated the cytochrome *P-448* system in human skin and have found that there is marked variability in the hydroxylase activity of different skin samples and that incubation of skin in tissue culture with the polycyclic hydrocarbon benzanthracene induces more enzyme activity. A skin biopsy is easy to do, so that assaying skin aryl hydroxylase in the presence or absence of polycyclic hydrocarbons may provide a convenient test of individual differences in the capacity to metabolize certain environmental carcinogens. The possibility that induction of aryl hydrocarbon hydroxylase is of considerable significance in chemical carcinogenesis is suggested by the results of some recent experiments performed by Gottfried Kellermann and his associates at the University of Texas. They found that the inducibility of such hydroxylase activity in lymphocytes is significantly greater in cigarette smokers who have lung cancer than it is in healthy nonsmokers.

It is clear that the liver's mixed-function oxidase system is implicated in a number of processes affecting human health. The ability of a drug or other foreign substance to stimulate the metabolism of another drug by the system may explain some of the adverse drug reactions observed in clinical practice. Drug interactions have been well documented in anticoagulant therapy; more research is required to explore other drug interactions, since most patients are given several drugs at the same time. The ability of environmental pollutants to modify drug action is now under active investigation. It is clear that insecticides, for example, stimulate drug-metabolizing enzymes and that heavy-metal substances such as lead and methyl mercury inhibit the enzymes; the clinical significance of these effects may be appreciable in populations that are exposed occupationally to such agents. Finally, recently acquired evidence seems to indicate that the induction of aryl hydrocarbon hydroxylase in human tissues by polycyclic hydrocarbons may play a significant role in chemical carcinogenesis. Research on this biological action of carcinogens may provide important leads toward predicting the susceptibility of individuals to certain kinds of chemically induced cancer.

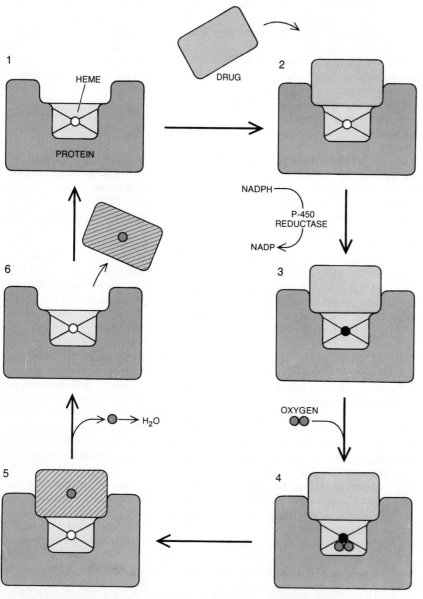

OXIDATION OF A DRUG by the enzyme cytochrome *P*-450 is visualized here as a sequential process. The enzyme (*1*) is a complex of a protein and the oxygen-binding compound heme, which contains an iron atom that is initially in the ferric (Fe^{+++}) form (*open circle*). The cytochrome binds the drug (*2*). Then (*3*) the enzyme cytochrome *P*-450 reductase, utilizing the coenzyme NADPH, reduces the iron of the heme to the ferrous (Fe^{++}) form (*black dot*), in which it can bind a molecule of oxygen (*4*). It supplies one atom of oxygen to oxidize the drug and one generally to form water, in the process reverting to its oxidized form (*5*). The drug, oxidized and in most cases inactive, is thereupon released (*6*).

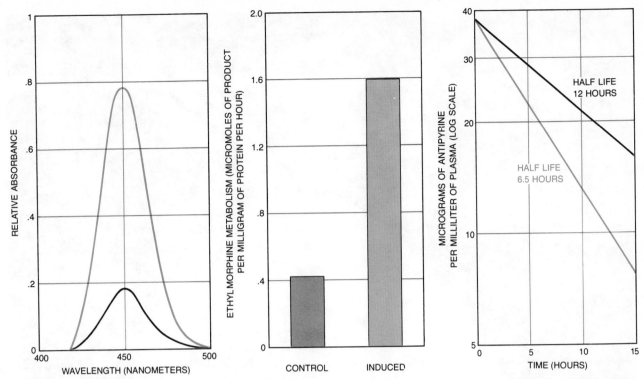

PHENOBARBITAL induces cytochrome *P*-450 and thus stimulates the metabolism of drugs oxidized by that enzyme. The curves (*left*) show the light-absorbance spectrum, with a peak at 450 nanometers, characteristic of reduced cytochrome *P*-450 complexed with carbon monoxide, the usual means of identifying and quantifying the enzyme. In a rat treated with phenobarbital the cytochrome *P*-450 content of microsomes (*colored curve*) is substantially higher than that of microsomes from an untreated rat (*black curve*). More of the test drug ethylmorphine is therefore metabolized (*center*) by liver tissue from a phenobarbital-treated rat (*colored bar*) than by the same amount of tissue from an untreated rat (*gray*). Similarly, phenobarbital speeds up metabolism of the drug antipyrine in man (*right*): antipyrine level in plasma falls faster after phenobarbital treatment (*colored curve*) than ordinarily (*black*).

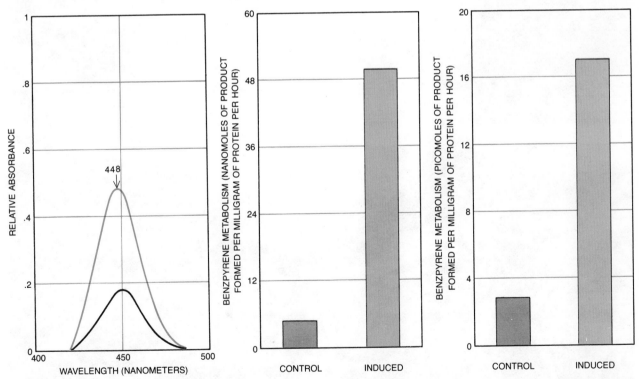

CARCINOGENS induce a different species of cytochrome, cytochrome *P*-448. The curves (*left*) compare the absorbance of microsomes from the liver of a rat treated with the carcinogen methylcholanthrene (*color*) with the absorbance of control microsomes (*black*); the induced cytochrome has a peak at 448 nanometers. Methylcholanthrene enhances metabolism of another carcinogen, benzpyrene (*center*): the microsomes of treated rats produce more metabolite (*color*) than control microsomes (*gray*) do. Similarly, treating human skin in tissue culture with the carcinogen benzanthracene stimulates skin-cell metabolism of benzpyrene (*right*).

CRUSHED EGG in the nest of a brown pelican off the California coast had such a thin shell that the weight of the nesting parent's body destroyed it. The concentration of DDE in the eggs of this 300-pair colony reached 2,500 parts per million; no eggs hatched.

7

Pesticides and the Reproduction of Birds

by David B. Peakall
April 1970

*High concentrations of chlorinated hydrocarbon
residues accumulate in such flesh-eaters as hawks and
pelicans. Among the results are upsets in normal
breeding behavior and eggs too fragile to survive*

The birds of prey have had an uneasy coexistence with man. Apart from the training of certain hawks for falconry and the veneration of the eagle as a symbol of fortitude, the predatory birds have been preyed on by the human species. In many parts of the world farmers, hunters and bird-lovers have waged unceasing warfare on the rapacious birds as pests, and egg collectors have further threatened their survival by raiding their nests for the beautifully pigmented eggs. Nevertheless, over the centuries the birds of prey on the whole survived well. The peregrine falcon, for example, is known to have maintained a remarkably stable population; records of aeries that have been occupied more or less continuously by peregrines go back in some cases to the Middle Ages.

About two decades ago, however, the peregrines in Europe and in North America suddenly suffered a crash in population. The peregrine is now rapidly vanishing in settled areas of the world, and in some places, particularly the eastern U.S., it is already extinct [*see illustration on page 72*]. The abrupt population fall of the peregrine (known in the U.S. as the duck hawk) has been paralleled by sharp declines of the bald eagle, the osprey and Cooper's hawk in the U.S. and of the golden eagle and the kestrel, or sparrow hawk, in Europe. The osprey, or fish hawk, has nearly disappeared from its haunts in southern New England and on Long Island; along the Connecticut River, where 150 pairs nested in 1952, only five pairs nested in 1969.

The population declines of all these raptorial birds are traceable not to the killing of adults but to a drastic drop in reproduction. It has been found that the reproduction failures follow much the same pattern among the various species: delayed breeding or failure to lay eggs altogether, a remarkable thinning of the shells and much breakage of the eggs that are laid, eating of broken eggs by the parents, failure to produce more eggs after earlier clutches were lost, and high mortality of the embryos and among fledglings.

Examination of the geographic patterns suggests a cause for the birds' reproductive failure. The regions of population decline coincide with areas where persistent pesticides—the chlorinated hydrocarbons such as DDT and dieldrin—are widely applied. Attrition of the predatory birds has been most severe in the eastern U.S. and in western Europe, where these pesticides first came into heavy use two decades ago. Analysis confirmed the suspicions about the pesticides: the birds were found to contain high levels of the chlorinated hydrocarbons. In areas such as northern Canada, Alaska and Spain, where the use of these chemicals has been comparatively light, the peregrine populations have remained normal or nearly normal. Recent studies show, however, that even in the relatively isolated North American arctic region the peregrines now have fairly high levels of chlorinated hydrocarbons and their populations apparently are beginning to decline.

The birds of prey are particularly vulnerable to the effects of a persistent pesticide such as DDT because they are the top of a food chain. As George M. Woodwell of the Brookhaven National Laboratory has shown, DDT accumulates to an increasingly high concentration in passing up a chain from predator to predator, and at the top of the chain it may be concentrated a thousandfold or more over the content in the original source [see "Toxic Substances and Ecological Cycles," by George M. Woodwell; SCIENTIFIC AMERICAN Offprint 1066]. The predatory birds, as carnivores, feed on birds that have fed in turn on insects and plants. Hence the birds of prey accumulate a higher dose of the persistent pesticides and are more likely to suffer the toxic effects than other birds.

The idea that the predatory birds' decline is due to an internal toxic effect, rather than to a change in their behavior or their habitat, has been verified by many experiments. One of the most interesting was a field test made by Paul Spitzer, now at Cornell University, working in cooperation with the Patuxent Wildlife Research Center in Maryland. He transferred eggs from nests of the failing osprey population in New England to nests of a successful population in the Chesapeake Bay area and placed the Chesapeake eggs in the New England nests. The Chesapeake eggs hatched as successfully in the New England nests as they would have at home with their own parents, whereas the New England eggs transferred to Chesapeake nests produced as few viable young as would have been expected if they had been incubated in their original nests in New England. The experiment thus indicated that the fate of the eggs was determined by an intrinsic factor in the egg itself.

The first clue to what was happening to the predatory birds' reproduction system came in the early 1960's when Derek Ratcliffe of the British Nature Conservancy, puzzled by the extraordinary number of broken eggs he found in peregrine nests, examined the shells of peregrine eggs that had been collected over a period of many years. He found that the eggs collected since the late 1940's show a sharp drop in thickness of the shell, averaging 19 percent. Similar findings were subsequently made on peregrine eggs in North America and on the eggs of other species of predatory birds whose populations were decreasing. It became apparent that something must be wrong with the birds' calcium metabolism and that the effects of the suspected pesticides would bear looking into.

Experiments were started in several laboratories. At the Patuxent Wildlife Research Center, Richard D. Porter and Stanley N. Wiemeyer, working with kestrels, found that a mixture of DDT and dieldrin in doses measured in a few parts per million brought about a significant decrease in the shell thickness of the birds' eggs. Robert G. Heath of the Patuxent center tested the effects of

DDE, the principal metabolic product of DDT, on mallard ducks. DDE is now a ubiquitous feature of the earth's environment; it is estimated that there are a billion pounds of the substance in the world ecosystem, and traces of it have been found in animals everywhere, from polar bears in the Arctic to seals in the Antarctic. Heath found that DDE caused the failure of mallard eggs in two ways: by increasing the fragility of the eggs, leading to increased breakage soon after laying, and by the death of the embryos in intact eggs toward the end of the period of incubation. James H. Enderson of Colorado College and his associate Daniel D. Berger, studying the eggs of prairie falcons in the Southwest

desert, established that the amount of thinning of the shells and the mortality rate for the embryos were related to the quantity of DDE in the egg. Enderson and Berger also found that when they fed starlings loaded with dieldrin to falcons, the falcons' eggs showed similar thinning.

The ultimate in thinness of birds' eggshells was discovered recently in colonies of the brown pelican off the California coast. The DDE content in the eggs of this wild population (as measured by Robert Risebrough of the University of California at Berkeley) ranged as high as 2,500 parts per million, and the eggshells were so thin that the eggs could not be picked up without denting the

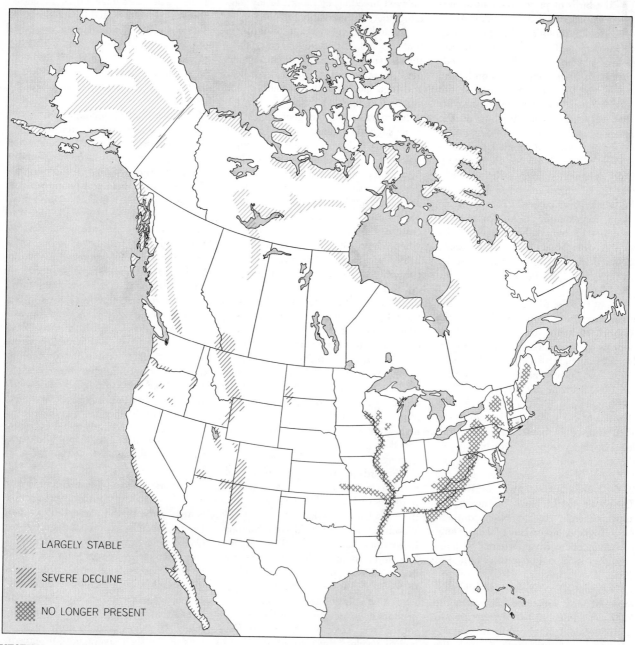

LARGELY STABLE

SEVERE DECLINE

NO LONGER PRESENT

NESTING AREAS of the peregrine falcon, or duck hawk, in the Northern Hemisphere of the New World are shown on this map.   Shades of color show the extent of interference with normal reproduction resulting from ingestion of pesticides by the birds.

shells [*see illustration on page 70*]. In a colony on the Anacapa Islands off the coast it was found that the 300 pairs of nesting pelicans had not produced a single viable egg. Their nests, visited shortly after the eggs were laid, contained many broken eggs.

Field studies and laboratory experiments suggest that the thinning of eggshells does not increase in direct proportion to the DDE dose. In fact, small doses can produce dramatic effects. A content of only 75 parts per million in the egg reduces the shell thickness by more than 20 percent; beyond that, as the dose increases the decrease in shell thickness is more gradual [*see illustration at right*]. In the case of the brown pelican very heavy doses may thin the shell to a mere film.

Studies of white pelicans and cormorants have implicated the polychlorinated biphenyls (PCB's), now widely used as plasticizers, as another threat to birds of prey. These compounds cause thinning of the eggshells, although not as effectively as DDT and its metabolites do. Preliminary laboratory studies show that PCB's are particularly effective, however, in delaying the onset of breeding. The PCB's are given off when plastic materials are burned, and they are widely distributed over the earth. They resemble DDT in molecular structure and produce similar physiological actions in animals.

Much interest has focused on the question of how the chlorinated pesticides produce their destructive effects in the predatory birds—a question that is of no small concern to man, who also is the top of a food chain. Oddly enough, the beginning of light on this question came about through an accidental discovery involving an animal totally unrelated to the birds: the laboratory rat. Larry G. Hart and James R. Fouts of the University of Iowa College of Medicine were investigating the effects of food deprivation on the metabolism of drugs in rats. The drug they were using was hexobarbital, and in one experiment they were startled to find that the rats' sleeping time after receiving a standard dose of the barbiturate was much shorter than it had been in previous tests. Reexamining the conditions of the experiment, they found that the only unusual factor was that the cages had been sprayed with chlordane to control bedbugs. Pursuit of this clue led to the finding that chlordane induced rat liver cells to synthesize enzymes that speeded up the metabolism of hexobarbital. The enzymes brought about hydroxylation of the barbiturate, thereby making it more soluble

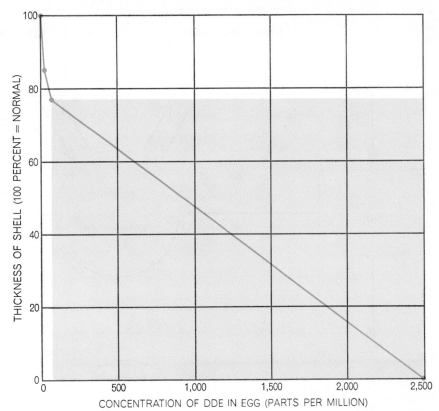

in water and hastening its excretion. Further experiments showed that these enzymes could hydroxylate a wide variety of substances, including the sex hormones: estrogen, testosterone and progesterone.

Because the investigators were interested primarily in drug research and their reports were published mainly in pharmacological journals, these discoveries did not come to the attention of workers studying the effects of pesticides on wildlife until several years later. I myself came on the published findings only incidentally in the course of preparing lectures for medical students. The fact that chlordane could change the balance of sex hormones in animals immediately suggested a possible explanation of the mechanism whereby the chlorinated pesticides inhibit reproduction in birds. It was capable of explaining their reproductive failure in general and the alteration of the calcium balance in the egg in particular.

My colleagues and I at Cornell University launched on a program of experiments designed to explore the interesting questions suggested by this

SEVERE EFFECT of the concentration of relatively small amounts of the persistent chlorinated hydrocarbon pesticides is evident in this graph. When the parent's concentration is enough to add as few as 25 parts per million of pesticide to the egg, the shell becomes 15 percent thinner than normal. Soon after the shells become more than 20 percent thinner than normal (*area of light color*) eggs are usually not found in nests because of breakage.

new aspect of the problem. To explain them I must briefly outline the complex chain of physiological events that characterizes breeding by birds. The cycle is initiated by a seasonal or climatic stimulus: the lengthening of daylight in spring in the northern Temperate Zone or rainfall in the arid and tropical regions. These signals cause an increase in the production of hormones in the nerve cells of the medial eminence of the bird's brain. The bloodstream carries these hormones to the anterior pituitary gland, which in turn dispatches to the gonads (the testes or ovaries) hormones that stimulate these organs to produce the sex hormones. The sex hormones not only generate physical changes in the reproductive organs and evoke breeding behavior but also promote the storage of a supply of calcium for the eggs.

Let us look first into the question of how a pesticide may affect the calcium supply. We carried out our experiments on the rather small Asian pigeon known as the ringdove, so that I shall describe the situation in this bird. The female forms the shell of the egg in the uterus within a period of 20 hours, and she needs 240 milligrams of calcium to pro-

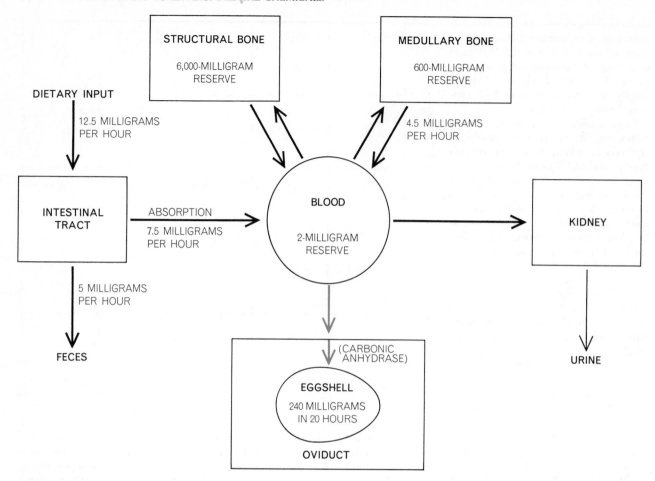

CALCIUM FOR EGGSHELL, which is formed around each egg in the last 20 hours before laying, is drawn in part from the bird's food supply and in part from calcium reserves in the bird's bones. The key to shell formation, however, is the enzyme carbonic an- hydrase, which makes the supply of calcium carried in the ring- dove's bloodstream available to the bird's oviduct at a rate of 12 milligrams per hour. When laying ringdoves are injected with DDE, the action of the enzyme is severely inhibited, causing thin shells.

duce a shell of normal thickness. Since the calcium content of the circulating blood, even at the time of ovulation, is only two milligrams (barely a 10-minute supply), the bird must draw on other sources to meet the demand. About 60 percent of the demand is supplied by the bird's food intake; the rest is provided by a store of calcium in the marrow of the bones [see "How an Eggshell Is Made," by T. G. Taylor; SCIENTIFIC AMERICAN Offprint 1171]. This calcium reserve is laid down in the bone cavities early in the breeding cycle, and the amount of the deposit is controlled by the levels of estrogen in the blood and tissues. Obviously, therefore, a defi- ciency of estrogen will reduce the bird's calcium reserve. It seemed unlikely, however, that the reduction of this re- serve alone could account for the drastic shell-thinning observed in eggs loaded with pesticides. If the *supply* of calcium were the sole problem, the birds could augment the supply by drawing on the calcium embodied in the skeleton; fur- thermore, birds on a very low calcium

diet have been found to cease egg-laying rather than laying eggs with abnormally thin shells. Was it possible, then, that the thinness of the eggshells was due less to the deficiency in supply than to a failure in delivery of calcium to the shell?

In our experiments we bred pairs of ringdoves in cages and delayed feed- ing the birds a pesticide until after they had completed at least one successful breeding cycle, thereby demonstrating their natural capability. For the experi- ment we separated the members of each pair, isolated them in individual cages where they had an eight-hour day in- stead of their normal 16-hour day and fed them a standard dose of DDT in their food. After three weeks we gave each bird an oral dose of radioactive cal- cium and returned the birds to cages with their original partners for pairing under long-day conditions. A number of days later we examined the birds, some before they laid eggs, others immediate- ly after they finished laying their clutch. In both cases the birds showed a consid-

erable rise of enzyme activity in the liver. A substantially lower level of es- trogen was found in the bloodstream of the birds that had not yet laid eggs. After the eggs had been laid, low estrogen levels were found in both experimental and control birds; this was to be ex- pected because the level of estrogen falls at the time of egg-laying. We found that less labeled calcium was stored in the bone marrow of the experimental birds than in the marrow of control birds that had not been fed the pesticide.

Eggs laid by the pesticide-treated birds were notably thin-shelled, as was to be expected. We proceeded to experi- ments designed to determine whether this was due simply to the shortage of stored calcium or to something that pre- vented calcium from reaching the shell. In order to resolve this question we re- sorted to the tactic of injecting pesticides into females within a period of hours before they laid their eggs. In that short interval there would not be time for any significant change in the supply of cal- cium by way of an alteration of the estrogen levels through the activity of liver enzymes; consequently if the pesticide produced an effect, it would be not on the stored supply but on the delivery of calcium to the eggshell, which as we have noted is laid down within 20 hours of the laying of the egg. And with regard to delivery it was known that an enzyme, carbonic anhydrase, plays an important role in making calcium available to the eggshell in the oviduct. One could therefore look for a possible effect on the activity of this enzyme.

We tried two chlorinated hydrocarbons: dieldrin and DDE. Dieldrin, when injected into a ringdove shortly before it laid its egg, did not produce any significant thinning of the eggshell or inhibit the activity of carbonic anhydrase in the oviduct. DDE, on the other hand, severely depressed the activity of the enzyme and brought about a marked decrease in the thickness of the eggshell.

Our experiments with ringdoves also showed that the chlorinated hydrocarbons cause a significant delay in breeding by birds. Females that were fed pesticides did not lay eggs until 21.5 days (on the average) and sometimes as long as 25 days after pairing, whereas the normal interval, as indicated by the control birds, is 16.5 days on the average. The delay evidently was caused by the depression of the estrogen level resulting from the induction of liver enzymes by the pesticide. It turned out that dieldrin and the polychlorinated biphenyls were more powerful inducers of these enzymes than DDT was.

Delayed breeding is another factor in the predatory birds' population decline. Most birds do their breeding in the season when food is most plentiful, thus giving their young an optimal chance for survival. An artificial delay in their breeding consequently reduces the chances for reproductive success, and it is most serious for large birds, with their long egg-incubation period and the slower growth of the fledglings to maturity. It was found that the now extinct peregrine colonies along the Hudson River, the declining cormorant rookery at Lake DuBay in Wisconsin and the failing pelican colonies in California were all notably late in breeding.

From this point of view it appears that dieldrin and the PCB's are greater threats to the predatory birds than DDT. Certain field and laboratory studies tend to bear out that deduction. Derek Ratcliffe and J. D. Lockie, in long-term observation of the nests of golden eagles in Scotland, found that although abnormal eggshell breakage began in 1952, about the time that DDT was introduced,

marked decline in the breeding success of these birds did not begin until 1960, after the introduction of dieldrin. In laboratory experiments on the bobwhite quail James B. DeWitt and John L. George of the U.S. Fish and Wildlife Service found that one part per million of dieldrin was effective in reducing the success in hatching and survival of chicks, whereas it took 200 parts per million of DDT to produce the same effect. Robert Heath found in his studies of mallard ducks, however, that DDE severely impaired reproductive success at doses as low as 10 parts per million. Thus there appears to be a considerable difference in the effect of DDT and its metabolites on different species of birds.

We come to the following conclusions concerning the physiological mechanisms responsible for the various harmful effects on bird breeding that are brought about by the persistent insecticides. Abnormally late breeding and the failure of birds to lay eggs after their early clutches have been lost can be explained in terms of the induction of liver enzymes that lower the estrogen levels in the birds. The failure, or apparent failure, of birds to lay any eggs at all may be due either to depression of the estrogen level or to the circumstance that the eggs were broken and eaten by the parents shortly after they were laid, so that observers found no eggs in the nest on visiting them. The reduction in clutch size may also be accounted for by early breakage and eating of some of the eggs, as this has been noted mainly in cases where the nests were not checked frequently. The thinning of eggshells and breakage of the eggs evidently is due largely to the inhibition of carbonic anhydrase by DDT and its metabolites. We are left with some phenomena that are still unexplained. Why does a low dose of pesticide produce relatively more thinning of the eggshell than larger doses do? What is the mechanism that kills embryos in the shell? These questions need further investigation.

The effect of the pesticides in disturb-

BREEDING SUCCESS in birds involves the five sequential responses to external stimuli shown in the illustration at right. Breeding failures, due to late breeding or an inability to lay more eggs after earlier clutches are destroyed, result from the action of pesticides on the fifth response. They stimulate the activity of enzymes in the breeding bird's liver; the enzymes cut the amount of estrogen in the system below the level that is needed for normal sexual behavior.

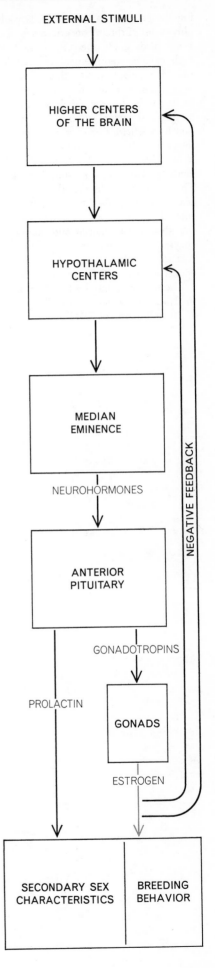

EXTERNAL STIMULI

HIGHER CENTERS OF THE BRAIN

HYPOTHALAMIC CENTERS

MEDIAN EMINENCE

NEUROHORMONES

ANTERIOR PITUITARY

GONADOTROPINS

PROLACTIN

GONADS

ESTROGEN

SECONDARY SEX CHARACTERISTICS   BREEDING BEHAVIOR

NEGATIVE FEEDBACK

ing the calcium balance of birds probably is not of direct concern to man, because birds are a special case in their high calcium requirement at breeding time. It seems, however, that we should be concerned about the pesticides' effects on the hormone balance and on other physiological systems. The induction by pesticides of liver enzymes that lower the estrogen levels has been found in a wide variety of vertebrates, including a primate, the squirrel monkey. There is little doubt that this effect applies to man as well. Moreover, the chlorinated hydrocarbons are known to alter the glucose metabolism and inhibit an enzyme (adenosine triphosphatase, or ATPase) that plays a vital role in the energy economy of the human body.

The recent finding by investigators at the National Cancer Institute that a dose of 46 milligrams of DDT per kilogram of body weight can produce a fourfold increase in tumors of the liver, lungs and lymphoid organs of animals indicates that DDT should be banned for that reason alone. Human cancer victims have been found to have two to two and a half times more DDT in their fat than occurs in the normal population. Investigators in the U.S.S.R. recently reported that DDD, another metabolite of DDT, reduces the islets of Langerhans, the site of insulin synthesis.

The peregrine population crash has prompted two international conferences of concerned investigators, in 1965 and again in 1969. It is encouraging to note that in Britain, where severe restrictions were imposed in 1964 on the use of chlorinated hydrocarbon pesticides, the peregrine population has increased in the past two years. The Canadian government recently announced licensing restrictions that are expected to reduce the use of these pesticides by 90 percent, and many states in the U.S. are also instituting or considering such restrictions. Environmental problems do not respect political boundaries, and in the long run it will do little good if restrictions on the use of these hazardous toxins are applied only to certain regions or parts of the globe.

The long-term effects of the chlorinated hydrocarbons in the environment on human beings are admittedly much more difficult to detect or assess than the spectacular effects that have been seen in the predatory birds. Still, the story told by the birds is alarming enough. It seems obvious that agents capable of causing profound metabolic changes in such small doses should not be broadcast through the ecosystem on a billion-pound scale.

Lead Poisoning

by J. Julian Chisolm, Jr.
February 1971

*Among the natural substances that man concentrates
in his immediate environment, lead is one of the most
ubiquitous. A principal cause for concern is the effect on
children who live in decaying buildings*

Lead has been mined and worked by men for millenniums. Its ductility, high resistance to erosion and other properties make it one of the most useful of metals. The inappropriate use of lead has, however, resulted in outbreaks of lead poisoning in humans from time to time since antiquity. The disease, which is sometimes called "plumbism" (from the Latin word for lead) or "saturnism" (from the alchemical term), was first described by the Greek poet-physician Nicander more than 2,000 years ago. Today our concerns about human health and the dissemination of lead into the environment are twofold: (1) there is a need to know whether or not the current level of lead absorption in the general population presents some subtle risk to health; (2) there is an even more urgent need to control this hazard in the several subgroups within the general population that run the risk of clinical plumbism and its known consequences. In the young children of urban slums lead poisoning is a major source of brain damage, mental deficiency and serious behavior problems. Yet it remains an insidious disease: it is difficult to diagnose, it is often unrecognized and until recently it was largely ignored by physicians and public health officials. Now public attention is finally being focused on childhood lead poisoning, although the difficult task of eradicating it has just begun.

Symptomatic lead poisoning is the result of very high levels of lead in the tissues. Is it possible that a content of lead in the body that is insufficient to cause obvious symptoms can nevertheless give rise to slowly evolving and long-lasting adverse effects? The question is at present unanswered but is most pertinent. There is much evidence that lead wastes have been accumulating during the past century, particularly in congested urban areas. Increased exposure to lead has been shown in populations exposed to lead as an air pollutant. Postmortem examinations show a higher lead content in the organs of individuals in highly industrialized societies than in the organs of most individuals in primitive populations. Although no population group is apparently yet being subjected to levels of exposure associated with the symptoms of lead poisoning, it is clear that a continued rise in the pollution of the human environment with lead could eventually produce levels of exposure that could have adverse effects on human health. Efforts to control the dissemination of lead into the environment are therefore indicated.

The more immediate and urgent problem is to control the exposure to lead of well-defined groups that are known to be directly at risk: young children who live in dilapidated housing where they can nibble chips of leaded paint, whiskey drinkers who consume quantities of lead-contaminated moonshine, people who eat or drink from improperly lead-glazed earthenware, workers in certain small-scale industries where exposure to lead is not controlled. Of these the most distressing group is the large group of children between about one and three to five years of age who live in deteriorating buildings and have the habit of eating nonfood substances including peeling paint, plaster and putty containing lead. (This behavior is termed pica, after the Latin word for magpie.) The epidemiological data are still scanty: large-scale screening programs now in progress in Chicago and New York City indicate that between 5 and 10 percent of the children tested show evidence of asymptomatic increased lead absorption and that between 1 and 2 percent have unsuspected plumbism. Small-scale surveys in the worst housing areas of a few other cities reveal even higher percentages.

There is little doubt that childhood lead poisoning is a real problem in many of the older urban areas of the U.S. and perhaps in rural communities as well. Current knowledge about lead poisoning and its long-term effects in children is adequate to form the basis of a rational attack on this particular problem. The ubiquity of lead-pigment paints in older substandard housing and the prevalence of pica in young children indicate, however, that any effective program will require the concerted and sustained effort of each community. Furthermore, the continued use of lead-pigment paints on housing surfaces that are accessible to young children and will at some future date fall into disrepair can only perpetuate the problem.

Traces of lead are almost ubiquitous in nature and minute amounts are found in normal diets. According to the extensive studies of Robert A. Kehoe and his associates during the past 35 years at the Kettering Laboratories of the University of Cincinnati, the usual daily dietary intake of lead in adults averages about .3 milligram. Of this, about 90 percent passes through the intestinal tract and is not absorbed. Kehoe's data indicate that the small amount absorbed is also excreted, so that under "normal" conditions there is no net retention of lead in the body. In addition the usual respiratory intake is estimated at between five and 50 micrograms of lead per day. These findings must be reconciled with postmortem analyses, which indicate that the concentration of lead in bone increases with age, although its concentration in the soft tissues is relatively stable throughout life. The physiological significance of increasing storage in bone is not entirely clear, but it has caused considerable concern. It is quite clear that as the level of intake of lead increases, the rate of absorption may exceed the rate at which lead can be excreted or stored in bone. And when the rates of excretion and storage are exceeded, the levels of lead in the soft tissues rise. Studies in adults indicate that as the sustained daily intake of lead rises above one milligram of lead per day, higher levels of lead in the blood result and metabolic, functional and clinical responses follow [*see illustration on pages 84 and 85*]. The reversible effects abate when the rate and amount of lead absorbed are reduced again to the usual dietary range.

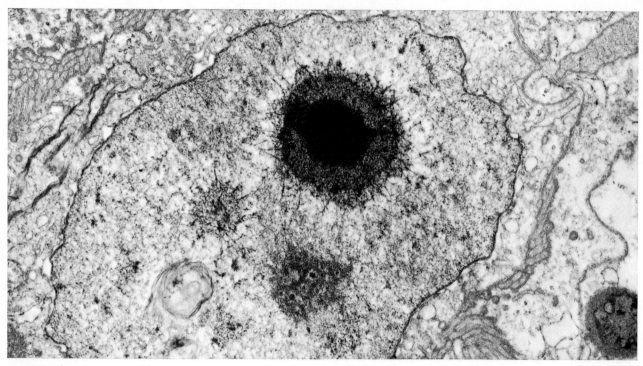

EXCESS LEAD complexed with protein forms inclusion bodies in the nuclei of certain cells in lead-poisoned animals and man. In an electron micrograph made by Robert A. Goyer and his colleagues at the University of North Carolina School of Medicine the nucleus of a cell from a proximal renal tubule of a lead-poisoned rat is enlarged 15,000 diameters. The large structure with a dense core and a filamentous outer zone is an inclusion body; below it to the left is a smaller one. The dark area below the large body is the nucleolus.

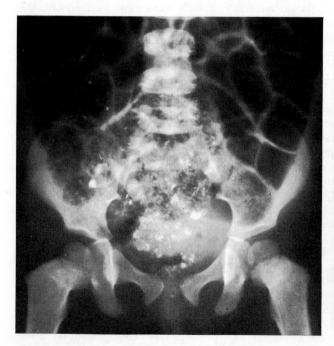 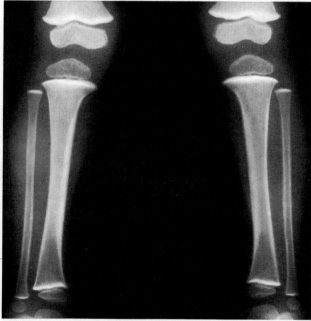

X-RAY PLATES may show evidence of lead ingestion or of an excessive body burden of the metal. The abdominal X ray (*left*) shows a number of bright opaque particles in the large intestine: bits of lead-containing paint that had been eaten by the 18-month-old subject. The X ray of the same child's legs (*right*) shows bright "lead lines": excess lead stored at the ends of the long bones.

As far as is known, lead is not a trace element essential to nutrition, but this particular question has not been adequately examined. Some of the adverse effects of lead on metabolism have nonetheless been studied in considerable detail. These effects are related to the concentration of lead in the soft tissues. At the level of cellular metabolism, the best-known adverse effect of lead is its inhibition of the activity of enzymes that are dependent on the presence of free sulfhydryl (SH) groups for their activity. Lead interacts with sulfhydryl groups in such a way that they are not available to certain enzymes that require them. In the living organism, under most conditions, this inhibition is apparently partial. Inhibitory effects of lead on other aspects of cellular metabolism have been demonstrated in the test tube. Such studies are preliminary. Most of the effects reported are produced with concentrations of lead considerably higher than are likely to be encountered in the tissues of man, so that speculation about such effects is unwarranted at this point.

The clearest manifestation of the inhibitory effect of lead on the activity of sulfhydryl-dependent enzymes is the disturbance it causes in the biosynthesis of heme. Heme is the iron-containing constituent that combines with protein to form hemoglobin, the oxygen-carrying pigment of the red blood cells. Heme is also an essential constituent of the other respiratory pigments, the cytochromes, which play key roles in energy metabolism. The normal pathway of heme synthesis begins with activated succinate (produced by the Krebs cycle, a major stage in the conversion of food energy to biological energy) and proceeds through a series of steps [see illustration below]. Two of these steps are inhibited by the presence of lead; two others may also be inhibited, but at higher lead concentrations.

Lead is implicated specifically in the metabolism of delta-aminolevulinic acid (ALA) and in the final formation of heme from iron and protoporphyrin. Both of these steps are mediated by enzymes that are dependent on free sulfhydryl groups for their activity and are therefore sensitive to lead. The two steps at which lead may possibly be implicated are the formation of ALA and the conversion of coproporphyrinogen to protoporphyrin. Although the exact mechanism is not known, coproporphyrin (an oxidized product of coproporphyrinogen) accumulates in the urine and the red cells in lead poisoning. Whatever the mechanisms, the increased excretion of ALA and coproporphyrin is almost always observed before the onset of symptoms of lead poisoning, and the presence of either is therefore important in diagnosis.

The enzyme that catalyzes ALA metabolism is ALA dehydrase. A number of investigators, including Sven Hernberg and his colleagues at the University of Helsinki and Abraham Goldberg's group at the University of Glasgow, have studied the extent to which varying levels of lead in the blood inhibit ALA-dehydrase activity in red blood cell preparations in the laboratory. They have shown a direct relation between the concentration of lead in blood and the activity of the enzyme. Moreover, they find that there seems to be no amount of lead so small that it does not to some extent decrease ALA-dehydrase activity; in other words, there appears to be no threshold for this effect [see top illustration page 80]. If that is so, however, one would expect to see a progressive increase in the urinary ex-

PATIENTS	LEAD OUTPUT (MILLIGRAMS PER 24 HOURS)		
	MEAN	MEDIAN	RANGE
UNEXPOSED CONTROLS	.132	.157	.012—.175
HOUSEHOLD CONTROLS	.832	.651	.087—1.93
INCREASED LEAD ABSORPTION, NO SYMPTOMS	2.16	1.11	.116—9.60
LEAD POISONING, WITH AND WITHOUT BRAIN DAMAGE DURING EXPOSURE: AFTER TREATMENT:	44.0 .362	27.0 .240	5.040—104.0 .062—0.850

EXCRETION OF LEAD in feces is an index of exposure to lead. These results of a study by the author and Harold E. Harrison illustrate the massive exposures seen in lead poisoning. Unexposed controls were children with no known exposure to lead. The other groups were children with increased lead absorption (high blood lead), children with lead poisoning and members of their households with neither high blood values nor overt symptoms.

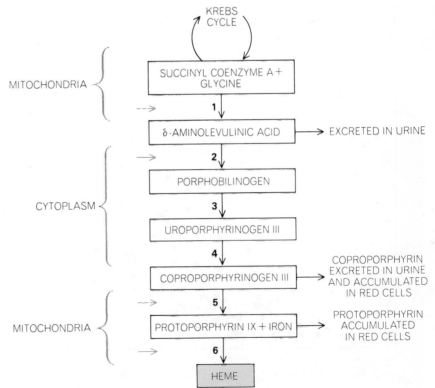

BIOSYNTHESIS OF HEME, a constituent of hemoglobin, is inhibited by lead, resulting in accumulation of intermediates in the synthetic pathway. Of six steps in the pathway, the first and the last two take place in mitochondria, the others elsewhere in the cell cytoplasm. Lead inhibits two steps (solid colored arrows) and may inhibit two others (broken arrows).

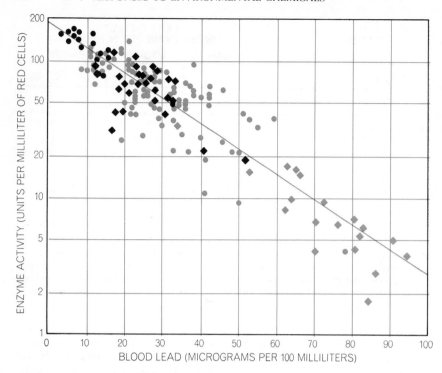

CORRELATION between blood lead and the activity of delta-aminolevulinic acid dehydrase, an enzyme inhibited by lead, was shown by Sven Hernberg and his colleagues at the University of Helsinki. The vertical scale is logarithmic. The values are well correlated, as indicated by the straight regression line, over a wide range of blood-lead levels in groups with different lead exposures: students (*black dots*), automobile repairmen (*black squares*), printshop employees (*colored dots*) and lead smelters and ship scrappers (*colored squares*).

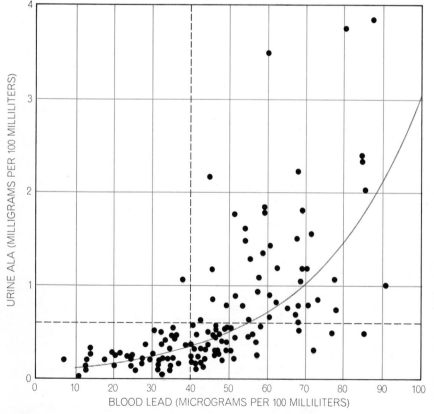

ENZYME SUBSTRATE, delta-aminolevulinic acid (ALA) accumulates in the urine when lead inhibits enzyme activity. Stig Selander and Kim Cramér found that a decrease in lead below about 40 micrograms does not produce a comparable decrease in ALA, suggesting that an enzyme reserve may be involved. Broken lines show presumed normal values.

cretion of the enzyme's substrate, ALA, beginning at very low blood-lead levels. This does not seem to be the case. Stig Selander and Kim Cramér in Sweden, correlating blood-lead and urine-ALA values, found that the first measurable increase in urine ALA is observed only after blood lead rises above approximately 30 micrograms of lead per 100 milliliters of whole blood [*see bottom illustration at left*]. The apparent inconsistency between the effect of lead on the activity of an enzyme in the test tube and the accumulation of the enzyme's substrate in the body might be explained by the presence of an enzyme reserve. This hypothesis is consistent with the functional reserve exhibited in many biological systems.

Almost all the information we have on the effect of lead on the synthesis of heme comes from observations of red blood cells. Yet all cells synthesize their own heme-containing enzymes, notably the cytochromes, and ALA dehydrase is also widely distributed in tissues. The observations in red blood cells may therefore serve as a model of lead's probable effects on heme synthesis in other organ systems. Even so, the degree of inhibition in a given tissue may vary and will depend on the concentration of lead within the cell, on its access to the heme synthetic pathway and on other factors. For example, J. A. Millar and his colleagues in Goldberg's group found that ALA-dehydrase activity is inhibited in the brain tissue of heavily lead-poisoned laboratory rats at about the same rate as it is in the blood [*see illustration on opposite page*]. When these workers used amounts of lead that produced an average blood-lead level of 30 micrograms per 100 milliliters of blood, the level of ALA-dehydrase activity in the brain did not differ significantly from the levels found in control rats that had not been given any added lead at all. It is now established experimentally that lead does interfere with heme synthesis in tissue preparations from the kidney, the brain and the liver as well as in red cells but the concentrations of lead that may begin to cause significant inhibition in these organs are not yet known.

Only in the blood is it as yet possible to see a direct cause-and-effect relation between the metabolic disturbance and the functional disturbance in animals or people. In the blood the functional effect is anemia. The decrease in heme synthesis leads at first to a decrease in the life-span of red cells and later to a decrease in the number of red cells and in the amount of hemoglobin per cell. In compensation for the shortage, the blood-forming tissue steps up its production of red cells; immature red cells, reticulocytes and basophilic stippled cells (named for their stippled appearance after absorbing a basic dye) appear in the circulation. The presence of stippled cells is the most characteristic finding in the blood of a patient with lead poisoning. The stippling represents remnants of the cytoplasmic constituents of red cell precursors, including mitochondria. Normal mature red cells do not contain mitochondria. The anemia of lead poisoning is a reversible condition: the metabolism of heme returns to normal, and the anemia improves with removal of the patient from exposure to excessive amounts of lead.

The toxic effect of lead on the kidneys is under intensive investigation but here the story is less clear. In acute lead poisoning there are visible changes in the kidney and kidney function is impaired. Again the mitochondria are implicated: their structure is visibly changed. Much of the excess lead is concentrated in the form of dense inclusions in the nuclei of certain cells, including those lining the proximal renal tubules. Robert A. Goyer of the University of North Carolina School of Medicine isolated and analyzed these inclusions and found that they consist of a complex of protein and lead [*see upper illustration on page 78*]. He has suggested that the inclusions are a protective device: they tend to keep the lead in the nucleus, away from the vulnerable mitochondria. Involvement of the mitochondria is also suggested by the fact that lead-poisoned kidney cells consume more oxygen than normal cells in laboratory cultures, which indicates that their energy metabolism is affected.

Kidney dysfunction, apparently due to this impairment in energy metabolism, is expressed in what is called the Fanconi syndrome: there is an increased loss of amino acids, glucose and phosphate in the urine because the damaged tubular cells fail to reabsorb these substances as completely as normal tubular cells do. The excessive excretion of phosphates is the important factor because it leads to hypophosphatemia, a low level of phosphate in the blood. There is some evidence that, when phosphate is mobilized from bone for the purpose of maintaining an adequate level in body fluids, lead that is stored with relative safety in the bones may be mobilized along with the phosphate and enter the soft tissues where it can do harm. The effect of acute lead poisoning on the kidney can be serious but, like the effect on blood cells, it is reversible with the end of abnormal exposure. Furthermore,

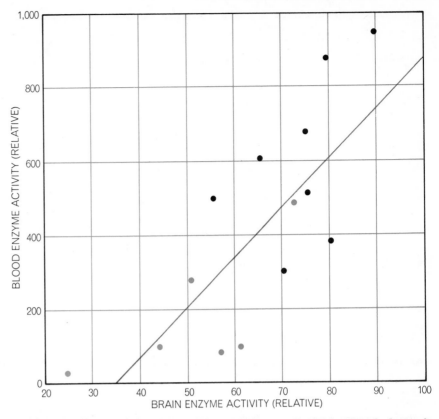

CORRELATION between the activity of ALA dehydrase in the blood and in the brain of normal rats (*black dots*) and lead-poisoned rats (*colored dots*) suggests that the enzyme may be implicated in brain damage, according to J. A. Millar and his colleagues. These data are for severely poisoned rats; in others with blood-lead values of about 30 micrograms per 100 milliliters of blood, brain enzyme activity was not significantly less than in controls.

the Fanconi syndrome is seen only at very high levels of lead in blood (greater than 150 micrograms of lead per 100 milliliters of blood) and only in patients with severe acute plumbism.

In the central nervous system the toxic effect of lead is least understood. Little is known at the metabolic level; most of the information comes from clinical observation of patients and from postmortem studies. Two different mechanisms appear to be involved in lead encephalopathy, or brain damage: edema and direct injury to nerve cells. The walls of the blood vessels are somehow affected so that the capillaries become too permeable; they leak, causing edema (swelling of the brain tissue). Since the brain is enclosed in a rigid container, the skull, severe swelling destroys brain tissue. Moreover, it appears that certain brain cells may be directly injured, or their function inhibited, by lead.

The effects I have been discussing are all those of acute lead poisoning, the result of a large accumulation of lead in a relatively short time. There are chronic effects too, either the aftereffects of acute plumbism or the result of a slow buildup

of a burden of lead over a period of years. The best-known effect is chronic nephritis, a disease characterized by a scarring and shrinking of kidney tissue. This complication of lead poisoning came to light in Australia in 1929, when L. J. J. Nye became aware of a pattern of chronic nephritis and early death in the state of Queensland. Investigation revealed that Queensland children drank quantities of rainwater that was collected by runoff from house roofs sheathed with shingles covered with lead-pigmented paint. In 1954 D. A. Henderson found that of 352 adults in Queensland who had had childhood lead poisoning 15 to 40 years earlier, 165 had died, 94 of chronic nephritis. Chronic lead nephropathy, which is sometimes accompanied by gout, is also seen in persistent, heavy moonshine drinkers and in some people who have had severe industrial exposure. In all these cases, however, the abnormal intake of lead persists for more than a decade or so before the onset of nephropathy. Most of the patients have a history of reported episodes of acute plumbism, which suggests that they have levels of lead in the tissues far above those found in the general popula-

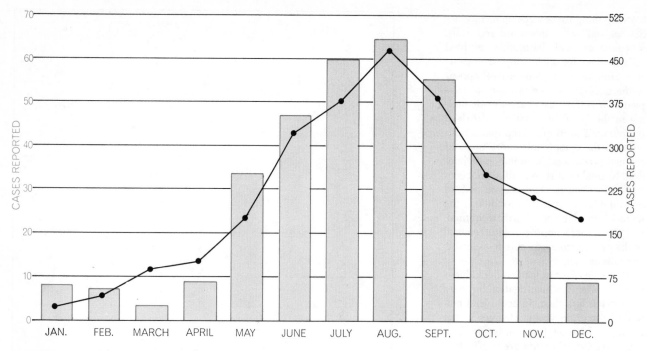

SEASONAL PATTERN of lead-poisoning cases is striking. The bars show the average number of cases reported monthly in Balti-more from 1931 through 1951 (*numbers at left*). Curve shows cases reported monthly in New York City last year (*numbers at right*).

tion. Furthermore, there is the suspicion that factors in addition to lead may be involved.

The other known result of chronic overexposure to lead is peripheral nerve disease, affecting primarily the motor nerves of the extremities. Here the tissue damage appears to be to the myelin sheath of the nerve fiber. Specifically, according to animal studies, the mitochondria of the Schwann cells, which synthesize the sheath, seem to be affected. Various investigators, including Pamela Fullerton of Middlesex Hospital in London, have found that conduction of

the nerve impulse may be impaired in the peripheral nerves of industrial workers who have had a long exposure to lead but who have no symptoms of acute lead poisoning.

These findings and others raise serious questions. It is clear that a single attack of acute encephalopathy can cause profound mental retardation and other forms of neurological injury that is permanent. Similarly, in young children repeated bouts of symptomatic plumbism can result in permanent brain damage ranging from subtle learning deficits to profound mental incompetence and epi-

lepsy. Can a level of absorption that is insufficient to cause obvious acute symptoms nevertheless cause "silent" brain damage? This question remains unanswered, in part because of the difficulty in recognizing mild symptoms of lead poisoning in children and in part because the experimental studies that might provide some answers have not yet been undertaken.

Classical plumbism—the acute disease—is seen today primarily in children with the pica habit. Before discussing these cases in some detail I shall

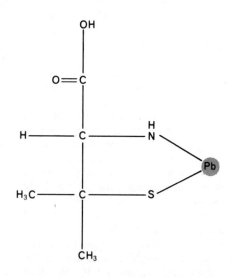

CHELATING AGENTS used in treating lead poisoning bind lead atoms (*Pb*) firmly in one or more five-member chelate rings. Dia-grams show lead chelates formed by EDTA (*left*), BAL (*middle*) and d-penicillamine (*right*). The last structure is still hypothetical.

briefly take up two other current environmental sources of lead: earthenware improperly glazed with lead and lead-contaminated alcoholic beverages.

Michael Klein and his colleagues at McGill University recently reported two cases of childhood lead poisoning, one of which was fatal, that they traced to an earthenware jug in which the children's mother kept a continuously replenished supply of apple juice. The slightly acidic juice was leaching lead out of the glaze, the thin layer of glassy material fused to the ceramic surfaces of the jug. The investigators thereupon tested 117 commercial earthenware food and beverage containers and 147 samples made with 49 different commonly used glazes in the McGill ceramics laboratory. Excessive amounts of lead—more than the U.S. maximum permissible amount for glazes of seven parts per million—were leached out of half the vessels. (The maximum permissible amount should probably be reevaluated, since past methods of testing have not taken account of such variables as the quantity of the food or beverage consumed, its acidity, the length of time it is stored and whether or not it is cooked in the pottery.) As the McGill report points out, the danger of poisoning from lead-glazed pottery has been rediscovered periodically since antiquity. The Greeks knew about the danger but the Romans did not; they made the mistake of storing wine in earthenware. James Lind, who in 1753 recommended lemon or lime juice as a preventive for scurvy, also warned that the juices should not be stored in earthenware jugs. Now the index of suspicion has fallen too low: one physician poisoned himself recently by drinking a cola beverage (and 3.2 milligrams of lead) every evening for two years from a mug his son had made for him. Do these cases represent isolated occurrences? How many other people are similarly exposed? Clearly the first step is the testing of earthenware and a reevaluation of its fabrication and use for food and drink.

In the manufacture of moonshine whiskey, lead solder is used in the tubing of distillation units. Moreover, discarded automobile radiators that contain lead often serve as condensers. Lead is therefore found in most samples of confiscated moonshine. Lead encephalopathy, nephritis with gout and other lead-related conditions have been reported in moonshine consumers, largely in the southeastern part of the U.S. The problem of diagnosis is complicated by the fact that the symptoms of acute alcoholism and acute lead poisoning are similar in many

POPULATION	EXPOSURE (MICROGRAMS PER CUBIC METER OF AIR)	MEAN BLOOD LEAD (MICROGRAMS PER 100 GRAMS)
RURAL U.S.	0.5	16
URBAN U.S.	1.0	21
DOWNTOWN PHILADELPHIA	2.4	24
CINCINNATI POLICEMEN	2.1	25
CINCINNATI TRAFFIC POLICEMEN	3.8	30
LOS ANGELES TRAFFIC POLICEMEN	5.2	21
BOSTON AUTOMOBILE-TUNNEL EMPLOYEES	6.3	30

RESPIRATORY EXPOSURE to lead is reflected in the mean blood-lead values of various groups, according to John R. Goldsmith and Alfred C. Hexter of the California Department of Public Health. Groups apparently exposed to more lead in the air have generally higher blood-lead values; whether these indicate higher body burdens of lead is not known.

ways. (Again there is a historical record. The McGill report noted that the Massachusetts Bay Colony forbade rum distillation in leaded stills in 1723 in an effort to prevent "dry gripes," an intestinal condition. In 1767 Sir George Baker blamed "the endemic colic of Devonshire" on the use of lead-lined troughs in the making of apple cider.)

Childhood lead poisoning in the U.S. is seen almost exclusively in children of preschool age who live in deteriorated housing built before 1940 (when titanium dioxide began to replace lead in the pigment of most interior paints). The causative factors are commonly a triad: a dilapidated old house, a toddler with pica and parents with inadequate resources (emotional, intellectual, informational and/or economic) to cope with the family's needs. The three factors interact to increase the likelihood that the child will eat chips of leaded paint. A chip of paint about the size of an adult's thumbnail can contain between 50 and 100 milligrams of lead, and so a child eating a few small chips a day easily ingests 100 or more times the tolerable adult intake of the metal! In one study conducted some years ago at the Baltimore City Hospitals and the Johns Hopkins Hospital, Harold E. Harrison and I found that the average daily fecal excretion of lead by children with severe plumbism was 44 milligrams. In a group of normal unexposed children we found a daily fecal lead excretion of less than .2 milligram of lead. In other words, pica for leaded paint results in genuinely massive exposures. And when the abnormal intake ceases, it may be several months

or years before blood-lead levels return to normal.

The repeated ingestion of leaded-paint chips for about three months or longer can lead to clinical symptoms and eventually to the absorption of a potentially lethal body burden of lead. During the first four to six weeks of abnormal ingestion there are no symptoms. After a few weeks minor symptoms such as decreased appetite, irritability, clumsiness, unwillingness to play, fatigue, headache, abdominal pain and vomiting begin to appear. These, of course, are all quite nonspecific symptoms, easily ignored as behavior problems or blamed on various childhood diseases. In a few weeks the lassitude may progress to intermittent drowsiness and stupor; the vomiting may become persistent and forceful; brief convulsions may occur. If the exposure to lead continues, the course of the disease can culminate abruptly in coma, intractable convulsions and sometimes death.

This picture of fulminating encephalopathy is commonest in children between 15 and 30 months of age; older children tend to suffer recurrent but less severe acute episodes and are usually brought to the hospital with a history of sporadic convulsions, behavior problems, hyperactivity or mental retardation. The symptoms tend to wax and wane, usually becoming more severe in summer. (Some 85 percent of all lead-poisoning cases are reported from May through October. This remarkably clear seasonal pattern is still not understood. It may be due at least in part to the fact that the ultraviolet component of sunlight increases the absorption of lead from the intestine.)

The symptoms of even acute encephalopathy are nonspecific, resembling those of brain abscesses and tumors and of viral and bacterial infections of the brain. Diagnosis depends, first of all, on a high level of suspicion. To make a positive diagnosis it is necessary to show high lead absorption as well as the adverse effects of lead. This requires the measurement of lead in blood and other specialized tests. Mild symptoms may be found in the presence of values of between 60 and 80 micrograms of lead per 100 milliliters of blood. As the blood-lead level rises above 80 micrograms the risk of severe symptoms increases sharply. Even in the absence of symptoms, in children blood-lead levels exceeding 80 micrograms call for immediate treatment and separation of the child from the source of lead.

Treatment is with potent compounds known as chelating agents (from the Greek *chēlē*, meaning claw): molecules that tend to bind a metal atom firmly, sequestering it and thus rendering it highly soluble [see "Chelation," by Harold F. Walton; SCIENTIFIC AMERICAN, June, 1953]. Chelating agents remove lead atoms from tissues for excretion through the kidney and through the liver. With chelating agents very high tissue levels of lead can be rapidly reduced to levels approaching normal, and the adverse metabolic effects can be promptly suppressed. Initially two agents are administered by injection: EDTA and BAL. (EDTA, or edathamil, is ethylenediaminetetraacetic acid; BAL is "British Anti-Lewisite," developed during World War II as an antidote for lewisite, an arsenic-containing poison gas.) After the lead level has been reduced another agent, d-penicillamine, may be administered orally as a follow-up therapy.

Before chelating agents were available about two-thirds of all children with lead encephalopathy died. Now the mortality rate is less than 5 percent. Unfortunately the improvement in therapy has not substantially reduced the incidence of brain damage in the survivors. Meyer A. Perlstein and R. Attala of the Northwestern University Medical School found that of 59 children who developed encephalopathy, 82 percent were left with permanent injury: mental retardation, convulsive disorders, cerebral palsy or blindness. This high incidence of permanent damage suggests that some of these children must have had recurrent episodes of plumbism; we have found that if a child who has been treated for acute encephalopathy is returned to the same hazardous environment, the risk of permanent brain damage rises to virtually 100 percent. In Baltimore, with the help of the Health Department and through the efforts of dedicated medical social workers, we are able to make it an absolute rule that no victim of lead poisoning is ever returned to a dangerous environment. The child goes from the hospital to a convalescent home and does not rejoin his family until all hazardous lead sources have been removed or the family has been helped to find lead-free housing. Cases of permanent brain damage nevertheless persist. It appears that even among children who suffer only one episode, are properly treated and are thereafter kept away from lead, at least 25 percent of the survivors of lead encephalopathy sustain lasting damage.

Clearly, then, treatment is not enough; the disease must be prevented. Children with increased lead absorption must be identified before they become poisoned. Going a step further, the sources of excessive lead exposure must be eliminated.

Baltimore has taken a "case-finding" approach to these tasks. Free diagnostic services were established by the city Health Department in the 1930's. Physicians took advantage of the services, and increasing numbers of cases

	I NO DEMONSTRABLE EFFECTS	II MINIMAL SUBCLINICAL EFFECTS DETECTABLE	III COMPENSATION	IV FUNCTIONAL INJURY (SHORT, INTENSE EXPOSURE)
METABOLIC EFFECTS	NORMAL	URINARY ALA MAY INCREASE	INCREASE IN SEVERAL METABOLITES IN BLOOD AND URINE	FURTHER INCREASE IN METABOLITES
FUNCTIONAL EFFECTS: BLOOD	NONE	NONE	REDUCED RED CELL LIFE-SPAN. INCREASED PRODUCTION	REDUCED RED CELL LIFE-SPAN WITH OR WITHOUT ANEMIA (REVERSIBLE)
KIDNEY FUNCTION	NORMAL	NORMAL	SOMETIMES MINIMAL DYSFUNCTION	FANCONI SYNDROME (REVERSIBLE)
CENTRAL NERVOUS SYSTEM	NONE	NONE	?	MINIMAL TO SEVERE BRAIN DAMAGE (PERMANENT)
PERIPHERAL NERVES	NONE	NONE	?	POSSIBLE DAMAGE
SYMPTOMS	NONE	NONE	SOMETIMES MILD, NON-SPECIFIC COMPLAINTS	ANEMIA, COLIC, IRRITABILITY, DROWSINESS; IN SEVERE CASES, MOTOR CLUMSINESS, CONVULSIONS AND COMA
RESIDUAL EFFECTS	NONE	NONE	NONE KNOWN	RANGE FROM MINIMAL LEARNING DISABILITY TO PROFOUND MENTAL AND BEHAVIORAL DEFICIENCY, CONVULSIVE DISORDERS, BLINDNES

EFFECTS OF LEAD are associated in a general way with five levels of exposure and rates of absorption of the metal. Level I is associated with blood-lead concentrations of less than 30 micrograms of lead per 100 milliliters and Level II with the 30–50 microgram range. Level III, at which compensatory mechanisms apparently minimize or prevent obvious functional injury, may be associated with concentrations of between 50 and 100 micrograms. Level IV is usually associated with concentrations greater than 80 micrograms but impairment may be evident at lower levels, particularly if compensatory responses are interfered with by some other disease state.

were discovered. Since 1951 the removal of leaded paint has been required in any dwelling where a child is found with a blood-lead value of more than 60 micrograms. The number of cases reported each year rose for some time as diagnostic methods and awareness improved, but recently it has leveled off. In order to reach children before they are poisoned, however, more is required than case-finding; what is needed is a screening program that examines entire populations of children in high-risk areas of cities. Chicago undertook that task in the 1960's. Last year New York City inaugurated a new and intensive screening program in which children are being tested for blood lead in hospitals and at a large number of neighborhood health centers; an educational campaign has been launched to bring lead poisoning and the testing facilities to public notice. As in Baltimore, a blood-lead finding of more than 60 micrograms results in an examination of the child's home. If any samples of paint and plaster contain

UNCTIONAL INJURY (CHRONIC
R RECURRENT INTENSE EXPOSURE)

NCREASE ONLY IN CASE OF
ECENT EXPOSURE

OSSIBLE ANEMIA (REVERSIBLE)

HRONIC NEPHROPATHY
PERMANENT)

EVERE BRAIN DAMAGE, PARTICULARLY
N CHILDREN (PERMANENT)

MPAIRED CONDUCTION
MAY BE CHRONIC)

MENTAL DETERIORATION.
EIZURES, COMA,
OOT OR WRIST DROP

MENTAL DEFICIENCY
OFTEN PROFOUND), KIDNEY
NSUFFICIENCY, GOUT (UNCOMMON),
OOT DROP (RARE)

What one can say is that the risk of functional injury increases as the concentration of lead in the blood exceeds 80 micrograms per 100 milliliters. The residual effects persist after blood-lead levels return to normal.

more than 1 percent of lead, the landlord is ordered to correct the condition by covering the walls with wallboard to a height of at least four feet and by removing all leaded paint from wood surfaces; if the landlord does not comply, the city undertakes the work and bills him. Before the new program was begun New York was screening about 175 blood tests a week; by the end of the year it was doing about 2,000 tests a week. Whereas 727 cases of lead poisoning were reported in the city in 1969, last year more than 2,600 were reported. As Evan Charney of the University of Rochester School of Medicine and Dentistry has put it, "the number of cases depends on how hard you look."

Screening is complicated by technical difficulties in testing both children and dwellings. The standard dithizone method of determining blood lead requires between five and 10 cubic centimeters of blood taken from a vein—a difficult procedure in very small children—and the analysis is time-consuming. What is needed is a dependable test that can be carried out on a drop or two of blood from a finger prick. A variety of approaches are now being tried in several laboratories in order to reach this goal; as yet no microtest utilizing a drop or two of blood has been proved practical on the basis of large-scale use in the field. Several appear to be promising in the laboratory, so that field testing in the near future can be anticipated. As for the checking of dwellings, the standard method is laborious primarily because it requires the collection of a large number of samples. Several different portable instruments are under development, including an X-ray fluorescence apparatus that gives a lead-content reading when it is pointed at a surface, but these devices have not yet been proved reliable in the field.

Since World War II the incidence of lead poisoning (usually in the form of lead palsy) among industrial workers, which was once a serious problem, has been reduced by various control measures. The danger is now limited primarily to small plants that are not well regulated and to home industries.

There is increasing concern over environmental lead pollution. Claire C.

Patterson of the California Institute of Technology has shown that the levels of lead in polar ice have risen sharply since the beginning of the Industrial Revolution. Henry A. Schroeder of the Dartmouth Medical School has shown that the burden of lead in the human body rises with age, and that this rise is due almost entirely to the concentration of lead in bone. Although man's exposure to lead in highly industrialized nations may come from a variety of sources, the evidence points to leaded gasoline as the principal source of airborne lead today. These observations have occasioned much speculation. It is nonetheless clear that a further rise in the dissemination of lead wastes into the environment can cause adverse effects on human health; indeed, concerted efforts to lower the current levels of exposure must be made, particularly in congested urban areas.

At the moment there is no evidence that any groups have mean blood levels that approach the dangerous range. Some, however, do have levels at which a minimal increase in urinary ALA, but nothing more, is to be expected. This includes people whose occupation brings them into close and almost daily contact with automotive exhaust. These observations emphasize the need to halt any further rise in the total level of exposure. A margin of safety needs to be defined and maintained. This will require research aimed at elucidating the effects of long-term exposure to levels of lead insufficient to cause symptoms or clear-cut functional injury. With regard to respiratory exposure, it is still not clear what fraction of the inhaled particles reaches the lungs and how much of that fraction is actually absorbed from the lung. Still another important question is the storage of lead in bone. Can any significant fraction of lead in bone be easily and quickly mobilized? If so, under what circumstances is it mobilized? There are more questions than answers to the problems posed by levels of lead only slightly higher than those currently found in urban man. Much research is required.

With regard to childhood lead poisoning, however, we know enough to act. It is impermissible for a humane society to fail to do what is necessary to eliminate a wholly preventable disease.

III

RESPONSES TO THE PHYSICAL ENVIRONMENT

RESPONSES TO THE PHYSICAL ENVIRONMENT   III

INTRODUCTION

The first two sections of this anthology were concerned with the responses of the body to chemicals, both nutrient and non-nutrient. The environment may also be characterized by its so-called "physical" parameters—energy (mechanical, thermal, electric, electromagnetic, and chemical) and force fields (gravitational, magnetic, and electric)—and this section introduces the reader to a few examples of the research about them. Certain problems, such as the adaptation to extremes of temperature and altitude, have been favorites of environmental physiologists for many years and have been extensively studied. Others, such as magnetic forces, have been much less studied in the past, but have rapidly attracted attention as biologists have come to recognize that every characteristic of the external environment is likely to have important effects on human function.

"The Diving Women of Korea and Japan," by Hong and Rahn, beautifully illustrates the complexities of environmental physiology (and the fun of being an environmental physiologist). These women, the ama, must solve simultaneously the problems of oxygen deprivation, carbon-dioxide excess, elevated external hydrostatic pressure (causing chest compression), rapidly changing nitrogen pressures, and cold. Hong and Rahn describe the various physiological adaptations which develop and enable the ama to cope with these challenges, but the mechanisms underlying many of them remain poorly understood.

The high-altitude environment offers yet another complex set of environmental stresses: oxygen lack, cold, increased ultraviolet radiation and ionization of the air, low humidity. The body's responses to these stresses are described in Hock's "The Physiology of High Altitude." These adaptations, which involve virtually every organ and tissue of the body, are very evident in native mountain-dwellers, but most of them also develop gradually in lowlanders sojourning in the mountains. Thus a major question has been whether any of the adaptations seen in mountain natives are due to genetic differences or whether all are the results of acclimatization. There are some findings which may perhaps indicate the existence of genetic differences: sojourners, even after many years, never exhibit quite as much change in certain physiological functions as mountain natives; and several of the mountain native's characteristics are not reversed by years of sea-level dwelling. However, recent studies have demonstrated that most, if not all, of these facts can be explained in terms of developmental acclimatization instead of genetic difference; i.e., exposure to high altitude during infancy and early childhood (the "critical periods" of development) caused phenotypic changes of greater magnitude and less reversibility than those caused by exposure later in life. At present, no genetic differences have been firmly established.

"The Effects of Light on the Human Body" by Wurtman illustrates how ubiquitous the effects of an environmental parameter may be, yet how little their potential importance may be known or appreciated. Even the well-known responses to solar radiation—tanning and the formation of vitamin D—have often been disregarded as trivial or unimportant (tanning as being merely an aesthetic concern, and vitamin D as being mainly found in food). The very name "vitamin D" is a tribute to misunderstanding, for "vitamin D" is really a hormone, whose formation is homeostatically regulated by the body as long as adequate sunlight is present. To call this substance a vitamin reinforces the (incorrect) concept that the body is "normally" incapable of synthesizing it and that one must depend on a dietary source for it; instead, this situation is abnormal, because we have so deprived ourselves of what should be our normal exposure to the sun. Moreover, because most food sources are deficient in vitamin D, we must resort to fortification with the result that ingestion of excessive and potentially toxic amounts becomes a possibility. To Wurtman's concluding plea that we need far more research into the biological effects of light so that we can more intelligently design interior lighting, we must add that such research is also required in order for us to appreciate the potential effects of altering the quantity and type of solar radiation reaching the earth (as, for example, would result from a decrease in the atmosphere's high-altitude ozone layer).

In contrast to solar radiation, ionizing radiations (which include both electromagnetic waves and high-energy particles) are of such high energy that when they interact with matter, they remove electrons from molecules, thus creating charged particles, or *ions*. Such ionization leads to the formation of free radicals and peroxides, which in turn react with larger molecules, altering and damaging their structures. Human beings have always been exposed to ionizing radiation from natural sources, but as our own actions have increased the amount of exposure, the question of its biological effects and potential hazards has been subjected to extensive study.

"Radiation and the Human Cell," by Puck, describes the extremely important tissue-culture experiments which documented that human cells, and especially the chromosomes, are far more sensitive to the killing effects of ionizing radiation than was originally suspected. We now recognize that the DNA molecules themselves are especially injured by ionizing radiation, and that this damage manifests itself in a variety of ways. First, if the DNA is so damaged that DNA replication and cell division are prevented, the total number of cells in the afflicted organ or tissue will decrease, as old cells die and are not replaced by new ones. This phenomenon will be most profound in those organs and tissues that have the highest normal rates of cell turnover; accordingly, the symptoms of acute exposure to intense ionizing radiation are anemia, bleeding, and massive infection (all due to failure of the blood-forming cells to divide) as well as diarrhea and gastrointestinal hemorrhage (due to loss of the lining of the gastrointestinal tract).

However, unless a nuclear war or accident occurs, acute radiation sickness is much less important to the population as a whole than the more subtle effects of chronic exposure to much lower doses of radiation. Here is where the other manifestations of DNA damage become prominent, in mutations, cancer, and acceleration of the aging process. There are immense problems involved in trying to measure the actual incidences of such toxic effects at low doses of radiation, and no trustworthy estimate is currently possible. The central stumbling block here is that we don't know whether there is a threshold for DNA damage; we can measure cancer and mutation rates caused in experimental animals by high doses of radiation, but we do not know whether the amounts of response to given doses will describe a straight line down to the lowest possible doses. Many believe that this question will never be answered, that it lies beyond the limits of toxicology.

Thus far we have dealt only with the pathological effects of ionizing radiation on the body. As with any type of environmental stress, we must also ask whether the body musters any adaptive responses to protect itself or counteract the radiation. So far as is known, humans possess no receptors capable of detecting the presence of radiation (outside the range of heat and visible light), so that no autonomic or behavioral responses which might prevent damage can occur. (One might speculate whether, should elevated levels of ionizing radiation persist, evolution might ultimately provide us with relevant receptors and responses.) In contrast, once the damage has been sustained, some degree of repair may be possible, for cells can repair damaged DNA molecules, at least to some extent. The mechanisms by which DNA is repaired are extremely complex, and we do not know how effective they are in different situations; their existence makes any purely theoretical consideration of threshold even more tenuous.

9

The Diving Women of Korea and Japan

by Suk Ki Hong and Hermann Rahn
May 1967

Some 30,000 of these breath-holding divers, called ama, are employed in daily foraging for food on the bottom of the sea. Their performance is of particular interest to the physiologist

Off the shores of Korea and southern Japan the ocean bottom is rich in shellfish and edible seaweeds. For at least 1,500 years these crops have been harvested by divers, mostly women, who support their families by daily foraging on the sea bottom. Using no special equipment other than goggles (or glass face masks), these breath-holding divers have become famous the world over for their performances. They sometimes descend to depths of 80 feet and can hold their breath for up to two minutes. Coming up only for brief rests and a few breaths of air, they dive repeatedly, and in warm weather they work four hours a day, with resting intervals of an hour or so away from the water. The Korean women dive even in winter, when the water temperature is 50 degrees Fahrenheit (but only for short periods under such conditions). For those who choose this occupation diving is a lifelong profession; they begin to work in shallow water at the age of 11 or 12 and sometimes continue to 65. Childbearing does not interrupt their work; a pregnant diving woman may work up to the day of delivery and nurse her baby afterward between diving shifts.

The divers are called ama. At present there are some 30,000 of them living and working along the seacoasts of Korea and Japan. About 11,000 ama dwell on the small, rocky island of Cheju off the southern tip of the Korean peninsula, which is believed to be the area where the diving practice originated. Archaeological remains indicate that the practice began before the fourth century. In times past the main objective of the divers may have been pearls, but today it is solely food. Up to the 17th century the ama of Korea included men as well as women; now they are all women. And in Japan, where many of the ama are male, women nevertheless predominate in the occupa-

tion. As we shall see, the female is better suited to this work than the male.

In recent years physiologists have found considerable interest in studying the capacities and physiological reactions of the ama, who are probably the most skillful natural divers in the world. What accounts for their remarkable adaptation to the aquatic environment, training or heredity or a combination of both? How do they compare with their nondiving compatriots? The ama themselves have readily cooperated with us in these studies.

We shall begin by describing the dive itself. Basically two different approaches are used. One is a simple system in which the diver operates alone; she is called *cachido* (unassisted diver). The other is a more sophisticated technique; this diver, called a *funado* (assisted diver), has a helper in a boat, usually her husband.

The *cachido* operates from a small float at the surface. She takes several deep breaths, then swims to the bottom, gathers what she can find and swims up to her float again. Because of the oxygen consumption required for her swimming effort she is restricted to comparatively shallow dives and a short time on the bottom. She may on occasion go as deep as 50 or 60 feet, but on the average she limits her foraging to a depth of 15 or 20 feet. Her average dive lasts about 30 seconds, of which 15 seconds is spent working on the bottom. When she surfaces, she hangs on to the float and rests for about 30 seconds, taking deep breaths, and then dives again. Thus the cycle takes about a minute, and the diver averages about 60 dives an hour.

The *funado* dispenses with swimming effort and uses aids that speed her descent and ascent. She carries a counterweight (of about 30 pounds) to pull her to the bottom, and at the end of her

dive a helper in a boat above pulls her up with a rope. These aids minimize her oxygen need and hasten her rate of descent and ascent, thereby enabling her to go to greater depths and spend more time on the bottom. The *funado* can work at depths of 60 to 80 feet and average 30 seconds in gathering on the bottom—twice as long as the *cachido*. However, since the total duration of each dive and resting period is twice that of the *cachido*, the *funado* makes only about 30 dives per hour instead of 60. Consequently her bottom time per hour is about the same as the *cachido*'s. Her advantage is that she can harvest deeper bottoms. In economic terms this advantage is partly offset by the fact that the *funado* requires a boat and an assistant.

There are variations, of course, on the two basic diving styles, almost as many variations as there are diving locations. Some divers use assistance to ascend but not to descend; some use only light weights to help in the descent, and so on.

By and large the divers wear minimal clothing, often only a loincloth, during their work in the water. Even in winter the Korean divers wear only cotton bathing suits. In Japan some ama have recently adopted foam-rubber suits, but most of the diving women cannot afford this luxury.

The use of goggles or face masks to improve vision in the water is a comparatively recent development—hardly a century old. It must have revolutionized the diving industry and greatly increased the number of divers and the size of the harvest. The unprotected human eye suffers a basic loss of visual acuity in water because the light passing through water undergoes relatively little refraction when it enters the tissue of the cornea, so that the focal point of the image is considerably behind the retina [*see top*

JAPANESE DIVING WOMAN was photographed by the Italian writer Fosco Maraini near the island of Hekura off the western coast of Japan. The ama's descent is assisted by a string of lead weights tiéd around her waist. At the time she was diving for aba- lone at a depth of about 30 feet. At the end of each dive a helper in a boat at the surface pulls the ama up by means of the long rope attached to her waist. The other rope belongs to another diver. The ama in this region wear only loincloths during their dives.

illustration, page 98]. Our sharp vision in air is due to the difference in the refractive index between air and the corneal tissue; this difference bends light sharply as it enters the eye and thereby helps to focus images on the retinal surface. (The lens serves for fine adjustments.) Goggles sharpen vision in the water by providing a layer of air at the interface with the eyeball.

Goggles create a hazard, however, when the diver descends below 10 feet in the water. The hydrostatic pressure on the body then increases the internal body pressures, including that of the blood, to a level substantially higher than the air pressure behind the goggles. As a result the blood vessels in the eyelid lining may burst. This conjunctival bleeding is well known to divers who have ventured too deep in the water with

only simple goggles. When the Korean and Japanese divers began to use goggles, they soon learned that they must compensate for the pressure factor. Their solution was to attach air-filled, compressible bags (of rubber or thin animal hide) to the goggles. As the diver descends in the water the increasing water pressure compresses the bags, forcing more air into the goggle space and thus raising the air pressure there in proportion to the increase in hydrostatic pressure on the body. Nowadays, in place of goggles, most divers use a face mask covering the nose, so that air from the lungs instead of from external bags can serve to boost the air pressure in front of the eyes.

The ama evolved another technique that may or may not have biological value. During hyperventilation before

their dives they purse their lips and emit a loud whistle with each expiration of breath. These whistles, which can be heard for long distances, have become the trademark of the ama. The basic reason for the whistling is quite mysterious. The ama say it makes them "feel better" and "protects the lungs." Various observers have suggested that it may prevent excessive hyperventilation (which can produce unconsciousness in a long dive) or may help by increasing the residual lung volume, but no evidence has been found to verify these hypotheses. Many of the Japanese divers, male and female, do not whistle before they dive.

Preparing for a dive, the ama hyperventilates for five to 10 seconds, takes a final deep breath and then makes the plunge. The hyperventilation serves to

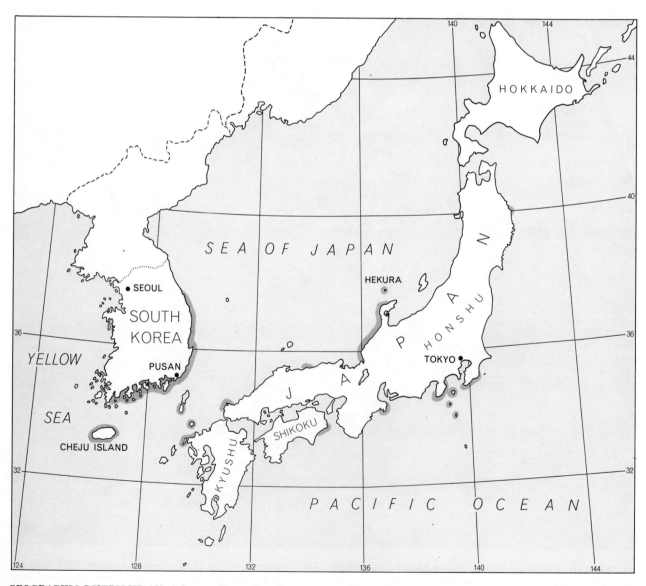

GEOGRAPHIC DISTRIBUTION of the ama divers along the seacoasts of South Korea and southern Japan is indicated by the colored areas. The diving practice is believed to have originated on the small island of Cheju off the southern tip of the Korean peninsula.

remove a considerable amount of carbon dioxide from the blood. The final breath, however, is not a full one but only about 85 percent of what the lungs can hold. Just why the ama limits this breath is not clear; perhaps she does so to avoid uncomfortable pressure in the lungs or to restrict the body's buoyancy in the water.

As the diver descends the water pressure compresses her chest and consequently her lung volume. The depth to which she can go is limited, of course, by the amount of lung compression she can tolerate. If she dives deeper than the level of maximum lung compression (her "residual lung volume"), she becomes subject to a painful lung squeeze; moreover, because the hydrostatic pressure in her blood vessels then exceeds the air pressure in her lungs, the pulmonary blood vessels may burst.

The diver, as we have noted, starts her dive with a lungful of air that is comparatively rich in oxygen and comparatively poor in carbon dioxide. What happens to the composition of this air in the lungs, and to the exchange with the blood, during the dive? In order to investigate this question we needed a means of obtaining samples of the diver's lung air under water without risk to the diver. Edward H. Lanphier and Richard A. Morin of our group (from the State University of New York at Buffalo) devised a simple apparatus into which the diver could blow her lung air and then reinhale most of it, leaving a small sample of air in the device. The divers were understandably reluctant at first to try this device, because it meant giving up their precious lung air deep under water with the possibility that they might not recover it, but they were eventually reassured by tests of the apparatus.

We took four samples of the diver's lung air: one before she entered the water, a second when she had hyperventilated her lungs at the surface and was about to dive, a third when she reached the bottom at a depth of 40 feet and a fourth after she had returned to the surface. In each sample we measured the concentrations and calculated the partial pressures of the principal gases: oxygen, carbon dioxide and nitrogen.

Normally, in a resting person out of the water, the air in the alveoli of the lungs is 14.3 percent oxygen, 5.2 percent carbon dioxide and 80.5 percent nitrogen (disregarding the rare gases and water vapor). We found that after hyperventilation the divers' alveolar air con-

KOREAN DIVING WOMAN from Cheju Island cooperated with the authors in their study of the physiological reactions to breath-hold diving. The large ball slung over her left shoulder is a float that is left at the surface during the dive; attached to the float is a net for collecting the catch. The black belt was provided by the authors to carry a pressure-sensitive bottle and electrocardiograph wires for recording the heart rate. The ama holds an alveolar, or lung, gas sampler in her right hand. The Korean ama wear only light cotton bathing suits even in the winter, when the water temperature can be as low as 50 degrees Fahrenheit.

5 SECONDS 15 SECONDS 5 SECONDS

UNASSISTED DIVER, called a *cachido*, employs one of the two basic techniques of ama diving. The *cachido* operates from a small float at the surface. On an average dive she swims to a depth of about 15 to 20 feet; the dive lasts about 25 to 30 seconds, of which 15 seconds is spent working on the bottom. The entire diving cycle takes about a minute, and the diver averages 60 dives per hour.

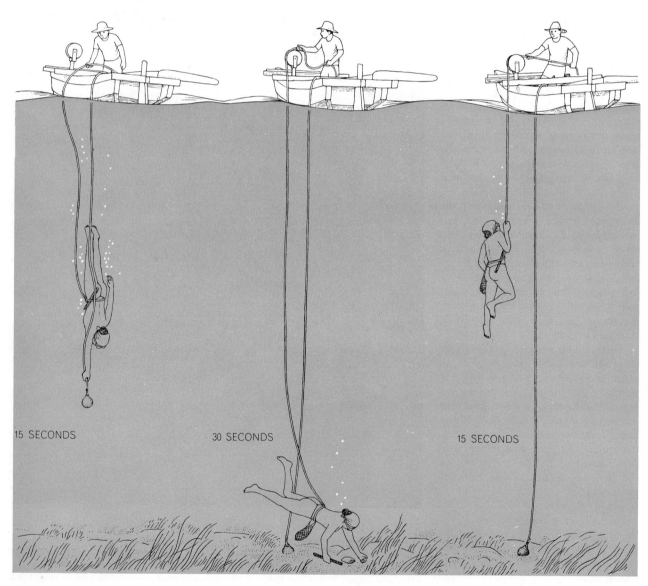

15 SECONDS 30 SECONDS 15 SECONDS

ASSISTED DIVER, called a *funado*, uses a counterweight to descend passively to a depth of 60 to 80 feet. She averages 30 seconds in gathering on the bottom but makes only about 30 dives per hour. At the end of each dive a helper in the boat pulls her up.

sists of 16.7 percent oxygen, 4 percent carbon dioxide and 79.3 percent nitrogen; translating these figures into partial pressures (in millimeters of mercury), the respective proportions are 120 millimeters for oxygen, 29 for carbon dioxide and 567 for nitrogen.

By the time the *cachido* (unassisted diver) reaches the bottom at a depth of 40 feet the oxygen concentration in her lungs is reduced to 11.1 percent, because of the uptake of oxygen by the blood. However, since at that depth the water pressure has compressed the lungs to somewhat more than half of their pre-dive volume, the oxygen pressure amounts to 149 millimeters of mercury—a greater pressure than before the dive. Consequently oxygen is still being transmitted to the blood at a substantial rate.

For the same reason the blood also takes up carbon dioxide during the dive. The carbon dioxide concentration in the lungs drops from 4 percent at the beginning of the dive to 3.2 percent at the bottom. This is somewhat paradoxical; when a person out of the water holds his breath, the carbon dioxide in his lungs increases. At a depth of 40 feet, however, the compression of the lung volume raises the carbon dioxide pressure to 42 millimeters of mercury, and this is greater than the carbon dioxide pressure in the venous blood. As a result the blood and tissues retain carbon dioxide and even absorb some from the lungs.

As the diver ascends from the bottom, the expansion of the lungs drastically reverses the situation. With the reduction of pressure in the lungs, carbon dioxide comes out of the blood rapidly. Much more important is the precipitous drop of the oxygen partial pressure in the lungs: within 30 seconds it falls from 149 to 41 millimeters of mercury. This is no greater than the partial pressure of oxygen in the venous blood; hence the blood cannot pick up oxygen, and Lanphier has shown that it may actually lose oxygen to the lungs. In all probability that fact explains many of the deaths that have occurred among sports divers returning to the surface after deep, lengthy dives. The cumulative oxygen deficiency in the tissues is sharply accentuated during the ascent.

Our research has also yielded a measure of the nitrogen danger in a long dive. We found that at a depth of 40 feet the nitrogen partial pressure in the compressed lungs is doubled (to 1,134 millimeters of mercury), and throughout the dive the nitrogen tension is sufficient to drive the gas into the blood. Lanphier has calculated that repeated dives to

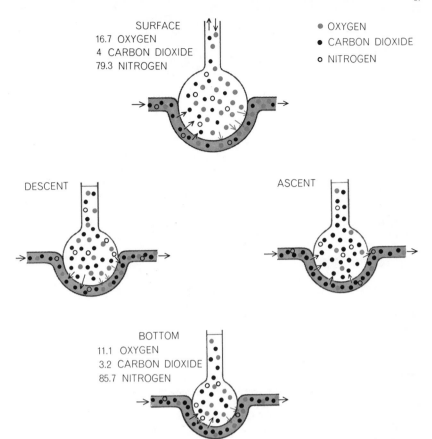

GASES EXCHANGED between a single alveolus, or lung sac, and the bloodstream are shown for four stages of a typical ama dive. The concentrations of three principal gases in the lung at the surface and at the bottom are given in percent. During descent water pressure on the lungs causes all gases to enter the blood. During ascent this situation is reversed.

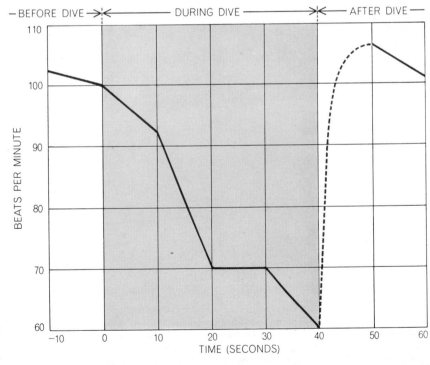

AVERAGE HEART RATE for a group of Korean ama was measured before, during and after their dives. All the dives were to a depth of about 15 feet. The average pattern shown here was substantially the same in the summer, when the water temperature was about 80 degrees Fahrenheit, as it was in winter, when water temperature was about 50 degrees F.

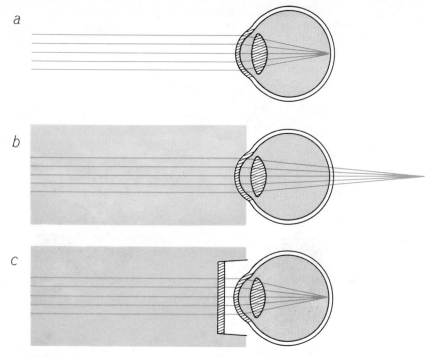

GOGGLES SHARPEN VISION under water by providing a layer of air at the interface with the eyeball (*c*). Vision is normally sharp in air because the difference in refractive index between air and the tissue of the cornea helps to focus images on the retinal surface (*a*). The small difference in the refractive index between water and corneal tissue causes the focal point to move considerably beyond the retina (*b*), reducing visual acuity under water.

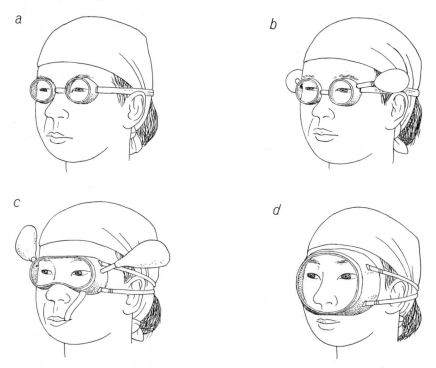

EVOLUTION OF GOGGLES has resulted in several solutions to the problem presented by the increase in hydrostatic pressure on the body during a dive. The earliest goggles (*a*) were uncompensated, and the difference in pressure between the blood vessels in the eyelid and the air behind the goggles could result in conjunctival bleeding. The problem was first solved by attaching air-filled, compressible bags to the goggles (*b*). During a dive the increasing water pressure compresses the bags, raising the air pressure behind the goggles in proportion to the increase in hydrostatic pressure on the body. In some cases (*c*) the lungs were used as an additional compensating gas chamber. With a modern face mask that covers the nose (*d*) the lungs provide the only source of compensating air pressure during a dive.

depths of 120 feet, such as are performed by male pearl divers in the Tuamotu Archipelago of the South Pacific, can result in enough accumulation of nitrogen in the blood to cause the bends on ascent. When these divers come to the surface they are sometimes stricken by fatal attacks, which they call *taravana*.

The ama of the Korean area are not so reckless. Long experience has taught them the limits of safety, and, although they undoubtedly have some slight anoxia at the end of each dive, they quickly recover from it. The diving women content themselves with comparatively short dives that they can perform again and again for extended periods without serious danger. They avoid excessive depletion of oxygen and excessive accumulation of nitrogen in their blood.

As far as we have been able to determine, the diving women possess no particular constitutional aptitudes of a hereditary kind. The daughters of Korean farmers can be trained to become just as capable divers as the daughters of divers. The training, however, is important. The most significant adaptation the trained diving women show is an unusually large "vital capacity," measured as the volume of air that can be drawn into the lungs in a single inspiration after a complete expiration. In this attribute the ama are substantially superior to nondiving Korean women. It appears that the divers acquire this capacity through development of the muscles involved in inspiration, which also serve to resist compression of the chest and lung volume in the water.

A large lung capacity, or oxygen intake, is one way to fortify the body for diving; another is conservation of the oxygen stored in the blood. It is now well known, thanks to the researches of P. F. Scholander of the Scripps Institution of Oceanography and other investigators, that certain diving mammals and birds have a built-in mechanism that minimizes their need for oxygen while they are under water [see "The Master Switch of Life," by P. F. Scholander; SCIENTIFIC AMERICAN, December, 1963]. This mechanism constricts the blood vessels supplying the kidneys and most of the body muscles so that the blood flow to these organs is drastically reduced; meanwhile a normal flow is maintained to the heart, brain and other organs that require an undiminished supply of oxygen. Thus the heart can slow down, the rate of removal of oxygen from the blood by tissues is reduced, and the animal can prolong its dive.

Several investigators have found recently that human subjects lying under water also slow their heart rate, although not as much as the diving animals do. We made a study of this matter in the ama during their dives. We attached electrodes (sealed from contact with the seawater) to the chests of the divers, and while they dived to the bottom, at the end of a 100-foot cable, an electrocardiograph in our boat recorded their heart rhythms. During their hyperventilation preparatory to diving the divers' heart rate averaged about 100 beats a minute. During the dive the rate fell until, at 20 seconds after submersion, it had dropped to 70 beats; after 30 seconds it dropped further to some 60 beats a minute [see *bottom illustration on page 97*]. When the divers returned to the surface, the heart rate jumped to slightly above normal and then rapidly recovered its usual beat.

Curiously, human subjects who hold their breath out of the water, even in an air pressure chamber, do not show the same degree of slowing of the heart. It was also noteworthy that in about 50 percent of the dives the ama showed some irregularity of heartbeat. These and other findings raise a number of puzzling questions. Nevertheless, one thing is quite clear: the automatic slowing of the heart is an important factor in the ability of human divers to extend their time under water.

In the last analysis the amount of time one can spend in the water, even without holding one's breath, is limited by the loss of body heat. For the working ama this is a critical factor, affecting the length of their working day both in summer and in winter. (They warm themselves at open fires after each long diving shift.) We investigated the effects of their cold exposure from several points of view, including measurements of the heat losses at various water temperatures and analysis of the defensive mechanisms brought into play.

For measuring the amount of the body's heat loss in the water there are two convenient indexes: (1) the increase of heat production by the body (through the exercise of swimming and shivering) and (2) the drop in the body's internal temperature. The body's heat production can be measured by examining its consumption of oxygen; this can be gauged from the oxygen content of the lungs at the end of a dive and during recovery. Our measurements were made on Korean diving women in Pusan harbor at two seasons of the year: in August,

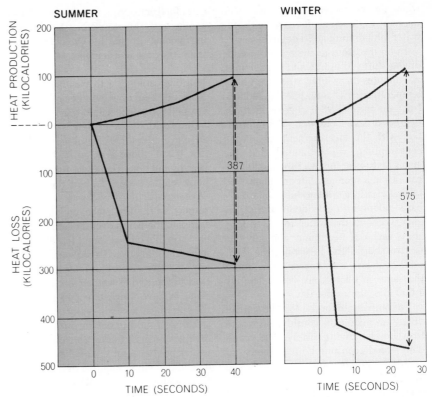

BODY HEAT lost by ama divers was found to be about 400 kilocalories in a summer shift (*left*) and about 600 kilocalories in a winter shift (*right*). The curves above the abscissa at zero kilocalories represent heat generated by swimming and shivering and were estimated by the rate of oxygen consumption. The curves below abscissa represent heat lost by the body to the water and were estimated by changes in rectal temperature and skin temperature.

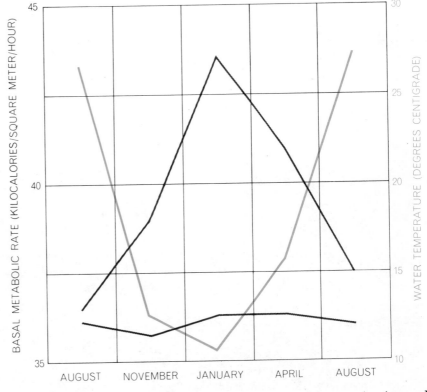

BASAL METABOLIC RATE of ama women (*top gray curve*) increases in winter and decreases in summer. In nondiving Korean women (*bottom gray curve*) basal metabolic rate is constant throughout the year. The colored curve shows the mean seawater temperatures in the diving area of Pusan harbor for the same period covered by the other measurements.

when the water temperature was 80.6 degrees F., and in January with the water temperature at 50 degrees.

In both seasons at the end of a single diving shift (40 minutes in the summer, 25 minutes in winter) the deep-body temperature was found to be reduced from the normal 98.6 degrees F. to 95 degrees or less. Combining this information with the measurements of oxygen consumption, we estimated that the ama's body-heat loss was about 400 kilocalories in a summer shift and about 600 kilocalories in a winter shift. On a daily basis, taking into consideration that the ama works in the water for three long shifts each day in summer and only one or two short shifts in winter, the day's total heat loss is estimated to be about the same in all seasons: approximately 1,000 kilocalories per day.

To compensate for this loss the Korean diving woman eats considerably more than her nondiving sisters. The ama's daily food consumption amounts to about 3,000 kilocalories, whereas the average for nondiving Korean women of comparable age is on the order of 2,000 kilocalories per day. Our various items of evidence suggest that the Korean diving woman subjects herself to a daily cold stress greater than that of any other group of human beings yet studied. Her

extra food consumption goes entirely into coping with this stress. The Korean diving women are not heavy; on the contrary, they are unusually lean.

It is interesting now to examine whether or not the diving women have developed any special bodily defenses against cold. One such defense would be an elevated rate of basal metabolism, that is, an above-average basic rate of heat production. There was little reason, however, to expect to find the Korean women particularly well endowed in this respect. In the first place, populations of mankind the world over, in cold climates or warm, have been found to differ little in basal metabolism. In the second place, any elevation of the basal rate that might exist in the diving women would be too small to have much effect in offsetting the large heat losses in water.

Yet we found to our surprise that the diving women did show a significant elevation of the basal metabolic rate—but only in the winter months! In that season their basal rate is about 25 percent higher than that of nondiving women of the same community and the same economic background (who show no seasonal change in basal metabolism). Only one other population in the world has been found to have a basal metabolic rate

as high as that of the Korean diving women in winter: the Alaskan Eskimos. The available evidence indicates that the warmly clothed Eskimos do not, however, experience consistently severe cold stresses; their elevated basal rate is believed to arise from an exceptionally large amount of protein in their diet. We found that the protein intake of Korean diving women is not particularly high. It therefore seems probable that their elevated basal metabolic rate in winter is a direct reflection of their severe exposure to cold in that season, and that this in turn indicates a latent human mechanism of adaptation to cold that is evoked only under extreme cold stresses such as the Korean divers experience. The response is too feeble to give the divers any significant amount of protection in the winter water. It does, however, raise an interesting physiological question that we are pursuing with further studies, namely the possibility that severe exposure to winter cold may, as a general rule, stimulate the human thyroid gland to a seasonal elevation of activity.

The production of body heat is one aspect of the defense against cold; another is the body's insulation for retaining heat. Here the most important factor (but not the only one) is the layer of fat

BETWEEN DIVES the ama were persuaded to expire air into a large plastic gas bag in order to measure the rate at which oxygen is consumed in swimming and diving to produce heat. The water temperature in Pusan harbor at the time (January) was 50 degrees F. One of the authors (Hong) assists. Data obtained in this way were used to construct the graph at the top of the preceding page.

under the skin. The heat conductivity of fatty tissue is only about half that of muscle tissue; in other words, it is twice as good an insulator. Whales and seals owe their ability to live in arctic and antarctic waters to their very thick layers of subcutaneous fat. Similarly, subcutaneous fat explains why women dominate the diving profession of Korea and Japan; they are more generously endowed with this protection than men are.

Donald W. Rennie of the State University of New York at Buffalo collaborated with one of the authors of this article (Hong) in detailed measurements of the body insulation of Korean women, comparing divers with nondivers. The thickness of the subcutaneous fat can easily be determined by measuring the thickness of folds of skin in various parts of the body. This does not, however, tell the whole story of the body's thermal insulation. To measure this insulation in functional terms, we had our subjects lie in a tank of water for three hours with only the face out of the water. From measurements of the reduction in deep-body temperature and the body's heat production we were then able to calculate the degree of the subject's overall thermal insulation. These studies revealed three particularly interesting facts. They showed, for one thing, that with the same thickness of subcutaneous fat, divers had less heat loss than nondivers. This was taken to indicate that the divers' fatty insulation is supplemented by some kind of vascular adaptation that restricts the loss of heat from the blood vessels to the skin, particularly in the arms and legs. Secondly, the observations disclosed that in winter the diving women lose about half of their subcutaneous fat (although nondivers do not). Presumably this means that during the winter the divers' heat loss is so great that their food intake does not compensate for it sufficiently; in any case, their vascular adaptation helps them to maintain insulation. Thirdly, we found that diving women could tolerate lower water temperatures than nondiving women without shivering. The divers did not shiver when they lay for three hours in water at 82.8 degrees F.; nondivers began to shiver at a temperature of 86 degrees. (Male nondivers shivered at 88 degrees.) It appears that the diving woman's resistance to shivering arises from some hardening aspect of their training that inhibits shiver-triggering impulses from the skin. The inhibition of shivering is an advantage because shivering speeds up the emission of body heat. L. G. Pugh, a British physiologist

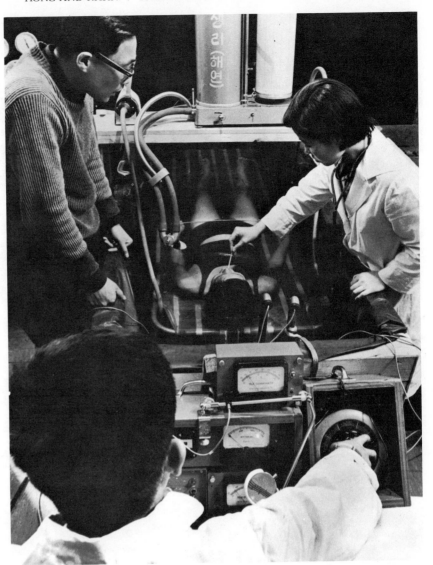

AMA'S THERMAL INSULATION (mainly fat) was measured by having the subjects lie in a tank of water for three hours. From measurements of the reduction of deep-body temperature and the body's heat production the authors were able to calculate the degree of the subject's overall thermal insulation. Once again Hong (*left*) keeps a close eye on the operation.

who has studied long-distance swimmers, discovered the interesting additional fact that swimmers, whether fat or thin, lose heat more rapidly while swimming than while lying motionless in the water. The whole subject of the body's thermal insulation is obviously a rather complicated one that will not be easy to unravel. As a general conclusion, however, it is very clearly established that women are far better insulated than men against cold.

As a concluding observation we should note that the 1,500-year-old diving occupation in Korea and Japan is now declining. The number of divers has dwindled during the past few decades, and by the end of this century the profession may disappear altogether, chiefly because more remunerative and less

arduous ways of making a living are arising. Nonetheless, for the 30,000 practitioners still active in the diving profession (at least in summer) diving remains a proud calling and necessary livelihood. By adopting scuba gear and other modern underwater equipment the divers could greatly increase their production; the present harvest could be obtained by not much more than a tenth of the present number of divers. This would raise havoc, however, with employment and the economy in the hundreds of small villages whose women daily go forth to seek their families' existence on the sea bottom. For that reason innovations are fiercely resisted. Indeed, many villages in Japan have outlawed the foam-rubber suit for divers to prevent too easy and too rapid harvesting of the local waters.

NUÑOA, at an altitude of 13,000 feet in the Andes Mountains of Peru, is the site of an experimental station operated by Pennsyl- vania State University for the study of human adaptation to high altitude. The native Indians herd llamas and alpacas, as shown here.

BARCROFT LABORATORY of the University of California is at an altitude of 12,500 feet in the White Mountains, a range just east of the Sierra Nevada in California. The environment appears un- inhabitable but is the home of a number of small animal species.

10

The Physiology of High Altitude

by Raymond J. Hock

February 1970

*To meet the stress of life at high altitude, notably
lack of oxygen, a number of changes in body processes
are required. What are these adaptations,
and are they inborn or the result of acclimatization?*

At altitudes above 6,000 feet the human organism leaves its accustomed environment and begins to feel the stresses imposed by an insufficiency of oxygen. Yet 25 million people manage to live and work in the high Andes of South America and the Himalayan ranges of Asia. More than 10 million of them live at altitudes above 12,000 feet, and there are mountain dwellers in Peru who daily go to work in a mine at an elevation of 19,000 feet. How does the human physiology contrive to acclimatize itself to such conditions? Over the past half-century, ever since the British physiologist Joseph Barcroft led an expedition to study the physiology of the mountain natives of Peru in the early 1920's, a small host of fascinated investigators has been exploring this puzzle. We now know many of the details of the body's remarkable ability to accommodate itself to life in an oxygen-poor environment.

The study has an intrinsic lure, akin to the challenge of an Everest for a mountain climber, and in these days of man's travels beyond the earth's atmosphere the subject of oxygen's relation to life has taken on added interest. The problem also has its practical aspects on our own planet. Already more and more people each year have recourse to mountain heights for recreations such as camping and skiing; for example, in 1968 there were five million visitor-days in the Inyo National Forest of California, nearly all at elevations of 7,000 feet or higher. From studies of the physiology of adjustment to oxygen deprivation we can expect some beneficial dividends, not only for the problems of living or vacationing in the high mountains but also for medical problems in diseases involving hypoxia.

The native mountain dwellers of Peru and the Sherpa people of Tibet in the Himalayas have served as very helpful subjects for the investigation of acclimatization to high-altitude life. British expeditions led by L. G. C. E. Pugh have conducted several important studies of the Sherpas. In the Andes the research has been centered principally in the Institute of Andean Biology, a permanent station founded in 1928 as a division of the University of San Marcos by Carlos Monge with Alberto Hurtado as director of research [see "Life at High Altitudes," by George W. Gray; SCIENTIFIC AMERICAN, December, 1955]. Recently Pennsylvania State University established another station in Peru at Nuñoa under the supervision of Paul T. Baker and Elsworth R. Buskirk. In the U.S. the University of California has a major center for high-altitude research on White Mountain, with Nello Pace as director. I was associated with the White Mountain Research Station for several years as resident physiologist, working primarily with experimental animals. There are four laboratories in the complex: the Barcroft Laboratory at 12,500 feet and others at 14,250 feet, 10,150 and 4,000 feet. Investigators from the U.S. Army Laboratory of the Fitzsimons Hospital have also been active in human high-altitude research in the Rockies, principally on Pikes Peak.

The High-Altitude Environment

One may wonder why people choose to live in the hostile environment of mountain heights, as the Quechua Indians of Peru, for example, have done for centuries. Life on the mountains is made rigorous not only by hypoxia but also by cold. Even in the equatorial Andes the air temperature decreases by one degree Celsius with each 640 feet of altitude. The winters are long, snowy and windy; summers are short and cool. At the 12,500-foot station on White Mountain in California temperature records over a period of 10 years showed that the mean temperature is below freezing during eight months of the year, and even in some of the summer months the nighttime minimum averages below freezing. Plant life at high altitudes has an extremely short growing season, and few animals can breed successfully at these altitudes. The relatively strong ultraviolet radiation, ionization of the air and other harsh factors no doubt affect life there. There are a few compensating factors. The intense sunlight in the thin atmosphere heats the rocks and provides warm niches for life. The heavy winter snowfall lays down a greater store of moisture than may be available in the surrounding lowlands. For a few months the highlands offer grazing for herdsmen's flocks (which are brought down to lower altitudes when the summer season ends).

Most of the people in the highlands live on herding or agriculture, raising short-season crops such as potatoes or some grains. In the Andes mining also is an important factor in the economy. The highest inhabited settlement in the world is a mining camp at 17,500 feet in Peru. The residents there work in the mine at 19,000 feet that I have already mentioned. The miners daily climb the 1,500 feet from their camp to the mine. Significantly, they rebelled against living in a camp that was built for them at 18,500 feet, complaining that they had no appetite, lost weight and could not sleep. It seems, therefore, that 17,500 feet is the highest altitude at which even acclimatized man can live permanently.

Notwithstanding the rigors of life on the heights, the Peruvian Indians of the "altiplano" have thrived in their environment. It is said that the Incas had

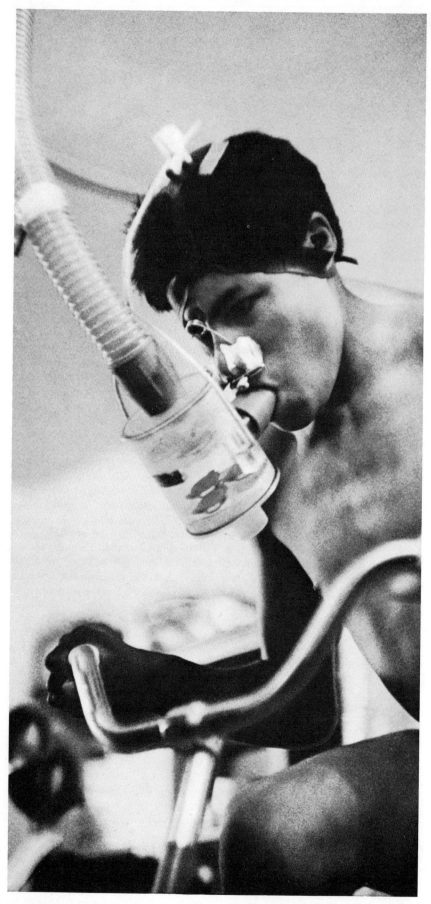

QUECHUA INDIAN breathes from a device that records oxygen intake and carbon dioxide output while he exercises on a bicycle ergometer, which measures the work performed.

two separate armies, one for the lowlands and one for high-mountain duty. Monge and Hurtado believe the high-altitude natives have become a distinctive breed ("Andean man") superbly fitted for life on the heights but probably incapable of surviving long in the lowlands. This is open to doubt. It may be that at sea level the highlanders would succumb to diseases to which they have not been exposed in the mountains, but it has not yet been demonstrated that they would be unable to adjust physiologically to the conditions at low altitudes.

Nevertheless, it is incontestable that high-altitude man, in the Andes and the Himalayas, does indeed possess unusual physiological capabilities. They are evidenced in his responses to hypoxia (defined here as a deficient supply of oxygen in the air). To see his special attributes in perspective, let us first examine the usual reactions of an unacclimatized person to hypoxia.

The Physiological Responses

The proportion of oxygen in the air is not reduced at high altitudes (it is constant at 21 percent throughout the atmosphere), but as the barometric pressure of the air as a whole declines with increasing altitude the partial pressure of the oxygen also declines correspondingly. Thus at 12,500 feet the barometric pressure drops to 480 millimeters of mercury (from 760 millimeters at sea level) and the partial pressure of oxygen is only 100 millimeters, as against 159 millimeters at sea level. That is to say, the number of oxygen and other molecules per cubic foot of air is reduced.

This decrease in oxygen tension, reducing the transfer of oxygen from inspired air to the blood in the lungs, calls forth several immediate reactions by the body. The breathing rate increases, in order to bring more air into the lungs. The heart rate and cardiac output increase, in order to enhance the flow of blood through the lung capillaries and the delivery of arterial blood to the body tissues. The body steps up its production of red blood cells and of hemoglobin to improve the blood's oxygen-carrying capacity. The hemoglobin molecule itself has a physicochemical property that enables it to take in and unload oxygen more readily when necessary at high altitudes. In a person who remains at high altitude these acclimatizing changes take place over a period of time. Investigators who measured them during a Himalayan expedition found that the hemoglobin content of the blood continued to increase for two or three months and then leveled off. As the climbers moved up from 13,000 feet to 19,000 feet and beyond, the number of red cells in the blood increased continuously for as long as 38 weeks.

The adjustments I have just recounted are not sufficient to enable a newcomer to high altitude to expend normal physical effort. Because of the interest stimulated by the holding of the 1968 Olympic Games in Mexico City (at an altitude of 7,500 feet) much study has recently been given to the effects of high altitude on the capacity for exercise. It has been found that at 18,000 feet, for instance, a man's capacity for performing exercise without incurring an oxygen debt is only about 50 percent of that at sea level. The tolerance of such a debt, and of the accumulation of lactic acid in the muscles, also is reduced. This accounts for the fact that mountain climbers at extreme altitudes can take only a few tortured steps at a time and must rest for a considerable period before going on. The limits on the capacity for work are set, of course, by the limits of the body's possible physiological adjustments to the high-altitude conditions. These limits affect the rate of ventilation of the lungs, the heart rate, the cardiac output and the blood flow to the exercising muscles. The limit for hyperventilation, for example, is a flow of 120 liters of air per minute through the lungs. This maximum, invoked at an altitude of about 16,400 feet, supplies two liters of oxygen per minute to the blood. At extreme altitudes the heart can speed up its beat during moderate exercise, but under the stress of maximum exercise the limit for both the

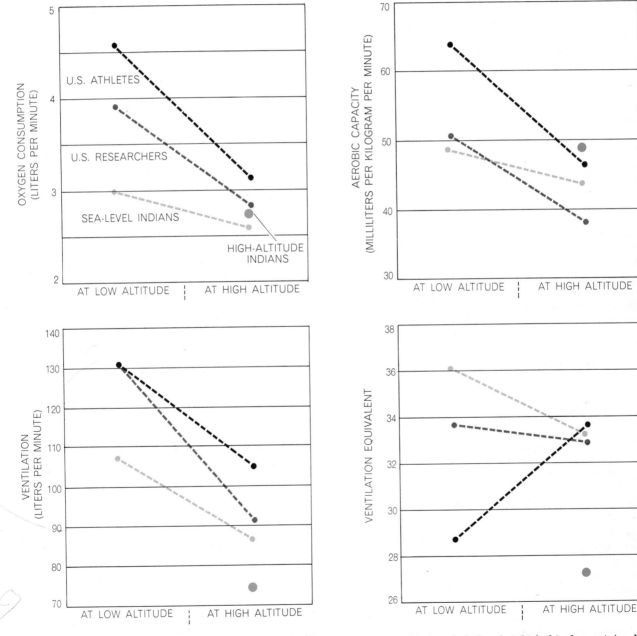

OXYGEN CONSUMPTION of Quechua Indians who were natives of Nuñoa, at 13,000 feet in the Andes (*solid color*), was compared with that of sea-level Quechua (*light color*), U.S. athletes (*black*) and U.S. research workers (*gray*) in a study made by Paul T. Baker of Pennsylvania State University. The data were collected during a bicycle exercise test. (The low-altitude subjects had been at Nuñoa at least four weeks before their high-altitude tests.) Aerobic capacity relates oxygen consumption to an individual's weight and therefore measures the success of his oxygen-transport system. Ventilation is the total volume of air breathed by the individual. Ventilation equivalent is ventilation divided by oxygen consumption; the lower the value, the greater the oxygen-extracting efficiency.

heart rate and the cardiac output is lower than at sea level.

Mountain Natives' Physiology

Let us now turn to the mountain natives' extraordinary adjustments for living at high altitudes. To begin with, the Quechua Indians of the Andes and the Sherpas of the Himalayas have developed an exceptionally large chest and lung volume, enabling them to take in a greater volume of air with each breath. Their breathing rate also is higher than that of dwellers at sea level, but they do not need to hyperventilate as much as lowlanders do when the latter go to high altitudes. The mountain natives also have a high concentration of red cells and hemoglobin in their blood, and their hemoglobin is geared to unload oxygen readily to the tissues.

In the high-altitude native the lung capillaries are dilated, so that the pulmonary circulation carries an unusually large percentage of the body's total blood volume. Moreover, the blood pressure in the lungs is higher than in the rest of the circulatory system. The heart is unusually large, apparently because of the heightened pressure in the pulmonary arteries. The heartbeat is slower than in sea-level dwellers. The mountain dwellers' metabolism also appears to be affected by the hypoxic conditions. Their basal metabolic rate is slightly higher than it is in lowlanders, and when this is considered in terms of body mass, it turns out that the rate of oxygen consumption per unit of metabolizable tissue is unusually high. That is, the hypoxic conditions exact a cost in lowered efficiency in the use of oxygen.

The mountain natives show their superior acclimatization most markedly in their capacity for exercise at high altitude. Sherpas show a smaller increase in ventilation, similar oxygen consumption and a greater heart-rate increase when performing the same exercise as low-altitude subjects who have become thoroughly acclimatized to a high altitude. The ability of mountain natives to perform physical labor daily at altitudes where even acclimatized visitors are quickly exhausted by exercise is itself obvious evidence of the mountaineers' extraordinary physiology.

In general, the physiological adjustments of the permanent mountain dwellers are similar in kind to those developed by sojourners in the mountains after a year of residence there. Furthermore, even mountain natives sometimes lose their acclimatization to high altitude and incur *soroche* (chronic mountain sickness), which is characterized by extreme elevation of the relative number and mass of red cells in the blood, pulmonary hypertension, low peripheral blood pressure, enlargement of the right lobe of the heart and ultimately congestive heart failure if the victim remains at high altitude. In general, the differences between mountain natives and sea-level natives are most apparent in the mountaineers' superior capacity for exercise at high altitude and their ability to produce children in that habitat; newcomers to the mountains, even after extended acclimatization, are much less successful in reproduction. The Spanish conquistadors who settled in the high Andes, for example, found themselves afflicted with relative infertility and a high rate of infant mortality.

Does the special physiology of the mountain people arise from genetic adaptation or is it acquired during their lifelong exposure (from the womb onward) to high altitude? One approach to answering that question has been through investigations with experimental animals. The laboratory studies of animals have also explored the physiological aspects of acclimatization much more exhaustively than is possible in man. Much of this animal work has been done at the White Mountain Research Station.

Hypoxia and Rats

Pace and an associate, Paola S. Timiras, carried out a series of investigations on rats at the White Mountain Research Station. Rats that had been bred at sea level were brought to the Barcroft Laboratory (at 12,500 feet), and the investigators examined the responses to hypoxia in these animals and in the second generation of offspring produced at the high altitude. The development of the animals exposed to the high altitude was compared with that of a control group of rats kept at sea level.

The rats at the Barcroft Laboratory exhibited acclimatizing reactions like those of human newcomers to high altitude. There was a marked increase, for example, in their red cells and hemoglobin: in the imported animals the red-cell concentration rose to 54.6 percent of the blood volume, and in their second-generation offspring it was 66.7 percent, as against 47.5 percent in the control rats of the same age at sea level. The rats also developed an enlargement of the heart like that of human mountain natives. After 10 months at the high altitude they had a 20 percent higher ratio of heart weight to body weight than the sea-level controls did, and in the second-generation rats born at the Barcroft station the increase in heart-weight ratio was 90 percent. An increase in the relative weight of the adrenal glands was also observed in the rats exposed to the high altitude.

The exposure to hypoxia stunted the rats' growth. Up to the age of about 120 days the rats brought to Barcroft (at age 30 days) gained weight at the same rate as the rats left at sea level, but thereafter their growth slowed, and their maximum weight (at about 300 days) was significantly lower than that of the sea-level controls. The growth rate of the second-generation rats at high altitude was lower still: at 130 days they weighed only 250 grams, whereas their parents and the sea-level controls attained this weight in 84 days.

Fenton Kelley at the Barcroft Laboratory investigated the rats' reproduction and high-altitude effects on their young. The hypoxic conditions did not impair the ability to conceive in young, healthy rats: more than 85 percent of the females that were mated after 30 days of acclimatization became pregnant. Their fetuses, however, suffered considerable attrition: by the 15th day of pregnancy 25 percent of the females had abnormally stunted fetuses. Whether this is due to inadequacy of the oxygen supply to the fetus, disturbance of hormone production, neurological anomalies or metabolic disturbances has not yet been determined. At all events, the females bred at high altitude bore substantially smaller litters of live young than those bred at sea level.

The offspring were generally normal in weight at birth, but by the age of 10 days their weight was 30 percent less than the sea-level norm. Their mortality rate in the first 10 days was about 20 percent, 10 times higher than in the sea-level control group. This was not attributable to lack of nursing ability in their mothers; in fact, the high-altitude infant rats had more milk in their stomachs than the sea-level young of the same age did. There were indications that the high-altitude young had metabolic defects that may account for their high mortality and for the slow rate of growth in those offspring that survived the postpartum period. The high-altitude young had a subnormal content of glycogen in the liver, apparently reflecting a defect in carbohydrate metabolism, and there is reason to believe the metabolism of fats and proteins also is affected by hypoxia. The dry atmosphere of high altitude may be another hazard for the newborn, affecting the

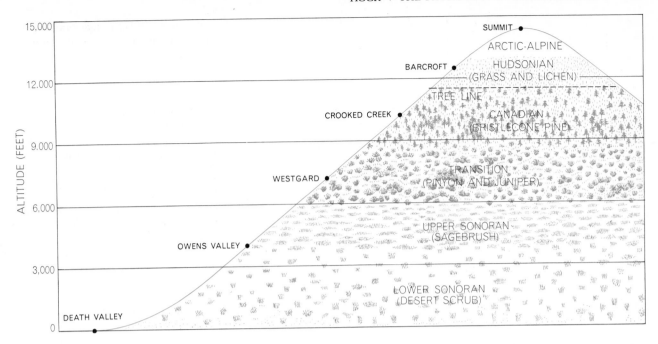

ALTITUDE RANGE of the deer mouse *Peromyscus maniculatus*, the animal the author studied, is remarkably large, extending from sea level to the summit of White Mountain in California. It en- compasses six of the seven "life zones" described some years ago by the American naturalist C. Hart Merriam, which are indicated, with their characteristic vegetation, on this schematic diagram.

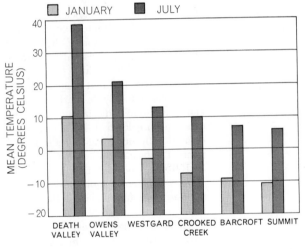

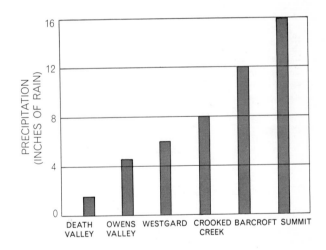

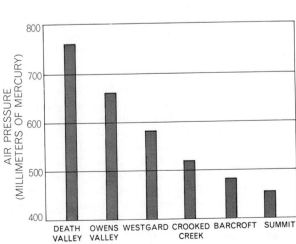

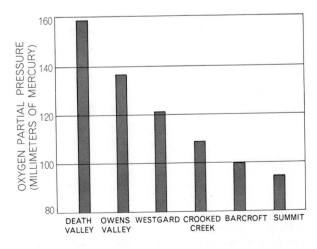

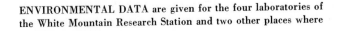

ENVIRONMENTAL DATA are given for the four laboratories of the White Mountain Research Station and two other places where the author worked, Death Valley and Westgard. Altitudes are in- dicated at the top of the page. (Summit precipitation is estimated.)

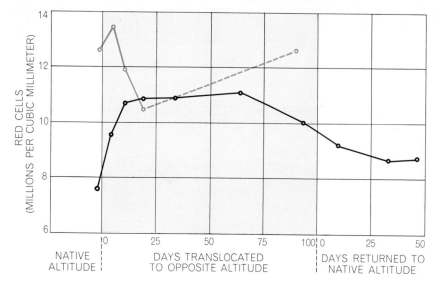

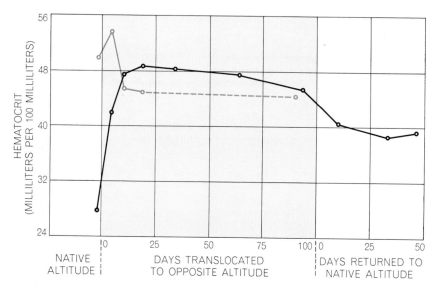

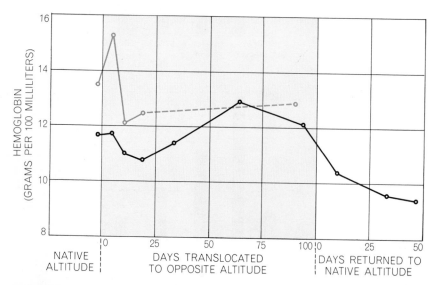

DEER MICE trapped at 12,500 feet (*color*) have more red blood cells, a higher hematocrit (red cells as a proportion of total blood volume) and more hemoglobin than mice trapped at sea level (*black*). As the curves indicate, these values increase in sea-level mice that are transported to the high altitude and decline again with their return to sea level. There is less long-term change, however, in high-altitude mice that are brought down to sea level.

body's water balance, temperature regulation, respiration and vulnerability to infection.

The Deer Mouse

My own investigations focused on the deer mouse (*Peromyscus maniculatus*), a small, white-footed species that is noted for its ubiquitous presence throughout North America and its ability to live in all climatic zones except extreme desert. Deer mice of various species are found inhabiting all altitudes from below sea level to about 15,000 feet. This one species is ideal for our studies not only because of its great variety of physiological responses to different conditions but also because a mouse spends its entire life in the same locality. A tiny deer mouse on White Mountain probably does not range more than a quarter of a mile from its birthplace during its lifetime; consequently we could be sure that a mouse trapped at 12,500 feet on White Mountain had been born at about that altitude (within 500 feet) and had been exposed to it throughout its life.

I trapped deer mice at seven different altitudes, ranging from below sea level to the 14,250-foot White Mountain summit, and for comparison of their native differences each population was examined at the altitude at which it was caught. It was apparent at once that the mice followed only in part the well-known rule that body size increases with exposure to cold: their body weight increased with increasing altitude up to a point. The heaviest weight was found at 10,150 feet; beyond that the oxygen scarcity apparently limits growth in deer mice as it was found to do in the experiments on rats.

There was also a clear progression, with increasing altitude, in the ratio of the heart size to the body weight. In the deer mice caught at 14,250 feet the relative heart weight was one and a half times greater than it was in mice living at an altitude of 4,000 feet. The relative number and mass of red cells and the amount of hemoglobin in the blood also increased with altitude in the deer mice as it does in rats and men. Contrary to what had been observed in rats, however, the mice's adrenal glands shrank with increasing altitude, both in absolute weight and in relation to body weight. Presumably this phenomenon in the mice reflects a different physiological response to the stresses of high altitude from that shown by the rats.

In an effort to determine whether the differences among the natives of various altitudes were genetic or simply the

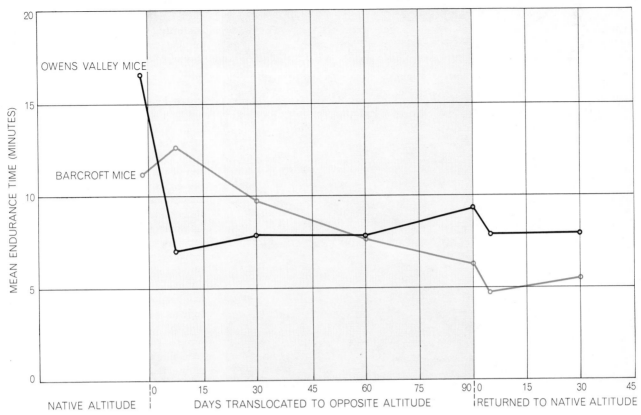

ENDURANCE of mice trapped at 4,000 feet (*black*) was greater than that of mice trapped at 12,500 feet (*color*) when both groups were in their home environments (*left*). The low-altitude mice showed a decided decrease in endurance when they were first taken to 12,500 feet but then improved somewhat. Surprisingly, the high-altitude natives did less well at 4,000 feet than at 12,500.

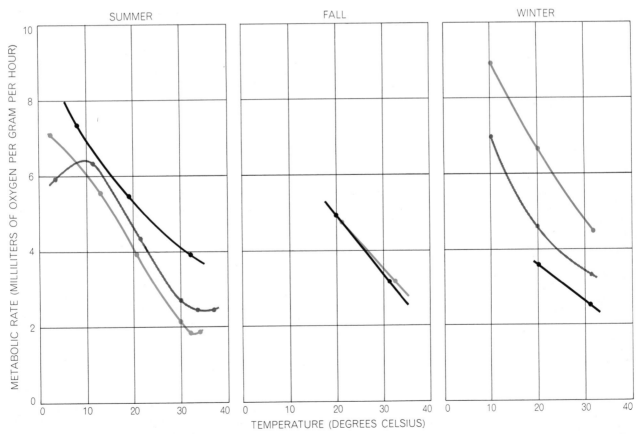

BASAL METABOLIC RATE (oxygen consumption at rest) of mice native to sea level (*black*), 4,000 feet (*gray*) and 12,500 feet (*color*) was measured at their native altitudes in the summer and winter over a wide range of ambient temperatures; two of the three groups were tested in the fall over a narrower temperature range. In the summer the sea-level mice had the highest rates and the 12,500-foot animals had the lowest rates; in the fall they were about equal, and in the winter the summertime findings were reversed (*see text*).

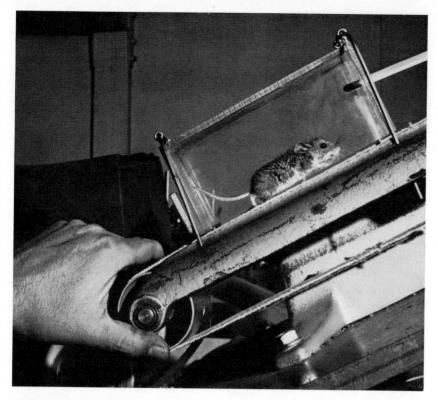

MINIATURE TREADMILL for testing the capacity of deer mice for strenuous exercise was improvised by mounting a linen belt on an old belt sander. The belt was moved at a speed of a mile and a half an hour. The mouse ran on it, restrained by the plastic barriers, until it was exhausted, and the length of time it ran was the measure of its endurance.

decreased with altitude. By the age of maturity (120 days) the average weight of mice born at 4,000 feet was 22.5 grams; of those born at 14,250 feet, 18.8 grams.

Exercise and Metabolism

In order to test the capacity of the mice for strenuous exercise I developed a miniature treadmill that could be run at various speeds. As was to be expected, the natives at low altitudes showed more endurance than those at higher levels. The mean time before exhaustion on the treadmill in one series of tests, for example, was 16.7 minutes for natives at 4,000 feet and only 11.2 minutes for natives at 12,500 feet. Was the difference in performance due to the handicap of hypoxia at the higher altitude or to some innate physical or physiological difference in the mice themselves? I examined this question by switching the environment for the mice, testing the low-altitude natives at the high station and the high-altitude natives at the lower station.

The 4,000-foot natives, when taken to the 12,500-foot level, at first showed a drop in endurance on the treadmill. Their performance slowly improved, however, as they became acclimatized over a period of 90 days. The transfer of 12,500-foot natives to tests at the 4,000-foot level, on the other hand, yielded a major surprise. Although they were now performing in an atmosphere richer in oxygen, their endurance on the treadmill declined instead of improving. At the end of 90 days of "acclimatization" they were able to run on the treadmill only half as long, before exhaustion, as they had done in their oxygen-poor native environment! The explanation may lie in abnormalities of the heart and certain other functions in the high-altitude mice and in the change to a new climate at the lower altitude.

In one study I compared low-altitude and high-altitude mice with regard to their consumption of oxygen during exercise. When sea-level natives were transferred to high altitude, they did not increase their oxygen consumption more while exercising than they had at sea level. High-altitude natives, on the other hand, showed a considerably greater increase in oxygen consumption during exercise than the low-altitude mice did, both at high and at low altitudes. In short, under both conditions the high-altitude mice paid a higher cost (that is, were less efficient) in the use of oxygen during exercise.

I measured the basal metabolic rate

result of acclimatization from birth, I then began to transfer mice from one altitude to another for study of their responses to the change. When mice from the sea-level colony were transferred to our 12,500-foot laboratory, their relative heart weight did not increase. They showed definite signs of acclimatization, however, in other responses. The adrenal glands diminished in size. The weight of the spleen decreased and the lung weight increased, indicating that circulatory and respiratory adjustments were taking place. The red-cell mass and the hemoglobin content of the cells increased to about the same values as in native high-altitude mice. When the surviving mice were later returned to sea level, they soon reverted to their original sea-level condition: the adrenal glands gained in weight and the red-cell mass and hemoglobin fell back to sea-level norms. It appears, therefore, that most of the adaptive mechanisms found in native high-altitude mice are actually adjustments acquired in the course of their exposure to the conditions of their environment.

The reverse experiment—transferring high-altitude mice to sea level—produced a mixed picture. In these mice the relative heart weight decreased and the spleen weight increased, but the adrenals and the lungs showed no change in relative weight. After 90 days of acclimatization to the low altitude the number of red cells in the mice's blood remained unchanged from what it had been at high altitude. This seems to suggest that the high red-cell count in high-altitude mice may represent a genetic adaptation, but it might be explainable on the basis that the translocated animals simply retained their original high red-cell count because there was nothing in the change to low-altitude conditions that would foster destruction of the cells. The total mass of the red cells in proportion to the blood volume did decrease to sea-level values; this may have been due to an increase in the amount of plasma. The concentration of hemoglobin, however, changed only slightly if at all.

The high-altitude natives showed no inferiority to low-altitude mice in fertility; in fact, the average litter size at 10,000 feet or above was six, as against five for females living at 4,000 feet. The high-altitude young, however, had a poor survival rate: mortality among them by the 30th day after birth was 23 percent, whereas all the low-altitude young survived beyond that age. I found that the rate of growth for the young was about the same at all levels during the first 45 days of life, but thereafter it

(as indicated by oxygen consumption at rest) of the mice at various altitudes under various temperature conditions. The stations used for comparison were at sea level, at 4,000 feet in the Owens Valley (which is hot in summer) and at 12,500 feet in the Barcroft Laboratory. The determinations were made during three different seasons and at temperatures ranging from near freezing to a maximum of about 99 degrees Fahrenheit. I found that in summer the high-altitude mice had the lowest metabolic rate, the 4,000-foot mice an intermediate rate and the sea-level mice the highest rate. (The highest temperature proved to be lethal for mice from the high altitude, and conversely lower-altitude mice tested at near-freezing temperatures became severely chilled.) In the fall (at temperatures between 68 and 90 degrees F.) the high-altitude mice and sea-level mice had about the same rate of metabolism. In winter (February) the situation was the reverse of the summer picture: now the high-altitude mice had the highest rate at all temperatures, the sea-level mice the lowest.

These observations could be interpreted as follows. At the high altitude the comparatively mild temperatures of summer enable the native mouse to adjust to the hypoxic environment by reducing oxygen consumption to the minimum required for nourishing the body tissues. Because of the mouse's small size it cannot grow a thick enough hair covering to insulate it effectively against the winter cold. Consequently the high-altitude mouse is forced to increase its metabolism in winter to maintain its body temperature. The sea-level mouse, on the other hand, is not subjected to extreme cold in winter. Hence it is adequately protected from the drop in the ambient temperature by a small increase in its furry insulation and by an adjustment in the form of "physiological insulation," that is, reduction of its body temperature, which cuts down heat loss by reducing the temperature difference between the body and the ambient air. (I found that the deep-body temperature of the sea-level mouse does indeed decrease in winter.) Moreover, the sea-level mouse in winter can afford to reduce its metabolic rate from the high rate associated with summer activities without stinting its tissues' needs for oxygen.

With two associates, Robert E. Smith and Jane C. Roberts, I looked into the metabolic response of mice at the cell and tissue levels. We found several marked differences in cell activities at low altitudes and at high altitudes. In most cases it was difficult to tell whether the observed differences were due to exposure to cold or to hypoxia. Studies of the brown fat in these animals, however, produced some significant findings.

Brown fat, found mainly between an animal's shoulder blades, is a heat-generating tissue [see "The Production of Heat by Fat," by Michael J. R. Dawkins and David Hull; SCIENTIFIC AMERICAN Offprint 1018]. In response to exposure to cold there is an increase in both the mass of brown fat and its heat production, which may multiply severalfold in a few weeks. We found that the mass of brown fat could be increased in deer mice by exposure to cold or to hypoxia (which also lowers the body temperature). It turned out, however, that when sea-level mice were transferred to high altitude, the respiration of their brown fat decreased, so that its heat production was reduced in spite of a clear increase in its mass. On the other hand, when high-altitude natives were brought down to sea level, both the mass and the respiration of their brown fat increased, and its heat production apparently equaled that of sea-level mice that had been acclimated to cold. Evidently the transfer to the oxygen abundance at sea level had improved the tissue's respiration so that it increased its production of heat. From these findings we concluded that hypoxia, although it gives rise to the growth of brown fat, may suppress heat production by limiting the tissue's respiration.

Heredity or Environment?

The extensive investigations leave unsettled the question of whether men native to high altitude are a race apart or merely human beings with a normal heredity who have adjusted to the conditions over a lifetime of habituation beginning in the uterus. In some respects the mountain natives, both animals and men, do seem to show innate physiological differences from their kindred species at sea level. There is a serious objection to considering them a separate strain, however, namely the lack of genetic isolation. There has been no barrier in this case to the pooling of genes, either for the deer mouse or for man. We know that Andean man has intermarried freely with lowlanders. Indeed, many of the mountain miners came from the lowlands and many highlanders have come down to live in the lowlands. It seems likely that the highlanders have derived their special qualities from acclimatization—in short, that their response to their environment is phenotypic rather than genotypic.

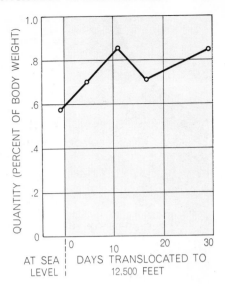

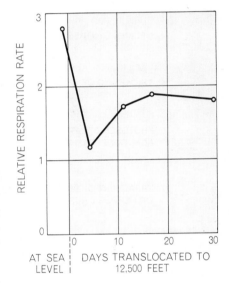

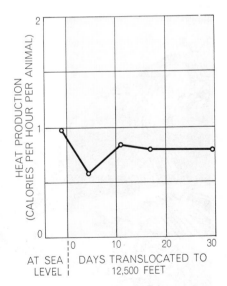

BROWN FAT, a tissue that generates heat, increased in quantity in sea-level deer mice transported to 12,500 feet (top). The rate of respiration of the tissue decreased, however, because of the relative lack of oxygen (middle). As a result the animal's total heat production was somewhat reduced (bottom).

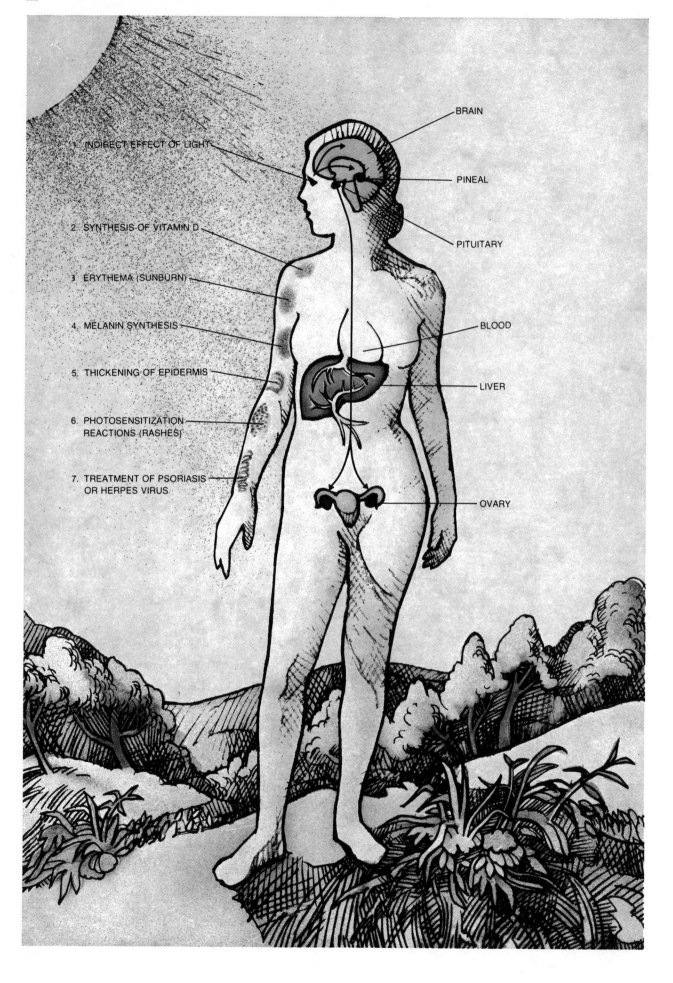

11

The Effects of Light
on the Human Body

by Richard J. Wurtman
July 1975

*Sunlight tans skin, stimulates the formation of
vitamin D and sets biological rhythms. Light is also
used in the treatment of disease. Such effects
now raise questions about the role of artificial light*

Since life evolved under the influence of sunlight, it is not surprising that many animals, including man, have developed a variety of physiological responses to the spectral characteristics of solar radiation and to its daily and seasonal variations. With the coming of summer in the Northern Hemisphere millions of people living in the North Temperate Zone will take the opportunity to darken the shade of their skin, even at the risk of being painfully burned. Coincidentally the sunbathers will replenish their body's store of vitamin D, the vitamin that is essential for the proper metabolism of calcium. Skin-tanning and subcutaneous synthesis of vitamin D from its precursors, however, are only the best-known consequences of exposure to sunlight.

Investigators are slowly uncovering subtler physiological and biochemical responses of the human body to solar radiation or its artificial equivalent. Within the past few years, for example, light has been introduced as the standard method of treatment for neonatal jaundice, a sometimes fatal disease that is common among premature infants. More recently light, in conjunction with a sen-sitizing drug, has proved highly effective in the treatment of the common skin inflammation psoriasis. It seems safe to predict that other therapeutic uses for light will be found.

At least equally significant for human well-being is the growing evidence that fundamental biochemical and hormonal rhythms of the body are synchronized, directly or indirectly, by the daily cycle of light and dark. For example, my co-workers at the Massachusetts Institute of Technology and I have recently discovered a pronounced daily rhythm in the rate at which normal human subjects excrete melatonin, a hormone synthesized by the pineal organ of the brain. In experimental animals melatonin induces sleep, inhibits ovulation and modifies the secretion of other hormones. In man the amount of the adrenocortical hormone cortisol in the blood varies with a 24-hour rhythm. Although seasonal rhythms associated with changes in the length of the day have not yet been unequivocally demonstrated in human physiology, they are well known in other animals, and it would be surprising if they were absent in man. The findings already in hand suggest that light has an important influence on human health, and that our exposure to artificial light may have harmful effects of which we are not aware.

The wavelengths of radiation whose physiological effects I shall discuss here are essentially those supplied by the sun after its rays have been filtered by the atmosphere, including the tenuous high-altitude layer of ozone, which removes virtually all ultraviolet radiation with a wavelength shorter than 290 nanometers. The solar radiation that reaches the earth's surface consists chiefly of the ultraviolet (from 290 to 380 nanometers), the visible spectrum (from 380 to 770 nanometers) and the near infrared (from 770 to 1,000 nanometers). About 20 percent of the solar energy that reaches the earth has a wavelength longer than 1,000 nanometers.

The visible spectrum of natural sunlight at sea level is about the same as the spectrum of an ideal incandescent source radiating at a temperature of 5,600 degrees Kelvin (degrees Celsius above absolute zero). The solar spectrum is essentially continuous, lacking only certain narrow wavelengths absorbed by elements in the sun's atmosphere, and at midday it has a peak intensity in the blue-green region from 450 to 500 nanometers [*see upper illustration on next page*]. The amount of ultraviolet radiation that penetrates the atmosphere varies markedly with the season: in the northern third of the U.S. the total amount of erythemal (skin-inflaming) radiation that reaches the ground in December is only about a fifteenth of the amount present in June. Otherwise there is little seasonal change in the spectral composition of the sunlight reaching the ground. The actual number of daylight hours, of course, can vary greatly, depending on the season and the distance north or south of the Equator.

SOME DIRECT AND INDIRECT EFFECTS OF LIGHT on the human body are outlined in the drawing on the opposite page. Indirect effects include the production or entrainment (synchronization) of biological rhythms. Such effects are evidently mediated by photoreceptors in the eye (*1*) and involve the brain and neuroendocrine organs. For example, excretion of melatonin, a hormone produced by the pineal organ, follows a daily rhythm. In animals melatonin synthesis is regulated by light. The hormone, acting on the pituitary, plays a role in the maturation and the cyclic activity of the sex glands. Ultraviolet radiation acts on the skin to synthesize vitamin D (*2*). Erythema, or reddening of the skin (*3*), is caused by ultraviolet wavelengths between 290 and 320 nanometers. In response melanocytes increase their synthesis of melanin (*4*), a pigment that darkens the skin. Simultaneously the epidermis thickens (*5*), offering further protection. In some people the interaction of light with photosensitizers circulating in the blood causes a rash (*6*). In conjunction with selected photosensitizers light can be used to treat psoriasis and other skin disorders (*7*). In infants with neonatal jaundice light is also used therapeutically to lower the amount of bilirubin circulating in the blood until infant's liver is mature enough to excrete the substance. The therapy prevents the bilirubin from concentrating in the brain and destroying brain tissue.

The most familiar type of artificial light is the incandescent lamp, in which the radiant source is a hot filament of tungsten. The incandescent filament in a typical 100-watt lamp has a temperature of only about 2,850 degrees K., so that its radiation is strongly shifted to the red, or long-wavelength, end of the spectrum. Indeed, about 90 percent of the total emission of an incandescent lamp lies in the infrared.

Fluorescent lamps, unlike the sun and incandescent lamps, generate visible light by a nonthermal mechanism. Within the glass tube of a fluorescent lamp ultraviolet photons are generated by a mercury-vapor arc; the inner surface of the tube is coated with phosphors, luminescent compounds that emit visible radiations of characteristic colors when they are bombarded with ultraviolet photons. The standard "cool white" fluorescent lamp has been designed to achieve maximum brightness for a given energy consumption. Brightness, of course, is a subjective phenomenon that depends on the response of the photoreceptive cells in the retina. Since the photoreceptors are most sensitive to yellow-green light of 555 nanometers, most fluorescent lamps are designed to concentrate much of their output in that wavelength region. It is possible, however, to make fluorescent lamps whose spectral output closely matches that of sunlight [see lower illustration on this page].

Since fluorescent lamps are the most widely used light source in offices, factories and schools, most people in industrial societies spend many of their waking hours bathed in light whose spectral characteristics differ markedly from those of sunlight. Architects and lighting engineers tend to assume that the only significant role of light is to provide adequate illumination for working and reading. The illumination provided at eye level in artificially lighted rooms is commonly from 50 to 100 footcandles, or less than 10 percent of the light normally available outdoors in the shade of a tree on a sunny day.

The decision that 100 footcandles or less is appropriate for indoor purposes seems to be based on economic and technological considerations rather than on any knowledge of man's biological needs. Fluorescent lamps could provide higher light intensities without excessive heat production, but the cost of the electric power needed for substantially higher light levels would probably be prohibitive. Nevertheless, the total amount of light to which a resident of Boston, say, is exposed in a conventionally lighted indoor environment for 16 hours a day is considerably less than would impinge on him if he spent a single hour each day outdoors. If future studies indicate that significant health benefits (for example better bone mineralization) might accrue from increasing the levels of indoor lighting, our society might, in a period of energy shortages, be faced with hard new choices.

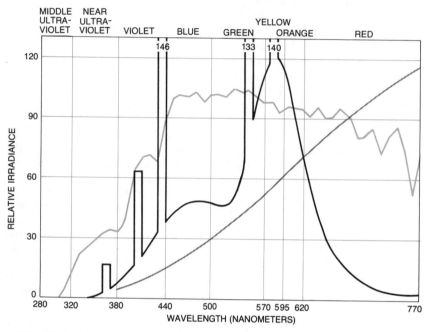

SPECTRUM OF SUN at sea level (*color*) is compared with the spectra of a typical incandescent lamp (*gray curve*) and of a standard "cool white" fluorescent lamp (*black curve*). The visible spectrum lies between the wavelengths of 380 and 770 nanometers. The peak of the sun's radiant energy falls in the blue-green region between 450 and 500 nanometers. Cool-white fluorescent lamps are notably deficient precisely where the sun's emission is strongest. Incandescent lamps are extremely weak in the entire blue-green half of the visible spectrum.

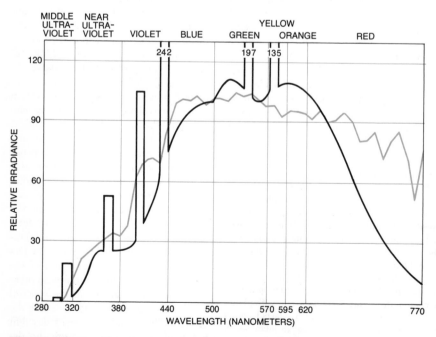

BROAD-SPECTRUM FLUORESCENT LAMP known as Vita-Lite (*black curve*) closely approximates spectral characteristics of sunlight (*color*). Wavelengths emitted by fluorescent lamps can be adjusted by selecting phosphors with which inner surface of lamp is coated.

Each of the various effects of light on mammalian tissues can be classified as direct or indirect, depending on whether the immediate cause is a photochemical reaction within the tissue or a neural or neuroendocrine signal generated by a photoreceptor cell. When the effect is direct, the molecule that changes may or may not be the one that actually absorbs the photon. For example, certain molecules can act as photosensitizers: when they are raised to transient high-energy states by the absorption of radiation, they are able to catalyze the oxidation of numerous other compounds before they return to the ground state. Photosensitizers sometimes present in human tissues include constituents of foods and drugs and of toxins produced in excess by some diseases.

In order to prove that a particular chemical change in a tissue is a direct response to light one must show that light energy of the required wavelength does in fact penetrate the body to reach the affected tissue. In addition the photoenergetic and chemical characteristics of the reaction must be fully specified, first in the test tube, then in experimental animals or human beings, by charting the reaction's "action spectrum" (the relative effectiveness of different spectral bands in producing the reaction) and by identifying all its chemical intermediates and products. Visible light is apparently able to penetrate all mammalian tissues to a considerable depth; it has even been detected within the brain of a living sheep.

Ultraviolet radiation, which is far more energetic than visible wavelengths, penetrates tissues less effectively, so that erythemal radiations barely reach the capillaries in the skin. The identification of action spectra for the effects of light on entire organisms presents major technical problems: few action spectra have been defined for chemical responses in tissues other than the skin and the eyes.

The indirect responses of a tissue to light result not from the absorption of light within the tissue but from the actions of chemical signals liberated by neurons or the actions of chemical messengers (hormones) delivered by circulation of the blood. These signals in turn are ultimately the result of the same process as the one that initiates vision: the activation by light of specialized photoreceptive cells. The photoreceptor transduces the incident-light energy to a neural signal, which is then transmitted over neural, or combined neural-endocrine, pathways to the tissue in which the indirect effect is observed. For example, when young rats are kept continuously under light, photoreceptive cells in their retina release neurotransmitters that activate brain neurons; these neurons in turn transmit signals over complex neuroendocrine pathways that reach the anterior pituitary gland,

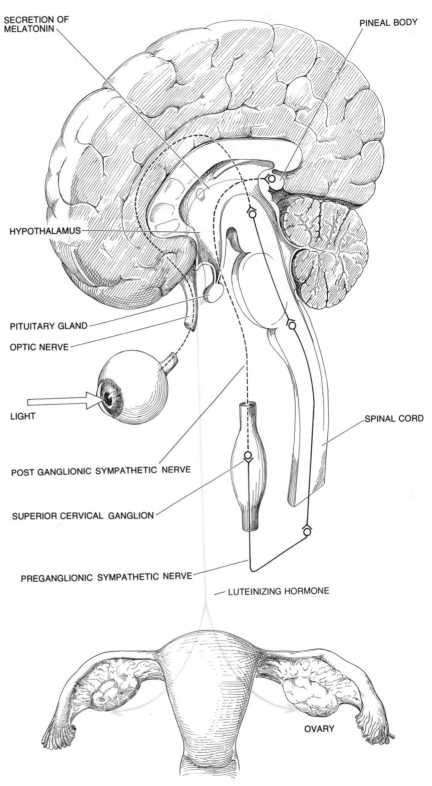

INDIRECT EFFECT OF LIGHT ON OVARIES OF RATS is shown schematically. Light activates receptors in the retina, giving rise to nerve impulses that travel via a chain of synapses through the brain, the brain stem and the spinal cord, ultimately decreasing the activity of neurons running to the superior cervical ganglion (in the neck) and of the sympathetic nerves that reenter the cranium and travel to the pineal organ. There the decrease in activity reduces both the synthesis and the secretion of melatonin. With less melatonin in blood or cerebrospinal fluid, less reaches brain centers (probably in hypothalamus) on which melatonin acts to suppress secretion of luteinizing hormone from anterior pituitary. Thus more hormone is released, facilitating ovarian growth and presumably ovulation.

where they stimulate the secretion of the gonadotropic hormones that accelerate the maturation of the ovaries [*see illustration on preceding page*].

That the ovaries are not responding directly to light can be shown by removing the eyes or the pituitary gland of the rat before exposing it to continuous light. After either procedure light no longer has any influence on ovarian growth or function. Various studies confirm that the effect of light on the ovaries is mediated by photoreceptive cells in the retina. It has not been possible to show, however, which of the photoreceptors in the eye release the neurotransmitters that ultimately affect the pituitary gland.

Natural sunlight acts directly on the cells of the skin and subcutaneous tissues to generate both pathological and protective responses. The most familiar example of a pathological response is sunburn; in susceptible individuals exposed over many years sunlight also causes a particular variety of skin cancer. The chief protective response is tanning. Ultraviolet wavelengths in the narrow band from 290 to 320 nanometers cause the skin to redden within a few hours of exposure. Investigators generally agree that the inflammatory reaction, which may persist for several days, results either from a direct action of ultraviolet photons on small blood vessels or from the release of toxic compounds from damaged epidermal cells. The toxins presumably diffuse into the dermis, where they damage the capillaries and cause reddening, heat, swelling and pain. A number of compounds have been proposed as the offending toxins, including serotonin, histamine and bradykinin. Sunburn is largely an affliction of industrial civilization. If people were to expose themselves to sunlight for one or two hours every day, weather permitting, their skin's reaction to the gradual increase in erythemal solar radiation that occurs during late winter and spring would provide them with a protective layer of pigmentation for withstanding ultraviolet radiation of summer intensities.

Immediately after exposure to sunlight the amount of pigment in the skin increases, and the skin remains darker for a few hours. The immediate darkening probably results from the photooxidation of a colorless melanin precursor and is evidently caused by all the wavelengths in sunlight. After a day or two, when the initial response to sunlight has subsided, melanocytes in the epidermis begin to divide and to increase their synthesis of melanin granules, which are then extruded and taken up into the adjacent keratinocytes, or skin cells [*see illustration at left*]. Concurrently accelerated cell division thickens the ultraviolet-absorbing layers of the epidermis. The skin remains tan for several weeks and offers considerable protection against further tissue damage by sunlight. Eventually the keratinocytes slough off and the tan slowly fades. (In the U.S.S.R. coal miners are given suberythemal doses of ultraviolet light every day on the theory that the radiation provides protection against the development of black-lung disease. The mechanism of the supposed protective effect is not known.)

In addition to causing sunburn and tanning, sunlight or its equivalent initiates photochemical and photosensitization reactions that affect compounds present in the blood, in the fluid space between the cells or in the cells themselves. A number of widely prescribed drugs (such as the tetracyclines) and constituents of foods (such as riboflavins) are potential photosensitizers. When they are activated within the body by light, they may produce transient intermediates that can damage the tissues in sensitive individuals. A typical response is the appearance of a rash on the parts of the body that are exposed to the sun.

In individuals with the congenital disease known as erythropoietic protoporphyria unusually large amounts of porphyrins (a family of photosensitizing chemicals) are released into the bloodstream as a result of a biochemical ab-

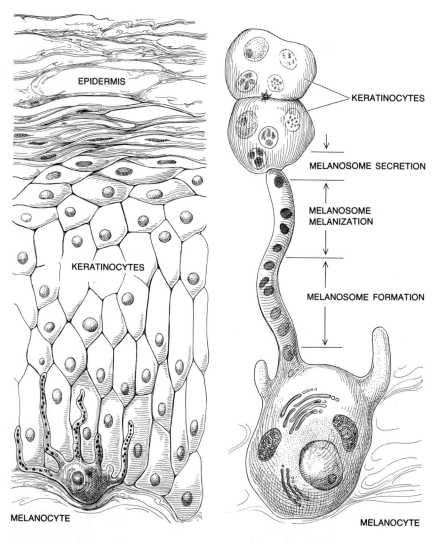

EPIDERMIS

KERATINOCYTES

KERATINOCYTES

MELANOCYTE

MELANOSOME SECRETION

MELANOSOME MELANIZATION

MELANOSOME FORMATION

MELANOCYTE

MECHANISM OF SUN-TANNING is an extension of the mechanism responsible for skin pigmentation. After exposure to the sun melanocytes begin to divide and increase their output of melanin granules, produced in the tiny intracellular bodies called melanosomes. The melanosomes are secreted into the adjacent keratinocytes, or skin cells, where the melanin causes the skin to take on a darker appearance. The tan fades as the keratinocytes slough off.

normality. The porphyrins absorb visible radiations and give rise to intermediates that are toxic to tissues. Patients with the disease complain at first of a burning sensation in areas of the skin that are exposed to sunlight; reddening and swelling soon follow.

Investigators can easily induce these typical symptoms without serious consequences in patients suffering from mild forms of erythropoietic protoporphyria, so that the disease is one of the few of its kind where the action spectrum for a direct effect of light has been studied in detail. The skin damage is caused by a fairly narrow band of wavelengths in the region of 400 nanometers. This band has also been shown to coincide with one of the absorption peaks of abnormal porphyrins. The symptoms of the disease can be ameliorated by administering photoprotective agents such as carotenoids, which quench the excited states of oxygen produced as intermediates in the photosensitization reactions.

In the past few years physicians have treated several skin diseases by deliberately inducing photosensitization reactions on the surface of the body or within particular tissues. The intent is to cause selective damage to invading organisms (such as the herpes virus), to excessively proliferating cells (as in psoriasis) or to certain types of malignant cells. The activated photosensitizers appear to be capable of inactivating the DNA in the viruses or in the unwanted cells. In treating herpes infections the photosensitizer (usually a dye, neutral red) is applied directly to the skin or to the mucous membrane under the ruptured blister; the area is then exposed to low-intensity white fluorescent light.

The treatment for psoriasis was devised by John A. Parrish, Thomas B. Fitzpatrick and their colleagues at the Massachusetts General Hospital. They administer a special photosensitizer (8-methoxypsoralen, or methoxalen) by mouth and two hours later expose the afflicted skin areas for about 10 minutes to the radiation from special lamps that emit strongly in the long-wave ultraviolet at about 365 nanometers. The sensitizing agent is present in small amounts in carrots, parsley and limes. It is derived commercially from an Egyptian plant (*Ammi majus* Linn.) that was used in ancient times to treat skin ailments. Scores of patients have responded successfully to the new light treatment, which will soon be generally available.

The formation of vitamin D_3, or cholecalciferol, in the skin and subcutaneous tissue is the most important of the beneficial effects known to follow exposure to sunlight. Vitamin D_3 is formed when ultraviolet radiation is absorbed by a precursor, 7-dehydrocholesterol. A related biologically active compound, vitamin D_2, can be obtained by consuming milk and other foods in which ergosterol, a natural plant sterol, has been converted to vitamin D_2 by exposure to ultraviolet radiation. Although vitamin D_2 can cure rickets in children who are deficient in vitamin D_3, it has not been demonstrated that vitamin D_2 is biologically as effective as the vitamin D_3 formed in the skin.

In a population of normal white adults living in St. Louis, studied by John G. Haddad, Jr., and Theodore J. Hahn of the Washington University School of Medicine, some 70 to 90 percent of the vitamin D activity in blood samples was found to be accountable to vitamin D_3 or its derivatives. The investigators concluded that sunlight was vastly more important than food as a source of vitamin D. (Although vitamin D_3 is also found in fish, seafood is not an important source in most diets.) In Britain and several other European countries the fortification of foods with vitamin D_2 has now been sharply curtailed because of evidence that in large amounts vitamin D_2 can be toxic, causing general weakness, kidney damage and elevated blood levels of calcium and cholesterol.

A direct study of the influence of light on the human body's ability to absorb calcium was undertaken a few years ago by Robert Neer and me and our coworkers. The study, conducted among elderly, apparently normal men at the Chelsea Soldiers' Home near Boston, suggests that a lack of adequate exposure to ultraviolet radiation during the long winter months significantly impairs the body's utilization of calcium, even when there is an adequate supply in the diet. The calcium absorption of a control group and an experimental group was followed for 11 consecutive weeks from the onset of winter to mid-March.

During the first period of seven weeks, representing the severest part of the winter, all the subjects agreed to remain indoors during the hours of daylight. Thus both groups were exposed more or less equally to a typical low level of mixed incandescent and fluorescent lighting (from 10 to 50 footcandles). At the end of the seven weeks the men in both groups were found to absorb only about 40 percent of the calcium they ingested. During the next four-week period, from mid-February to mid-March, the lighting was left unchanged for the control subjects, and their ability to ab-

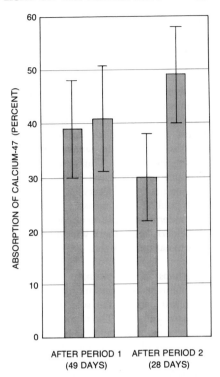

CALCIUM ABSORPTION was increased by a daily eight-hour exposure to broad-spectrum artificial light in a study made by the author and his colleagues at a veterans' home. During the first seven weeks after the beginning of winter, control subjects (*gray bars*) and experimental subjects (*colored bars*) were equally exposed to the same low levels of typical indoor lighting. The bars at the left show their ability to absorb calcium at the end of the initial period. During the next four-week period conditions for the control subjects were unchanged; their ability to absorb calcium fell about 25 percent. The experimental subjects, who were exposed to 500 footcandles of broad-spectrum fluorescent light for eight hours per day for four weeks, showed an average increase of about 15 percent in their calcium absorption.

sorb calcium fell by about 25 percent. The men in the experimental group, however, were exposed for eight hours per day to 500 footcandles of light from special fluorescent (Vita-Lite) lamps, which simulate the solar spectrum in the visible and near-ultraviolet regions. In contrast with the control subjects' loss of 25 percent of their capacity to absorb calcium, the experimental group exhibited an increase of about 15 percent [*see illustration above*]. The additional amount of ultraviolet radiation received by the experimental subects was actually quite small: roughly equivalent to what they would get during a 15-minute lunchtime walk in the summer.

Our study indicates that a certain amount of ultraviolet radiation, whether it is from the sun or from an artificial

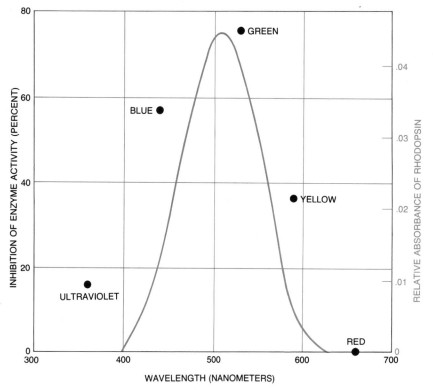

PINEAL ACTIVITY OF RATS can be suppressed by exposing the animals continuously to light. As in the case of the daily temperature rhythm, green light is more effective than light of other spectral colors in suppressing the organ's enzyme activity, as is shown by the labeled dots. The enzyme that is measured is the melatonin-forming enzyme hydroxyindole-O-methyltransferase. Presumably the suppression is mediated by rhodopsin (curve in color).

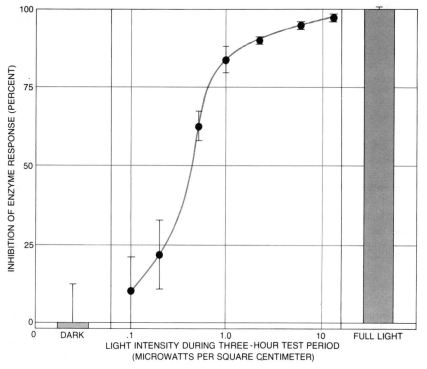

GRADATION IN LIGHT INTENSITY leads to proportional inhibition in the activity of the rat's pineal gland, indicating that light controls synthesis of the pineal hormone, melatonin. Rats that had been kept in constant light for 48 hours were exposed to various light intensities for the next three hours. Their pineals were then analyzed for serotonin-N-acetyltransferase, an enzyme that participates in melatonin synthesis, with the results plotted here.

of the pineal gland of rats [see top illustration at left].

Cycles in environmental lighting can interact with biological rhythms in two ways. The light cycle may directly induce the rhythm, in which case either continuous light or darkness should rapidly abolish it, or the cycle may simply entrain the biological rhythm so that all animals of a given species exhibit maximums or minimums at about the same time of day or night. In the latter case the rhythmicity itself may be generated by a cyclic input other than light, either exogenous (for example food intake) or endogenous (a biological clock). If the cycle is simply entrained by light, an environment of continuous light or darkness might not extinguish it. In human beings psychosocial factors are probably of greater importance than light cycles in generating or synchronizing biological rhythms. The biological utility of even so dramatic a rhythm as that of sleep and wakefulness, for example, remains to be discovered.

Annual rhythms in sexual activity, hibernation and migratory behavior are widespread among animals. The rhythms enable members of a species to synchronize their activities with respect to one another and to the exigencies of the environment. For example, sheep ovulate and can be fertilized only in the fall, thus anticipating the spring by many months, when food will be available to the mother for nursing the newborn. In man no annual rhythms have been firmly established, except, of course, those (such as in sun-tanning and vitamin D_3 levels) that are directly correlated with exposure to summer sunlight.

The best-characterized indirect effect of light on any process other than vision is probably the inhibition of melatonin synthesis by the pineal organ of mammals. Although melatonin seems to be the major pineal hormone, its precise role has not yet been established. When melatonin is administered experimentally, it has several effects on the brain: it induces sleep, modifies the electroencephalogram and raises the levels of serotonin, a neurotransmitter. In addition melatonin inhibits ovulation and modifies the secretion of other hormones from such organs as the pituitary, the gonads and the adrenals, probably by acting on neuroendocrine control centers in the brain.

Experiments performed on rats and other small mammals during the past decade provide compelling evidence that the synthesis of melatonin is suppressed by nerve impulses that reach the pineal over pathways of the sympathetic nervous system. These impulses in turn vary inversely with the amount of visible light impinging on the retina. In rats the pineal function is depressed to half its maximum level when the animals are subjected to an amount of white light only slightly greater than that shed by the full moon on a clear night [*see bottom illustration on opposite page*]. A multisynaptic neuronal system mediates the effects of light on the pineal. The pathway involved, which is apparently unique to mammals, differs from the route taken by the nerve impulses responsible for vision.

Quite recently Harry Lynch, Michael Moskowitz and I have found a daily rhythm in the rate at which normal human subjects excrete melatonin. During the third of the day corresponding to the bedtime hours, 11:00 P.M. to 7:00 A.M., the level of melatonin in the urine is much higher than it is in any other eight-hour period [*see illustration at right*]. It remains to be determined whether the rhythm in melatonin excretion in humans is induced by light or is simply entrained by it.

In some birds and reptiles the pineal responds directly to light, thereby serving as a photoreceptive "third eye" that sends messages about light levels to the brain. In the pineal organ of mammals any trace of a direct response to light is lost. Evidently photoreceptors in the retina mediate the control of the pineal by light. Since, as I have noted, the function of the pineal in rats is influenced most strongly by green light, corresponding to the peak sensitivity of the rod pigment rhodopsin, the retinal photoreceptor would seem to be a rod cell, at least in this species.

Light levels and rhythms influence the maturation and subsequent cyclic activity in the gonads of all mammals and birds examined so far. The particular response of each species to light seems to depend on whether the species is monestrous or polyestrous, that is, on whether it normally ovulates once a year (in the spring or fall) or at regular intervals throughout the year. Examples of polyestrous species are rats (ovulation every four or five days), guinea pigs (every 12 to 14 days) and humans (every 21 to 40 days). The gonadal responses also seem to depend on whether the members of the species are physically active during the daylight hours or during the night. Recently Leona Zacharias and I had the opportunity to examine more than a score of girls and women (members of a diurnally active polyestrous species) who had become blind in the

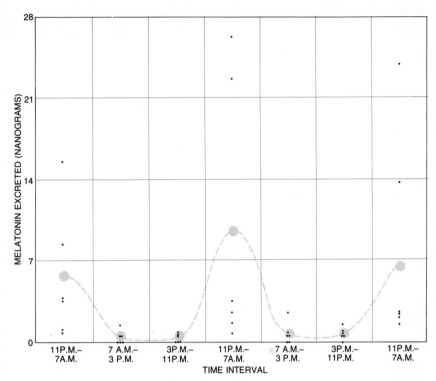

RHYTHM IN MELATONIN SECRETION in human beings has been found by the author and his colleagues. The black dots show the melatonin content of urine samples from six subjects during consecutive eight-hour periods. The colored circles and broken curve correspond to the mean values. High values that were recorded for the 11:00-P.M.-to-7:00-A.M. samples suggest that synthesis of melatonin in man, as in rats, increases with onset of darkness.

first year of life. We observed that gonadal maturation had in general occurred earlier in this group than in normal girls. In contrast, in rats (a polyestrous species that is active at night) blindness delays maturation, and continuous illumination accelerates the maturation of weanlings with normal vision.

The gonads of most birds and of most diurnally active, monestrous animals (the ferret, for instance) mature in the spring, in response to the gradual increase in day length. Ovulation can be accelerated in such animals by exposing them to artificially long days. The annual gonadal activity in domestic sheep, on the other hand, occurs in the fall, in response to the decrease in day length. The mechanisms that cause some species to be monestrous and others polyestrous, or that cause some animals to sleep by day and others by night, are entirely unknown, as are the factors that cause the gonadal responses of various species to light to vary as widely as they do.

The multiple and disparate effects of light I have described support the view that the design of light environments should incorporate considerations of human health as well as visual and aesthetic concerns. We have learned that the chemical constituents of the environ-

ment in the form of food, drugs and pollutants must be monitored and regulated by agencies with suitable powers of enforcement. A major part of their responsibility is to see that nothing harmful is put into food or drugs and that nothing essential is left out of food. The food and drug industries, for their part, look to public and private research organizations, including their own laboratories, for intellectual guidance in creating wholesome and beneficial (as well as profitable) products.

In contrast, only minuscule sums have been expended to characterize and exploit the biological effects of light, and very little has been done to protect citizens against potentially harmful or biologically inadequate lighting environments. Both government and industry have been satisfied to allow people who buy electric lamps—first the incandescent ones and now the fluorescent—to serve as the unwitting subjects in a long-term experiment on the effects of artificial lighting environments on human health. We have been lucky, perhaps, in that so far the experiment has had no demonstrably baneful effects. One hopes that this casual attitude will change. Light is potentially too useful an agency of human health not to be more effectively examined and exploited.

12 Radiation and the Human Cell

by Theodore T. Puck
April 1960

The exposure of single human cells to X-rays indicates that they are far more sensitive to radiation than had been thought. This explains why the relatively small dose of 400 roentgens is lethal to the body

Ever since the discovery of X-rays by Wilhelm Konrad Roentgen in 1895, man has been exposed to steadily increasing doses of high-energy radiation. Beyond their well-established functions in medicine, these radiations have been finding progressively wider use in agricultural and industrial research and in industrial processes. But the great energy that makes them so useful also makes them a hazard to living organisms. They are frequently called "ionizing" radiations because they strip electrons from (that is, ionize) atoms and thus break chemical bonds. The large and elaborately structured molecules of the living cell appear to be especially sensitive to damage by radia-

tion. Such damage may find expression in gross harm and death to the organism exposed, or it may be transmitted as a hereditary defect to the offspring. Such a defect may not be revealed until the appropriate mating occurs, an event that may not take place for many generations.

A major objective of investigation into the biological action of ionizing radiation has been to determine the primary site of radiation damage in the cell. In the case of hereditary defects, *i.e.*, those transmitted to the progeny, it seemed logical to look to the genetic structures: the genes and chromosomes and their attendant apparatus, which are localized in the nucleus. But in the case of pathological damage to the irradiated organ-

ism itself, the picture was not so clear-cut. Ionizing radiation will disrupt any chemical bond in any molecule of any cell that happens to absorb energy from an incident ray. Injuries as diverse as a drop in the level of white cells in the blood; severe skin damage; ulcerating, nonhealing wounds; loss of hair; gastrointestinal disturbance; internal bleeding due to capillary fragility; cancer; death—all follow as recognized consequences of the exposure of mammalian organisms to large doses of radiation. No readily recognizable target site within the individual cells has been demonstrated to be the seat of all these actions. Some investigators have postulated the nucleus to be the principal sensitive re-

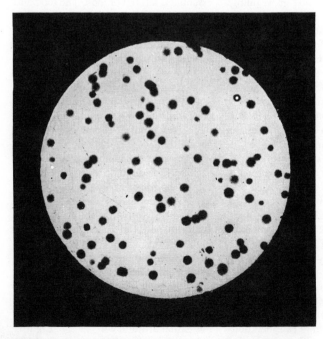
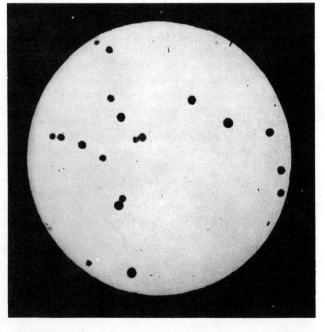

REPRODUCTIVE DEATH in human-tissue cells caused by ionizing radiation is demonstrated by single-cell culture technique. Spots in culture plate at left, shown roughly actual size, are colonies grown from unirradiated control cells. Smaller number of colonies grown from equal number of cells that had been exposed to radiation (*right*) shows how percentage surviving is measured.

gion, while others have held that the basic cellular damage might just as well occur in the structures outside the nucleus, that is, in the cytoplasm, the principal seat of the cell's nonhereditary physiological functions. The issue has been an important one in radiation biology, because a great many other questions have awaited its resolution. A finding that the cell nucleus is an important site of damage would afford new insight into the role of the nucleus in the maintenance of normal function in the body cells of humans and other mammals. It would also indicate the dose range at which genetic damage might be expected to occur.

A ray of high-energy radiation, striking an atom, may knock an electron from a shell of electrons close to the atomic nucleus or from the outermost shell of electrons. The effect on the target atom, however, is the same in the end. Any hole in an inner shell is quickly filled by electrons falling inward from outer shells until the deficiency is transferred to the outermost shell. Since this shell contains the valence electrons that establish the chemical bonds between the atoms in a molecule, the loss of an electron here disrupts the dynamic balance of forces that holds the molecule together. The molecular fragments, produced by the absorption of energy far exceeding that of ordinary chemical bonds, are chemically unsatisfied. They will attack almost any molecule with which they collide in an attempt to regain the lost electron and achieve new stable bonds. The process goes on until all the excited atoms have attained a lower energy-state. In such a simple substance as water the bonds are re-established in the stable configuration H_2O, and all the absorbed energy is converted into heat. Thus there is no permanent change in chemical constitution. A more complex system, such as that represented by a living cell, is not in its lowest possible energy-state initially and possesses countless alternative atomic combinations with similar energy states. These configurations, most of which are incompatible with normal life processes, are usually excluded by the existence of energy barriers. Absorption of high-energy radiation causes these barriers to be surmounted so that the system comes to rest with new chemical bonds established. As a result the chemical constitution of the system is changed by the exposure to radiation; the chemical bonds, broken at random, have been reconstituted in new and bizarre combina-

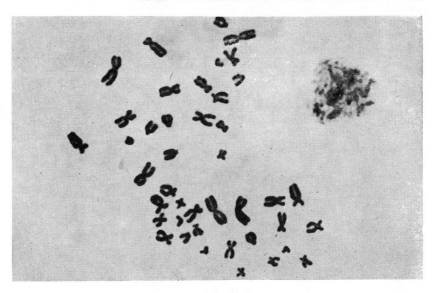

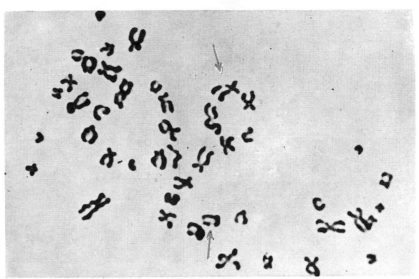

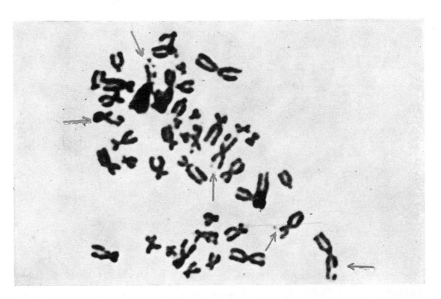

BREAKS IN CHROMOSOMES of human-tissue cells caused by radiation are marked by arrows in two lower photomicrographs. Chromosomes from an unirradiated control cell (*top*) show no breaks; two breaks are marked in chromosomes from cell irradiated with 50 roentgens (*middle*) and five breaks are marked in chromosomes from cell exposed to 75 roentgens (*bottom*). The photomicrographs on this and next page were made by the author.

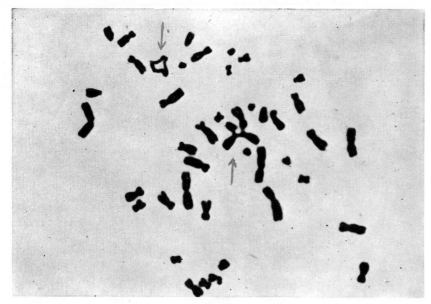

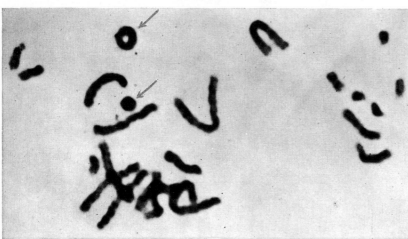

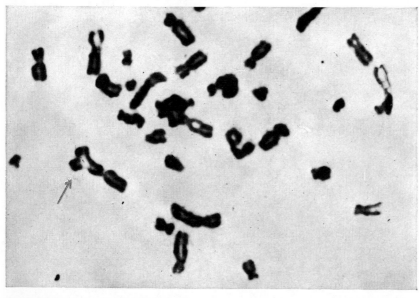

ABNORMAL RECOMBINATIONS that result when several chromosomes are simultaneously broken in the same cell indicate random nature of the healing process. Arrows in photomicrograph at top show recombination of several chromosomes (*center of photomicrograph*) and linkage of two chromosomes at two points (*upper left in photomicrograph*). In middle photomicrograph arrows point to rings formed by recombination of broken ends of the same chromosome. In bottom photomicrograph the arrow points to an aberration involving union of two broken arms of one chromosome and of one break in another.

tions. Since the cell is made up to a large degree of giant molecules whose architecture is vital to their function, the rupture of a few bonds may have far-reaching consequences.

The earliest studies of nonmammalian cells had indicated that the nucleus is far more sensitive to radiation than is the cytoplasm. In the cells of the corn plant and the fruit fly, biologists could observe such aberrations as the breakage of chromosomes, the fusion of broken ends of chromosomes that were not previously joined and the deletion of large areas of chromosome structure. Even where no such obvious damage could be seen, the investigators could demonstrate single-gene mutations by appropriate tests of the progeny. Many of these changes result in the ultimate loss of reproductive capacity, even though the cell may multiply for a few generations after irradiation. Investigators also found that the lethality of chromosomal damage depends upon a variety of factors: the total number of chromosomes, the number of duplicate chromosomes, the number of genes essential to reproduction and the degree to which the function of a given gene involves its association with particular neighbors.

Because the same pattern of chromosomal damage and effects upon function was observed in many different kinds of organism, a number of workers, including the distinguished geneticist H. J. Muller, postulated that like processes underlie the action of ionizing radiations upon mammals as well. In their view damage to the reproductive power of individual body-cells would explain many of the pathological symptoms observed in mammals exposed to radiation.

Against this conclusion, however, there stood some significant evidence that at first appeared convincing. The dose of radiation needed to destroy the reproductive capacity of certain cells, such as yeast and the common bacterium *Escherichia coli*, could be accurately measured; it was found to lie in the range of 5,000 to 15,000 roentgens. The paramecium, a protozoon that in many ways resembles human-tissue cells more closely, showed even greater resistance to exposure: radiation in the range from 20,000 to 100,000 roentgens was required to bring about reproductive death. In contrast, the mean lethal dose for whole-body radiation in man is only about 400 roentgens. It did not seem likely that the pathological consequences of radiation exposure in man could be attributed to reproductive death in human-tissue cells, such as that observed in one-celled organisms. Irradiation of human and other mammalian cells seeded in large numbers in tissue cultures seemed to settle the question. These experiments showed that doses in the neighborhood of a few hundred roentgens apparently produced no more than a temporary lag in growth; doses of 10,000 to 100,000 roentgens were required to achieve permanent inhibition of cellular reproduction by these techniques.

Many investigators accordingly turned to other parts of the cellular machinery in the search for the site most sensitive to radiation in mammalian cells. They tested the effect of radiation on many systems of enzymes—the large molecules that catalyze the chemical reactions of the cell—and demonstrated the existence of many new kinds of radiation damage. But none of these easily fitted the specifications of a truly primary site. In summing up the work on the effects of radiation upon the chemical mechanisms of the cell in 1958, the United Nations Subcommittee on the Effects of Atomic Radiation concluded: "The nature of the initial step of radiation damage remains to be determined."

Meanwhile our group in the biophysics laboratory at the University of Colorado was making progress in the development of a new technique for mammalian tissue-culture, a line of investigation we had undertaken in 1955. From each cell in a sample of tissue cells, added to nutrient medium in a glass dish, we learned to culture a large isolated colony of daughter cells that could easily be recognized and counted [see "Single Human Cells in Vitro," by Theodore T. Puck; SCIENTIFIC AMERICAN Offprint 33]. This technique has made it possible to measure with great accuracy the effect of radiation upon the reproductive capacity of a single cell. From a "clone," or colony, of cells cultured from an original single cell we take a sample for irradiation. We then seed the plates with the irradiated cells and with a sample of unirradiated cells as a control. By counting the colonies that grow from the two samples we find the percentage of the cell population that has lost its ability to reproduce.

This procedure separates the effect of a lag in the rate of reproduction, which is often induced by radiation, from the irreversible cessation of reproduction in a given cell. In an ordinary tissue-culture containing a mass of cells the two effects cannot be distinguished, and the recovery of reproductive capacity by some cells masks the reproductive death of others. In addition, our technique has made it possible to detect individual mutant cells. We can thus measure the mutation-inducing effect of given doses of radiation upon human and other mammalian cells. Tested in this fashion, human-tissue cells from a wide variety of normal organs have shown survival curves of a simple type.

Determination of many such curves shows that the lethal dose for reproduction in human-tissue cells is only 50 roentgens. This figure is obviously far smaller than those indicated by experiments that measured outgrowth from larger chunks of tissue. What this finding means is that a human cell exposed to X-rays suffers reproductive death when it has absorbed from the X-ray beam an amount of energy equivalent to a temperature rise of less than .001 degree centigrade. Cells with an abnormally large number of chromosomes, such as turn up occasionally in tissue cultures, proved to be about twice as resistant to X-rays, but they were still extremely susceptible to doses hitherto considered quite small [see illustration on page 127].

These experiments have also yielded

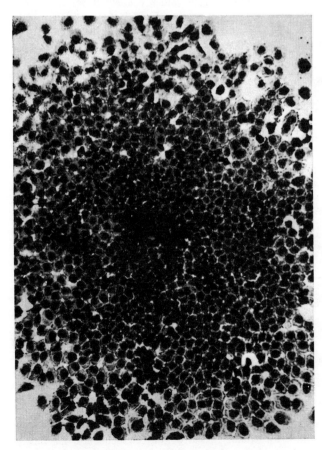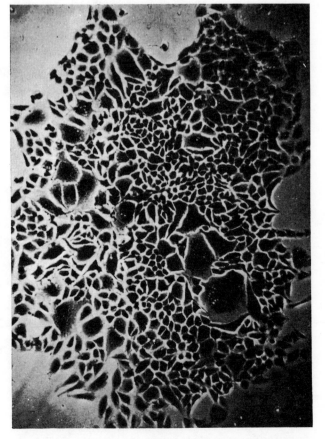

MUTATION OF CELLS, induced by irradiation, is indicated by the appearance of large "monster" cells in colony at right, as compared to cells grown from unirradiated cell in the colony at left. The cells are from a standard HeLa strain of human cancer tissue.

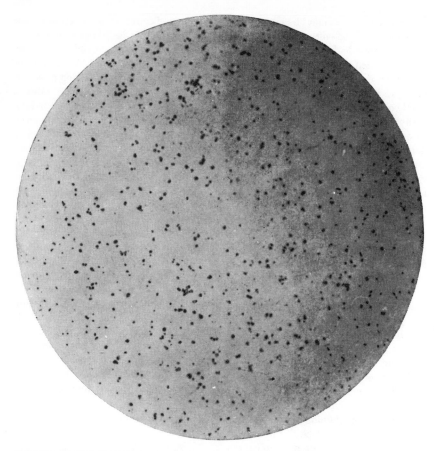

SINGLE GIANT CELLS, shown here at a magnification of only two diameters, demonstrate reproductive death from irradiation. The metabolic processes of the cells continue despite destruction of reproductive capacity; the cells become giants instead of dividing.

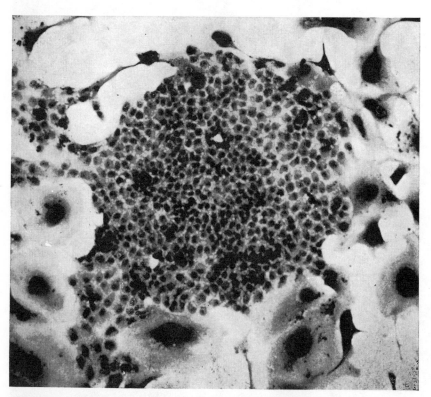

COLONY OF CELLS grown from single-cell survivor of irradiation is here contrasted with radiation-inactivated cells, surrounding the periphery of the colony, which are forming giants. The diameters of some giant cells approach a third the diameter of entire colony.

a variety of different lines of evidence indicating that damage to the genetic apparatus, specifically the chromosomes, occurs at sufficiently low doses to account for the destruction of reproductive capacity in the cells. Even the more indirect evidence is persuasive. For example, if the damage is done in the nucleus rather than in the cytoplasm, one would expect the cells that survived exposure to a lethal dose to bear evidence of this experience in the form of mutations. Colonies grown from such survivors frequently do show changes in their appearance and nutritional requirements which are transmitted to succeeding generations like other genetic characteristics. Similarly, except for the few cells actually dividing, chromosome damage should be relatively constant for a given dose of radiation; on the other hand, damage inflicted upon the physiological mechanisms in the cytoplasm might vary greatly, depending on the metabolic activity of the cells. When we cultured cells from colonies that had been irradiated under conditions of maximal growth and of no growth, we found practically identical X-ray survival curves.

In contrast to the great sensitivity of the genetic apparatus of mammalian cells, other cell structures have been found to be much more resistant to radiation. These studies, carried on in many laboratories, have demonstrated radiation damage to a large number of cellular enzyme systems. In every case the necessary X-ray dose was larger than the chromosome-damaging dose, except of course in functions dependent on chromosomal integrity. Cells irradiated with a dose 10 or 100 times that required to destroy their reproductive capacity can still carry on many metabolic functions. They can take in sugars and utilize their stored chemical energy, synthesize specific proteins and nucleic acids, and take up substances from their culture medium in a fashion virtually identical with that of an unirradiated cell. Cells in this condition may become giants, reaching a diameter as great as a millimeter when adhering to a glass plate, and so are readily visible to the naked eye [top illustration on this page]. A reproductively dead, but metabolically alive, cell can be infected with a virus, and it will proceed to synthesize large numbers of virus particles. The production of these intricate particles with their vast number of specific chemical bonds certainly requires the active functioning of a great deal of the cell's chemical machinery. Large regions of the cell's nongenetic apparatus must remain in working order.

Until recently it would have been difficult to go beyond this indirect evidence. The numerous small chromosomes in mammalian cells do not lend themselves to inspection so readily as those in other forms of life; thus it was not possible to correlate the effects of radiation with observed damage to these vital structures. New techniques have been contributed by several investigators, including T. C. Hsu of the University of Texas and J. H. Tjio in our laboratory, which have made it possible to delineate the chromosomes of mammalian cells. When the cells are placed in a salt solution whose osmotic pressure is lower than that of the cellular fluid, they swell to many times their normal size, so that the chromosomes become separated from one another and show up handsomely after staining. In cells that have been irradiated by doses in the range of 50 roentgens the chromosomal damage thus made visible appears sufficient to account for their reproductive death. As nearly as present experiments can determine the extent of primary damage observed varies directly with the dosage.

Since the chromosomes possess power of self-repair, the number of breaks observed also depends upon the time interval between the irradiation and the treatment of the cell for inspection of its chromosomes. The mechanism of this exceedingly important restitution process is now being thoroughly explored by a number of workers, particularly by Sheldon Wolff of the Oak Ridge National Laboratory. If the two ends of a broken chromosome rejoin, there may remain no visible evidence of the damage, although a gene at the site of the break conceivably might have suffered mutation. When more than one chromosome is broken, however, the chances are appreciable that the ends of different chromosomes will make connections with one another. Some of these abnormal junctions are recognizable under the microscope. Moreover, some of them may cause reproductive death of the cell. For example, the two centromeres that are supposed to separate the replicated chromosomes from one another during cell division may become joined to the two ends of a single chromosome. They will then pull the two ends to the opposite poles of the cell, producing a chromosomal bridge that prevents completion of the process of division. Even if the reproductive capacity survives, the random nature of the radiation process may introduce muta-

tions in the genes that happen to be located at the site of the breaks. Still other genetic changes will reflect the subtle working of the so-called position effect, which modifies a gene's potentiality when it becomes juxtaposed to different neighboring genes on the chromosome.

Careful counting of the chromosomal breaks has shown that the dose needed to cause a single break per cell is only about 40 roentgens. This figure, however, includes a discount for the healing of breaks, since the counting is ordinarily done two or three days after irradiation. If the breaks are scored immediately after irradiation, the average dose needed to produce one break per cell falls to 20 to 25 roentgens. This figure shows that significant chromosomal changes follow doses well within the lethal limit.

The genetic apparatus thus shows extraordinary sensitivity to radiation. Calculation indicates that the ionization of not more than a few hundred atoms within or around the space occupied by the chromosome may be sufficient to produce a visible break. It seems probable that an even greater number of lesions may occur and remain invisible, either

because they are submicroscopic or because they become resealed before they are examined. The biochemical nature of this reaction, resulting from so small an initiating stimulus, will afford a fascinating area of investigation.

How well do these experiments, performed in laboratory glassware, reflect events that occur in intact human tissue under irradiation? Several lines of experimentation indicate that the *in vitro* model is useful for understanding at least some events *in vivo*. H. B. Hewitt and C. W. Wilson of the Westminster Hospital in London have studied the X-ray survival curves of leukemia cells in mice. They irradiate leukemic animals and then test the capacity of cells from these animals to induce leukemia in other mice. This is a reliable test, because a single cell from an unirradiated animal suffices to carry the disease to a new animal. The survival curve for single irradiated cells established by this experiment corresponds closely to the curve for single cells growing in tissue culture.

Perhaps an even more striking demonstration is afforded by the results of

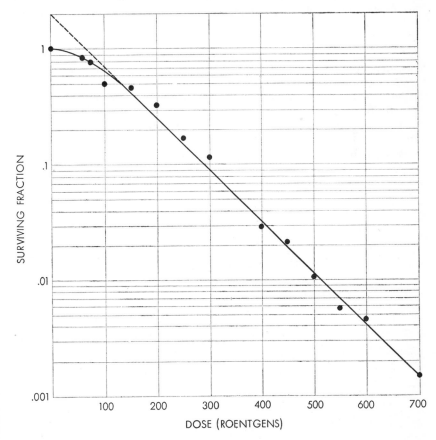

SURVIVAL CURVE for human cells originating from the standard HeLa cancer-cell strain shows correlation between exposure to radiation and reproductive death in the cells. Cells from normal human tissue show same curve but a somewhat higher sensitivity to radiation.

an experiment involving a situation in the living organism that closely resembles that in the tissue-culture experiments. This is the fertilized egg, a single cell whose history can be readily traced following an episode of irradiation. At the Oak Ridge National Laboratory W. L. Russell and his wife Liane Brauch Russell have exposed female mice to a standard dose of 200 roentgens soon after mating, while the egg is still presumably in the one-celled stage. Only 20 per cent of the embryos survived; this finding agrees well with the figure predicted from experiments with single-cell cultures. When irradiation occurred at later, multicelled stages of development, the survival rate increased, reflecting the capacity of the uninjured cells to replace those destroyed.

With the new measurement of the mean lethal dose for reproductive function of mammalian cells it is now possible to explain the relatively low mean lethal dose of 400 to 500 roentgens for the entire body. Such a dose leaves only about .5 per cent of the body's reproducing cells still able to multiply. Death, however, will not be immediate. The cells have each absorbed an almost infinitesimal amount of radiation energy. Though they have suffered an appreciable amount of chromosomal damage, their enzymatic machinery is, by and large, still active. Each such cell continues to perform its physiological functions in reasonably normal fashion until the time comes for it to reproduce. But at the next division, or at the next one or two divisions, reproduction will fail. One of the most characteristic features of radiation injury is thus explained: the relatively long lag that usually occurs between even severe irradiation and the

development of pathological symptoms.

The same line of reasoning helps explain some aspects of the typical course of radiation disease. One would deduce that the tissue functions first embarrassed are those most dependent upon rapid cell-multiplication. This is indeed the case, as is observed in the well-known generalization that the most rapidly dividing cells of the body are the most sensitive to radiation. Our experiments suggest, however, that exposure to X-rays may distribute chromosomal damage quite impartially among all cell types. The damage merely shows up first in those tissues that multiply most rapidly. Thus genetic damage to cells can cause physiologic damage to the entire organism through failure of functions requiring steady cell-multiplication.

Since the blood-cell-forming tissues of the bone marrow have the highest normal rate of division, the first symptom of whole-body irradiation in the lethal range is usually a depression in the white blood-cell count. Experimental animals that have suffered such exposure can often be saved if new, viable bone-marrow cells are injected into them. The new cells will colonize the bone marrow and restore the necessary rate of cell production. But the dose given may be high enough to stop the reproduction of so many cells in other more slowly reproducing tissues that the survivors may not be able to restore the needed numbers in time to maintain physiological function. The injection of marrow will no longer save the animal.

In experiments with frogs Harvey M. Patt and his associates at the Argonne National Laboratory have found that a reduction in the body temperature of the animals will delay the onset of the

symptoms of radiation disease. This is quite understandable in light of present knowledge. At lower temperatures both the rate of cell division and the body's normal dependence upon it are depressed. The damage remains latent even in tissues that multiply rapidly. Upon restoration of normal body temperature the usual sequence of events resumes, and the symptoms of radiation damage manifest themselves.

The new picture of radiation damage thus locates its primary site in the genetic apparatus and shows that the damage consists in the destruction of the capacity to reproduce in some cells and the induction of mutations in others. On this basis a great many otherwise inexplicable phenomena have become understandable. However, much still remains to be learned. The factors that govern the healing of chromosome breaks and the degree to which these may differ in different cell types are still largely unknown; the biochemistry of this process will contribute much to fundamental understanding and to the more effective use of X-rays in the treatment of cancer. The study of the effect of chromosomal damage upon metabolic activities will illuminate many aspects of the complex mechanisms by which the genetic apparatus regulates the physiology of the cell. Single-cell techniques also offer promise in exploration of possible radiation damage to extranuclear cell structures. Finally these techniques permit accurate measurement of mutation rates in body cells, and so should contribute to unraveling the difficult problems involved in setting permissible levels of human exposure to ionizing radiation and other mutagenic agents.

IV

RESPONSES TO PSYCHOSOCIAL STRESS

RESPONSES TO PSYCHOSOCIAL STRESS   IV

INTRODUCTION

Just as previous sections dealt with our ability to adapt to perturbations in the chemical and physical environment, the articles in this section ask how capable human beings are of adjusting to stresses in the psychosocial environment. Given the vagueness of the very concept of "stress," it is impossible to prove that modern Westernized people are subjected to more psychological stress than people in earlier societies were. Yet, if *change itself* is taken to be the preeminent stress (a view supported by much recent research), then such a claim might be true; accordingly, investigations into the body's normal responses to psychosocial stress and into the possibility that these responses contribute to certain diseases should receive a high priority.

In fact, it turns out that the body's responses to psychosocial stress are essentially quite similar to its responses to physical stress. It must be emphasized that any given stress (increased external temperature, lowered external oxygen tension, etc.) sets into motion *specific* responses, which maintain homeostasis in the face of the challenge; however, in addition to triggering their specific responses, many quite different types of physical and psychological stress also elicit a relatively stereotyped set of adaptive neuroendocrine reactions collectively known as the "general stress response."* Most prominent among the reactions are increased secretion of the adrenocortical hormone, cortisol, and increased activity of the sympathetic nervous system, including the adrenal medulla.

Many situations are known to elicit the general stress response; these include physical trauma, prolonged heavy exercise, infection, shock, decreased oxygen supply, prolonged exposure to cold, pain, and fright. What many of these stresses have in common is that they are all harmful or potentially harmful to the body, and it is mainly in this context that the adaptive role of the general stress response has been analyzed. Walter B. Cannon's concept that the increase in sympathetic activity prepares the body for "fight or flight" has been very productive, because the net effect of these responses is to mobilize nutrients (glucose from the liver and fatty acids from adipose tissue), "arouse" the central nervous system, provide more oxygen and nutrients to skeletal muscles (by increasing

---

* There is much current controversy about just how stereotyped the stress response really is; it is almost certain that its components do, to some extent, depend on the nature of the initiating stress.

ventilation, cardiac output, blood pressure, and blood flow to skeletal muscles), increase contractility of skeletal muscles, and reduce any potential blood loss from the body by increasing blood coagulability.

The increased cortisol released during stress would also favor survival in a fight-or-flight situation, since the dominant physiological effects of this hormone are to mobilize nutrients (by stimulating protein breakdown and conversion of the amino acids into glucose, and by facilitating fat mobilization from adipose tissue) and to assure the normal functioning of the cardiovascular system.

We can also usefully apply this sort of analysis to the other hormones whose secretion rates usually increase during the stress response. Glucagon and growth hormone help to mobilize organic nutrients, and aldosterone and antidiuretic hormone stimulate sodium and water retention by the kidneys, an important adaptation in the face of real or potential fluid loss from hemorrhage or sweating. This list of hormonal changes is by no means complete, for the secretion of almost every hormone may be altered during stress; however, the adaptive significance of many of these changes in terms of "fight or flight" remains unclear.

To reiterate, the "fight or flight" concept has proven quite useful in analyzing the adaptive value of the neuroendocrine responses to stress. However, many of the situations which elicit the stress response are psychosocial, not physical. For certain of them (seeing one's enemy approaching with a gun, for example) the model still holds, in that an anticipatory response, preparing the body for a likely "fight or flight," would be adaptive; but the psychosocial stimuli which elicit the stress response are so varied that one becomes uncomfortable in pushing the analysis too far. Moreover, many of them are so benign (mere exposure to a novel situation, for example) that the very term "stress" seems inappropriate.

It was largely for these reasons that physiologists and psychologists began looking for other ways in which the hormonal response to psychosocial stimuli might be adaptive. The logical place to look was in the realm of behavior, and the exciting first fruits of this search are described in Levine's "Stress and Behavior." From these studies emerges an entirely new adaptive role for cortisol* (and for ACTH†, the anterior pituitary hormone which controls the release of cortisol): by acting directly on the brain, these hormones may facilitate learning and other behavior appropriate for coping with the psychosocial stimulus which elicited the hormonal response.

The implications of Levine's article go beyond its specific topic. The field of neuroendocrinology has mainly been concerned with the effects of the nervous system on hormone secretion; now it has become clear that the hormones, in turn, may have profound effects on the nervous system and thus on behavior. Perhaps the term psychoneuroendocrinology would be more appropriate.

---

* Cortisol and hydrocortisone are synonyms.

† While reading this article, one must keep in mind that, so far as is known, the increased cortisol secretion which occurs during psychosocial stress is triggered by an increased secretion of ACTH (which in turn is stimulated by a neuroendocrine pathway acted on by the stress). Therefore, ACTH and cortisol *both* normally increase during stress, and the fact (described by Levine) that these two hormones sometimes have opposing actions on behavior makes interpretation of their roles difficult. Endocrinologists have previously assumed that the only important action of ACTH is to control cortisol secretion, but this view may require revision should physiological levels of ACTH be shown to have direct effects on the brain.

The second article by Levine, "Stimulation in Infancy," deals with several other critical aspects of the relationships among stress, pituitary and adrenal secretions, and behavior, and provides another excellent illustration of the "critical periods" concept. Perhaps its most striking message is that relatively trivial "stresses" early in life can have profound and possibly irreversible effects, not only on the body's future responses to stress, but also on over-all body growth and emotional behavior. Again, it must be emphasized that these studies were performed on rats, and that we don't really know whether, or how much, their results are true for human beings. Yet, many other conclusions about critical periods and development that were originally derived from animal experiments have proven relevant for human beings, and one cannot fail to recognize the analogies between these rat studies and the observations on human infants described in "Deprivation Dwarfism" by Gardner: the failure to grow, altered emotionality, decreased responsiveness of the pituitary-adrenal system, to name just a few. The precise mechanisms which lead to stunting of growth in these emotionally deprived children are still not understood; the decreased secretion of growth hormone suggested by Gardner does probably explain the defect in some children, but current evidence indicates that other hormonal imbalances triggered by the psychosocial environment are also important.

Given these basic ideas about normal responses to psychosocial stress, we now turn to their possible pathophysiological implications and to the field of psychosomatic medicine. This last term, which is often misinterpreted, might better be replaced by the more cumbersome title, "psychological factors in the causation of disease." Based on the recognition that all diseases have many causes, this field asks the question: are there physical diseases in which psychological factors help generate the observable anatomical and chemical changes in the body that are typical for the diseases?

The evidence that there are, in fact, many such diseases comes from two general sources: epidemiology and animal experiments. The former has established associations between certain diseases and personality types (for example, between heart disease and so-called "type A" personality) and between the presence of stress and the onset of disease. Of course, epidemiological correlations can never prove causality; for this reason, animal experiments are of critical importance. One of the most extensive and convincing of such experiments is described by Weiss in "Psychological Factors in Stress and Disease." This experiment conclusively demonstrates that psychological stress can contribute to the formation of stomach ulcers, a disease long suspected by physicians to have a psychosomatic component. Moreover, Weiss' experiments suggest general principles which provide a more sophisticated and quantitative way to evaluate the role of psychological factors in causing human ulcers. The concept that feedback about the effectiveness of one's actions is the major factor provides a fascinating analogy to other physiological control systems, the effectiveness of which also depends on continuous feedback.

Given that psychological stress can contribute to physical disease, a critical physiological question to be posed is what the efferent and effector mechanisms are which actually mediate the damage. In the specific example of gastric ulceration, one would like to know the answers to a host of questions: Do the gastric glands hypersecrete acid or hyposecrete mucus? Does blood flow to the gastric lining decrease, thereby causing ischemic damage? Have the lining cells been damaged, thus becoming less resistant to the effects of gastric acid? Are the smooth-muscle coatings of the gastric wall contracting abnormally? What hormonal or neural pathways are causing any observed changes in these glandular, muscular, and epithelial structures? In other words, once physiologists have accepted the psychological stress as the initiator, they wish to know just what the intervening links are between the stress and the actual physiological malfunction.

One theory holds that the damage might be due to the generalized stress response triggered by the psychological stress. This concept of psychosomatic causation holds that the increase in cortisol (and other hormones) and in sympathetic nerve activity which is characteristic of the general stress response may, under certain conditions, actually cause damage to some tissue or organ (in this case, the stomach). To take other examples, a hyperstimulated sympathetic nervous system might, given enough time, damage the heart (by increasing its oxygen requirement and excitability) or cause hypertension (by reducing the diameter of arterioles).

How might increased cortisol contribute to disease? Very large amounts of cortisol produce an array of effects which are quite different from the physiological effects described earlier. These are collectively known as cortisol's "pharmacological effects," and include gastric ulceration, inhibition of immune responses (which would lower one's resistance to infection and cancer), hypertension, atherosclerosis, sterility, and personality changes. The correspondence of this list to the major causes of mortality in industrialized countries makes it tantalizing to speculate about the possible role of psychological stress, acting by means of increased cortisol, in all these diseases. However, we do not know whether the levels of cortisol which occur spontaneously during stress are large enough to produce these "pharmacological effects," nor do we have any solid information about long-term patterns of blood cortisol (or any other hormone) in persons suffering from these diseases. Clearly no meaningful conclusions can be drawn at present, but the subject is worthy of intensive research. If participation of the general stress response in the diseases were ultimately shown, it would provide yet another example of how a set of responses, highly adaptive in one environment, might be maladaptive in a very different environment, as in modern life when psychological stress usually occurs in a setting in which neither fight nor flight is appropriate.

In contrast to (but not mutually exclusive from) the nonspecific general stress concept of psychosomatic disease are those theories which emphasize the highly specific character of psychosomatic disorders and explain them in terms of learned or conditioned responses to stress. The recent evidence that "involuntary" autonomic functions can be influenced by operant conditioning has supplied the theoretical base for these theories and is described by DiCara in "Learning in the Autonomic Nervous System." It should be pointed out that the validity of the specific experiments described by DiCara is still the subject of considerable controversy among researchers in this field as is the conclusion that an individual can control autonomic function *directly* rather than *indirectly* (through the mediation of some nonautonomic response). Nonetheless, despite the tenuousness of the theoretical underpinning, researchers in this area (dubbed "biofeedback") have accumulated considerable data indicating that many autonomic functions are, at least to some extent, subject to influence by operant conditioning. The role of such processes in the etiology of psychosomatic disorders and the possible contribution of biofeedback to their treatment remain to be determined.

The article "Physiology of Meditation" by Wallace and Benson presents another method, quite different from biofeedback, for attempting to control physiological reactions to the psychosocial environment. Whereas the techniques of operant conditioning are designed to elicit (or repress) highly specific responses, meditation aims rather to produce a widespread integrated state of relaxation (or "hypometabolism") mediated by the central nervous system. As Wallace and Benson point out, the condition might be viewed as opposite to the fight-or-flight response; however, to date, there are no experiments reporting *direct* measurements of sympathetic nervous system activity or blood hormone levels. Of far greater significance even than the documented acute changes which may occur during this particular form of meditation (known as transcendental meditation or TM) are the possible long-term effects. A large number of studies have appeared, claiming that the practice of TM has beneficial effects on many parameters of physical and mental health. Unfortunately, many of these studies are poorly controlled and subject to other interpretations; yet there are enough reports which do seem to have validity to warrant careful further study of the matter. It might also provide an excellent tool for studying the entire question of whether the level of present-day stress contributes significantly to the incidence of our most serious diseases.

13 Stress and Behavior

by Seymour Levine
January 1971

*The chain of pituitary and adrenal hormones that
regulates responses to stress plays a major role in learning
and other behaviors. It may be that effective behavior
depends on some optimum level of stress*

Hans Selye's concept of the general "stress syndrome" has surely been one of the fruitful ideas of this era in biological and medical research. He showed that in response to stress the body of a mammal mobilizes a system of defensive reactions involving the pituitary and adrenal glands. The discovery illuminated the causes and symptoms of a number of diseases and disorders. More than that, it has opened a new outlook on the functions of the pituitary-adrenal system. One can readily understand how the hormones of this system may defend the body against physiological insult, for example by suppressing inflammation and thus preventing tissue damage. It is a striking fact, however, that the system's activity can be evoked by all kinds of stresses, not only by severe somatic stresses such as disease, burns, bone fractures, temperature extremes, surgery and drugs but also by a wide range of psychological conditions: fear, apprehension, anxiety, a loud noise, crowding, even mere exposure to a novel environment. Indeed, most of the situations that activate the pituitary-adrenal system do not involve tissue damage. It appears, therefore, that these hormones in animals, including man, may have many functions in addition to the defense of tissue integrity, and as a psychologist I have been investigating possible roles of the pituitary-adrenal system in the regulation of behavior.

The essentials of the system's operation in response to stress are as follows. Information concerning the stress (coming either from external sources through the sensory system or from internal sources such as a change in body temperature or in the blood's composition) is received and integrated by the central nervous system and is presumably delivered to the hypothalamus, the basal area of the brain. The hypothalamus secretes a substance called the corticotropin-releasing factor (CRF), which stimulates the pituitary to secrete the hormone ACTH. This in turn stimulates the cortex of the adrenal gland to step up its synthesis and secretion of hormones, particularly those known as glucocorticoids. In man the glucocorticoid is predominantly hydrocortisone; in many lower animals such as the rat it is corticosterone.

The entire mechanism is exquisitely controlled by a feedback system. When the glucocorticoid level in the circulating blood is elevated, the central nervous system, receiving the message, shuts off the process that leads to secretion of the stimulating hormone ACTH. Two experimental demonstrations have most clearly verified the existence of this feedback process. If the adrenal gland is removed from an animal, the pituitary puts out abnormal amounts of ACTH, presumably because the absence of the adrenal hormone frees it from restriction of this secretion. On the other hand, if crystals of glucocorticoid are implanted in the hypothalamus, the animal's secretion of ACTH stops almost completely, just as if the adrenal cortex were releasing large quantities of the glucocorticoid.

Now, it is well known that a high level of either of these hormones (ACTH or glucocorticoid) in the circulating blood can have dramatic effects on the brain. Patients who have received glucocorticoids for treatment of an illness have on occasion suffered severe mental changes, sometimes leading to psychosis. And patients with a diseased condition of the adrenal gland that caused it to secrete an abnormal amount of cortical hormone have also shown effects on the brain, including changes in the pattern of electrical activity and convulsions.

Two long-term studies of my own, previously reported in *Scientific American* [see "Stimulation in Infancy," page 143 this volume and "Sex Differences in the Brain," Offprint 498], strongly indicated that hormones play an important part in the development of behavior. One of these studies showed that rats subjected to shocks and other stresses in early life developed normally and were able to cope well with stresses later, whereas animals that received no stimulation in infancy grew up to be timid and deviant in behavior. At the adult stage the two groups differed sharply in the response of the pituitary-adrenal system to stress: the animals that had been stimulated in infancy showed a prompt and effective hormonal response; those that had not been stimulated responded slowly and ineffectively. The other study, based on the administration or deprivation of sex hormones at a critical early stage of development in male and female rats, indicated that these treatments markedly affected the animals' later behavior, nonsexual as well as sexual. It is noteworthy that the sex hormones are steroids rather similar to those produced by the adrenal cortex.

Direct evidence of the involvement of the pituitary-adrenal system in overt behavior was reported by two groups of experimenters some 15 years ago. Mortimer H. Appley, now at the University of Massachusetts, and his co-workers were investigating the learning of an avoidance response in rats. The animals were placed in a "shuttle box" divided into two compartments by a barrier. An electric shock was applied, and if the animals crossed the barrier, they could avoid or terminate the shock. The avoidance response consisted in making the move across the barrier when a conditioned stimulus, a buzzer signaling the onset of the shock, was sounded. Appley found that when the pituitary gland was removed surgically from rats, their learning of the avoidance response was severely retarded. It turned out that an injection of ACTH in pituitary-deprived rats could restore the learning ability to normal. At about the same time Robert E. Miller and Robert Murphy of the University of Pittsburgh reported experiments showing that ACTH could affect extinction of the avoidance response. Normally if the shocks are discontinued, so that the animal receives no shock when it fails to react to the conditioned stimulus (the buzzer in this case), the avoidance response to the buzzer is gradually extinguished. Miller and Murphy found that when they injected ACTH in animals during the learning period, the animals continued to make the avoidance response anyway, long after it was extinguished in animals that had not received the ACTH injection. In short, ACTH inhibited the extinction process.

These findings were not immediately followed up, perhaps mainly because little was known at the time about the details of the pituitary-adrenal system and only rudimentary techniques were available for studying it. Since then purified preparations of the hormones involved and new techniques for accurate measurement of these substances in the circulating blood have been developed, and the system is now under intensive study. Most of the experimental investigation is being conducted at three centers: in the Institute of Pharmacology at the University of Utrecht under David de Wied, in the Institute of Physiology at the University of Pecs in Hungary under Elemér Endroczi and in our own laboratories in the department of psychiatry at Stanford University.

The new explorations of the pituitary-adrenal system began where the ground had already been broken: in studies of the learning and extinction of the avoidance response, primarily by use of the shuttle box. De Wied verified the role of ACTH both in avoidance learning and in inhibiting extinction of the response. He did this in physiological terms by means of several experiments. He verified the fact that removal of the pituitary gland severely retards the learning of a conditioned avoidance response. He also removed the adrenal gland from rats and found that the response was then not extinguished, presumably because adrenal hormones were no longer present to re-

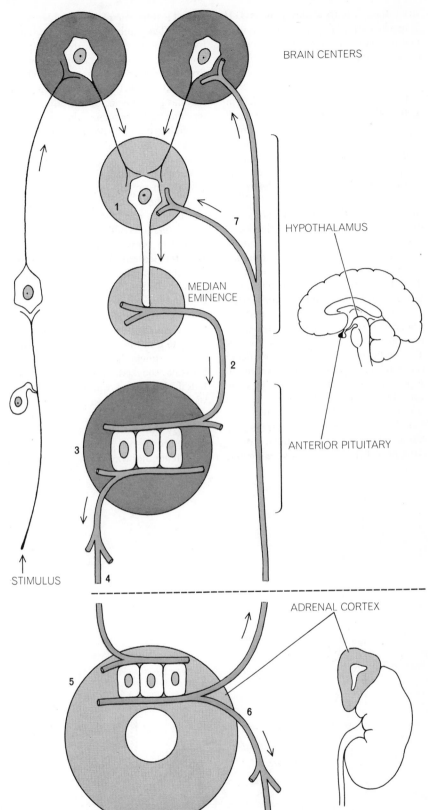

BRAIN CENTERS

HYPOTHALAMUS

MEDIAN EMINENCE

ANTERIOR PITUITARY

STIMULUS

ADRENAL CORTEX

PITUITARY-ADRENAL SYSTEM involves nerve cells and hormones in a feedback loop. A stress stimulus reaching neurosecretory cells of the hypothalamus in the base of the brain (1) stimulates them to release corticotropin-releasing factor (CRF), which moves through short blood vessels (2) to the anterior lobe of the pituitary gland (3). Pituitary cells thereupon release adrenocorticotrophic hormone (ACTH) into the circulation (4). The ACTH stimulates cells of the adrenal cortex (5) to secrete glucocorticoid hormones (primarily hydrocortisone in man) into the circulation (6). When glucocorticoids reach neurosecretory cells or other brain cells (it is not clear which), they modulate CRF production (7).

strict the pituitary's output of ACTH. When he excised the pituitary, thus eliminating the secretion of ACTH, the animals returned to near-normal behavior in the extinction of the avoidance response.

In further experiments De Wied injected glucocorticoids, including corticosterone, the principal steroid hormone of the rat's adrenal cortex, into animals that had had the adrenal gland, but not the pituitary, removed; as expected, this had the effect of speeding up the extinction of the avoidance response. Similarly, the administration to such animals of dexamethasone, a synthetic glucocorticoid that is known to be a potent inhibitor of ACTH, resulted in rapid extinction of the avoidance response; the larger the dose, the more rapid the extinction. Curiously, De Wied found that corticosterone and dexamethasone promoted extinction even in animals that lacked the pituitary gland, the source of ACTH. This indicated that the glucocorticoid can produce its effect not only through suppression of ACTH but also, in some way, by acting directly on the central nervous system. It has recently been found, on the other hand, that there may be secretions from the pituitary other than ACTH that can affect learning and

inhibit extinction of the avoidance response. The inhibition can be produced, for example, by a truncated portion of the ACTH molecule consisting of the first 10 amino acids in the sequence of 39 in the rat's ACTH—a molecular fragment that has no influence on the adrenal cortex. The same fragment, along with other smaller peptides recently isolated by De Wied, can also overcome the deficit in avoidance learning that is produced by ablation of the pituitary.

With an apparatus somewhat different from the shuttle box we obtained further light in our laboratory on ACTH's effects on behavior. We first train the animals to press a bar to obtain water. After this learning has been established the animal is given an electric shock on pressing the bar. This causes the animal to avoid approaching the bar (called "passive avoidance") for a time, but after several days the animal will usually return to it in the effort to get water and then will quickly lose its fear of the bar if it is not shocked. We found, however, that if the animal was given doses of ACTH after the shock, it generally failed to return to the bar at all, even though it was very thirsty. That is to say, ACTH suppressed the bar-pressing response, or, to put it another way, it strengthened the

passive-avoidance response. In animals with the pituitary gland removed, injections of ACTH suppressed a return to bar-pressing after a shock but injections of hydrocortisone did not have this effect.

The experiments I have described so far have involved behavior under the stress of fear and anxiety. Our investigations with the bar-pressing device go on to reveal that the pituitary-adrenal system also comes into play in the regulation of behavior based on "appetitive" responses (as opposed to avoidance responses). Suppose we eliminate the electric shock factor and simply arrange that after the animal has learned to press the bar for water it fails to obtain water on later trials. Normally the animal's bar-pressing behavior is then quickly extinguished. We found, however, that when we injected ACTH in the animals in these circumstances, the extinction of bar-pressing was delayed; the rats went on pressing the bar for some time although they received no water as reinforcement. Following up this finding, we measured the corticosterone levels in the blood of normal, untreated rats both when they were reinforced and when they were not reinforced on pressing the

WARNING LIGHT

SPEAKER

SPEAKER

ELECTRIC GRID

PHOTOELECTRIC CELL

"SHUTTLE BOX" used for studying avoidance behavior is a two-compartment cage. The floor can be electrically charged. A shock is delivered on the side occupied by the rat (detected by the photocell). The rat can avoid the shock by learning to respond to the conditioned stimulus: a light and noise delivered briefly before the shock. The avoidance response, once learned, is slowly "extinguished" if the conditioned stimulus is no longer accompanied by a shock. Injections of ACTH inhibited the extinction process.

bar. The animals that received no water reinforcement, with the result of rapid extinction of bar-pressing, showed a marked rise in activity of the pituitary-adrenal system during this period, whereas in animals that received water each time they pressed the bar there was no change in the hormonal output. In short, the extinction of appetitive behavior in this case clearly involved the pituitary-adrenal system.

Further investigations have now shown that the system affects a much wider range of behavior than learning and extinction. One of the areas that has been studied is habituation: the gradual subsidence of reactions that had appeared on first exposure to a novel stimulus when the stimulus is repeated. An organism presented with an unexpected stimulus usually exhibits what Ivan Pavlov called an orientation reflex, which includes increased electrical activity in the brain, a reduction of blood flow to the extremities, changes in the electrical resistance of the skin, a rise in the level of adrenal-steroid hormones in the blood and some overt motor activity of the body.

If the stimulus is repeated frequently, these reactions eventually disappear; the organism is then said to be habituated to the stimulus. Endroczi and his co-workers recently examined the influence of ACTH on habituation of one of the reactions in human subjects—the increase of electrical activity in the brain, as indicated by electroencephalography. The electrical activity evoked in the human brain by a novel sound or a flickering light generally subsides, after repetition of the stimulus, into a pattern known as electroencephalogram (EEG) synchronization, which is taken to be a sign of habituation. Endroczi's group found that treatment of their subjects with ACTH or the 10-amino-acid fragment of ACTH produced a marked delay in the appearance of the synchronization pattern, indicating that the hormone inhibits the process of habituation.

Experiments with animals in our laboratory support that finding. The stimulus we used was a sudden sound that produces a "startle" response in rats, which is evidenced by vigorous body movements. After a number of repetitions of the sound stimulus the startle response fades. It turned out that rats deprived of the adrenal gland (and consequently with a high level of ACTH in their circulation) took significantly longer than intact animals to habituate to the sound stimulus. An implant of the adrenal hormone hydrocortisone in the hy-

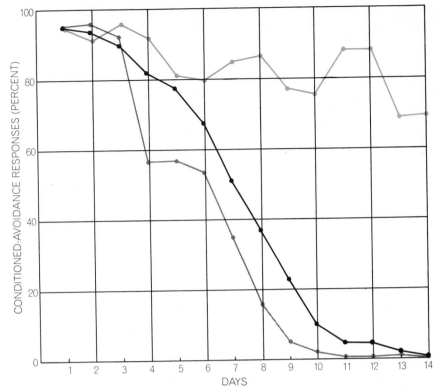

EXTINCTION of the avoidance response was studied by David de Wied of the University of Utrecht. Removal of the adrenal gland inhibited extinction (*color*); the rats responded to the conditioned stimulus in the absence of shock, presumably because adrenal hormones were not available to restrict ACTH output. When the pituitary was removed, the rate of extinction (*gray*) was about the same as in rats given only a sham operation (*black*).

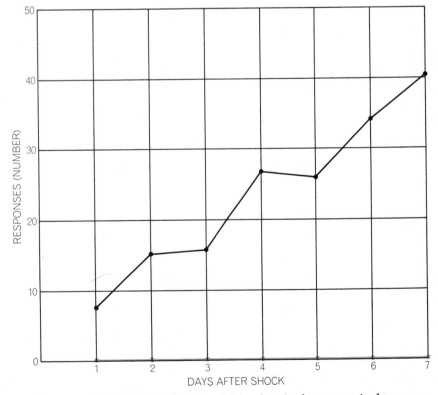

PASSIVE AVOIDANCE BEHAVIOR is studied by observing how rats, trained to press a bar for water, avoid the bar after they get a shock on pressing it. Before being shocked rats pressed the bar about 75 times a day. After the shock the control animals returned to the bar and, finding they were not shocked, gradually increased their responses (*black curve*). Rats injected with ACTH stayed away (*color*): ACTH strengthens the avoidance response.

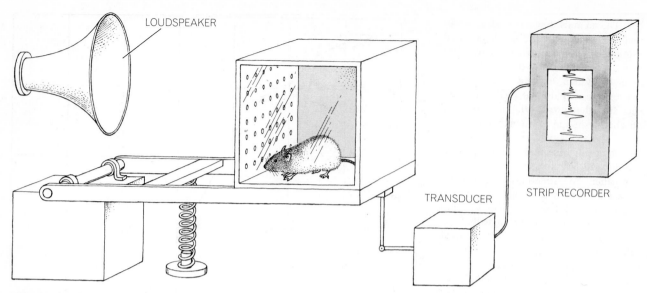

LOUDSPEAKER

TRANSDUCER

STRIP RECORDER

"STARTLE" RESPONSE is measured by placing a rat in a cage with a movable floor and exposing it to a sudden, loud noise. The rat tenses or jumps, and the resulting movement of the floor is transduced into movement of a pen on recording paper. After a number of repetitions of the noise the rat becomes habituated to it and the magnitude of the animal's startle response diminishes.

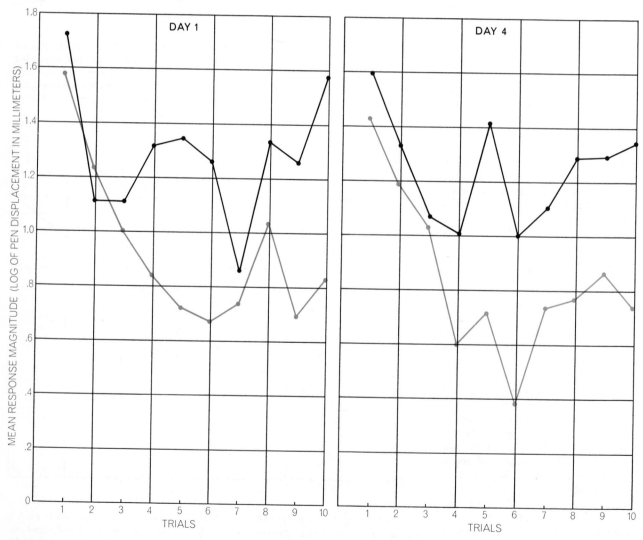

HABITUATION is affected by the pituitary-adrenal system. If a crystal of the adrenal hormone hydrocortisone is implanted in a rat's hypothalamus, preventing ACTH secretion, habituation is speeded up, as shown here. The mean startle response (shown as the logarithm of the recording pen's movement) falls away more rapidly in implanted rats (*color*) than in control animals (*black*).

pothalamus, on the other hand, speeded up habituation.

A series of studies by Robert I. Henkin of the National Heart Institute has demonstrated that hormones of the adrenal cortex play a crucial role in the sensory functions in man. Patients whose adrenal gland has been removed surgically or is functioning poorly show a marked increase in the ability to detect sensory signals, particularly in the senses of taste, smell, hearing and proprioception (sensing of internal signals). On the other hand, patients with Cushing's syndrome, marked by excessive secretion from the adrenal cortex, suffer a considerable dulling of the senses. Henkin showed that sensory detection and the integration of sensory signals are regulated by a complex feedback system involving interactions of the endocrine system and the nervous system. Although patients with a deficiency of adrenal cortex hormones are extraordinarily sensitive in the detection of sensory signals, they have difficulty integrating the signals, so that they cannot evaluate variations in properties such as loudness and tonal qualities and have some difficulty understanding speech. Proper treatment with steroid hormones of the adrenal gland can restore normal sensory detection and perception in such patients.

Henkin has been able to detect the effects of the adrenal corticosteroids on sensory perception even in normal subjects. There is a daily cycle of secretion of these steroid hormones by the adrenal cortex. Henkin finds that when adrenocortical secretion is at its highest level, taste detection and recognition is at its lowest, and vice versa.

In our laboratory we have found that the adrenal's steroid hormones can have a truly remarkable effect on the ability of animals to judge the passage of time. Some years ago Murray Sidman of the Harvard Medical School devised an experiment to test this capability. The animal is placed in an experimental chamber and every 20 seconds an electric shock is applied. By pressing a bar in the chamber the animal can prevent the shock from occurring, because the bar resets the triggering clock to postpone the shock for another 20 seconds. Thus the animal can avoid the shock altogether by appropriate timing of its presses on the bar. Adopting this device, we found that rats learned to press the bar at intervals averaging between 12 and 15 seconds. This prevented a majority of the shocks. We then gave the animals glucocorticoids and found that they became significantly more efficient!

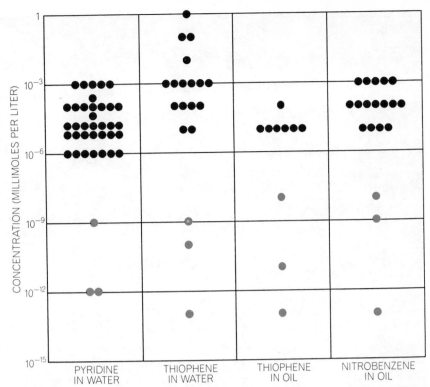

SENSORY FUNCTION is also affected by adrenocortical hormones. Robert I. Henkin of the National Heart Institute found that patients whose adrenal-hormone function is poor are much more sensitive to odor. Placing various chemicals in solution, he measured the detection threshold: the concentration at which an odor could be detected in the vapor. The threshold was much lower in the patients (*color*) than in normal volunteers (*black*).

They lengthened the interval between bar presses and took fewer shocks. Evidently under the influence of the hormones the rats were able to make finer discriminations concerning the passage of time. Monkeys also showed improvement in timing performance in response to treatment with ACTH.

The mechanism by which the pituitary-adrenal hormones act to regulate or influence behavior is still almost completely unknown. Obviously they must do so by acting on the brain. It is well known that hormones in general are targeted to specific sites and that the body tissues have a remarkable selectivity for them. The uterus, for instance, picks up and responds selectively to estrogen and progesterone among all the hormones circulating in the blood, and the seminal vesicles and prostate gland of the male select testosterone. There is now much evidence that organs of the brain may be similarly selective. Bruce Sherman McEwen of Rockefeller University has recently reported that the hippocampus, just below the cerebral cortex, appears to be a specific receptor site for hormones of the adrenal cortex, and other studies indicate that the lateral portion

of the hypothalamus may be a receptor site for gonadal hormones. We have the inviting prospect, therefore, that exploration of the brain to locate the receptor sites for the hormones of the pituitary-adrenal system, and studies of the hormones' action on the cells of these sites, may yield important information on how the system regulates behavior. Bela Bohun in Hungary has already demonstrated that implantation of small quantities of glucocorticoids in the reticular formation in the brain stem facilitates extinction of an avoidance response.

Since this system plays a key role in learning, habituation to novel stimuli, sensing and perception, it obviously has a high adaptive significance for mammals, including man. Its reactions to moderate stress may contribute greatly to the behavioral effectiveness and stability of the organism. Just as the studies of young animals showed, contrary to expectations, that some degree of stress in infancy is necessary for the development of normal, adaptive behavior, so the information we now have on the operations of the pituitary-adrenal system indicates that in many situations effective behavior in adult life may depend on exposure to some optimum level of stress.

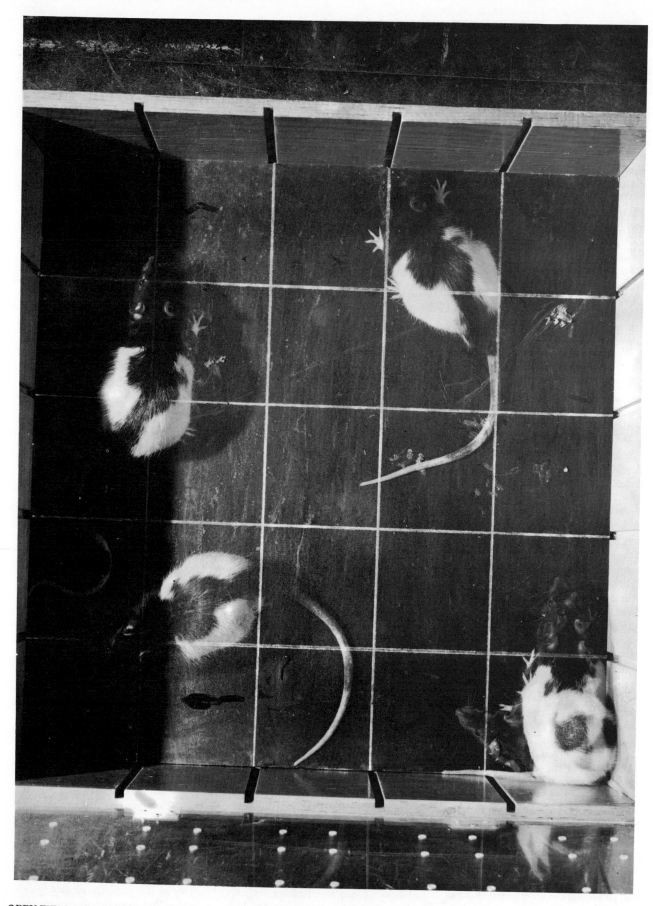

OPEN-FIELD EXPERIMENT illustrates how the behavior of a full-grown rat stimulated during infancy differs from that of a previously nonstimulated one. This multiple-exposure photograph shows how a nonstimulated rat (*lower right*) cowers in a corner when placed in an unfamiliar environment; the stimulated animal is much more willing to run about and explore his surroundings.

Stimulation in Infancy

by Seymour Levine

May 1960

*Both painful shocks and gentle handling enhance
the development of normal stress responses in infant
animals. The absence of such treatment leads to
behavioral disorders when the animal matures*

When the Emperor of Lilliput accepted Lemuel Gulliver into favor, His Most Sublime Majesty first secured Gulliver's solemn oath upon an agreement to observe certain rules of etiquette. The fourth article of the agreement stipulated that Gulliver should not take any Lilliputian subjects into his hands without their consent. Gulliver learned later to appreciate the sentiments behind this article in an intensely subjective way. In the country of Brobdingnag he was himself picked up in the huge hand of a Brobdingnagian. He recalled his reactions: "All I ventured was to raise my eyes towards the sun, and place my hands together in a supplicating posture, and to speak some words in an humble melancholy tone, suitable to the condition I then was in."

What Jonathan Swift describes here is the essence of an experience that befalls children and small animals every day. It happens whenever a parent picks up a baby, or a child tussles with his puppy. Almost all experiences of infancy involve some handling by a parent or some other larger and supremely powerful figure. Even the tenderest handling must at times be the occasion of emotional stress. Perhaps the only children insulated from such experience are those reared in orphanages and other institutions, and the only animals those that live in laboratories. Certainly the laboratory animal must find a minimum of stress and little stimulation of any other kind in an environment controlled for temperature, humidity, light and so on. In the ordinary world the infant must grow under the changing pressures and sudden challenges of an inconstant environment. One may well wonder how the stressful experiences of infancy affect the behavior and physiology of the adult organism later on.

When in 1954 we began our investigations into the broad area defined by this question, we naturally turned first to the presumably more obvious effects of early painful or traumatic experience. We subjected a group of infant rats to mild electric shocks, scheduled at the same hour each day. For control purposes we routinely placed the members of another group in the shock cage for the same length of time each day but did not give them shocks. A third group of infant rats was left in the nest and not handled at all. We expected that the shocked rats would be affected by their experience, and we looked for signs of emotional disorder when they reached adulthood. To our surprise it was the second control group—the rats we had not handled at all—that behaved in a peculiar manner. The behavior of the shocked rats could not be distinguished from that of the control group which had experienced the same handling but no electric shock. Thus the results of our first experiment caused us to reframe our question. Our investigation at the Columbus Psychiatric Institute and Hospital of Ohio State University has since been concerned not so much with the effects of stressful experience—which after all is the more usual experience of infants—as with the effects of the absence of such experience in infancy.

We have repeated our original experiment many times, subjecting the infant animals to a variety of stresses and degrees of handling. Invariably it is the nonmanipulated "controls" that exhibit deviations of behavior and physiology when they are tested as adults. Significantly these deviations involve the organism's response to stress, and they show up in most of the diverse aspects of that response. In a standard behavioral test, for example, the animal is placed in the unfamiliar, but otherwise neutral, surroundings of a transparent plastic box. The nonmanipulated animals crouch in a corner of the box; animals that have been handled and subjected to stress in infancy freely explore the space. The same contrast in behavior may be observed and recorded quantitatively in the "open field": an area three feet square marked off into smaller squares. In terms of the number of squares crossed during a fixed time period, shocked and manipulated animals show a much greater willingness to run about and explore their surroundings. In both situations the nonmanipulated animals, cowering in a corner or creeping timidly about, tend to defecate and urinate frequently. Since these functions are largely controlled by the sympathetic nervous system, and since certain responses to stress are principally organized around the sympathetic nervous system, this behavior is a sure sign of reactivity to stress.

Another objective and quantitative index of stress response is provided by the hormones and glands of the endocrine system. Under stress, in response to prompting by the central nervous system, the pituitary releases larger quantities of various hormones, one of the principal ones being the adrenal-corticotrophic hormone (ACTH). Stimulation by ACTH causes the outer layer, or cortex, of the adrenal gland to step up the release of its several steroids; distributed by the bloodstream, these hormones accelerate the metabolism of the tissues in such a way as to maintain their integrity under stress. The activity of the endocrine system may be measured conveniently in a number of ways: by the enlargement of the adrenal glands, by the volume of adrenal steroids in circulation or by the depletion of ascorbic acid (vitamin C) in the adrenals. By some of these measurements the nonstimulated animals showed a markedly higher reactivity when subjected to a variety of stresses, including toxic injection of glucose, conditioning to avoid a painful stimulus and swimming in a water maze.

The conclusion that these animals are hyperreactive to stress is, however, an oversimplification that conceals an even more important difference in their stress response. Recently we measured the steroids in circulation in both stimulated and nonstimulated animals during the period immediately following stress by electric shock. Whereas the two groups showed the same volume of steroids in circulation before shock, the animals that had been exposed to stress in infancy showed a much higher output of steroids in the first 15 minutes after shock. The nonstimulated animals achieve the same output but more slowly, and appear to maintain a high level of steroid secretion for a longer period of time. There is thus a distinct difference in the pattern of the stress response in the two kinds of animal.

This observation acquires its full significance when it is considered in the light of the biological function of the stress response. The speed and short duration of the response in the stimulated animal obviously serve the useful purpose of mobilizing the resources of the organism at the moment when it is under stress. The delay in the endocrine response of the nonstimulated animal

MILD STIMULATION consisted of picking up the infant rats, removing them from their breeding cage and enclosing them in a small compartment for three minutes a day. The rats were then returned to their nests. The rats shown here are about 11 days old.

is thus, by contrast, maladaptive. Moreover, the prolongation of the stress response, as observed in these animals, can have severely damaging consequences: stomach ulcers, increased susceptibility to infection and eventually death due to adrenal exhaustion.

The maladaptive nature of the stress response in the nonmanipulated animal is further manifested in the fact that it may be elicited in such a neutral situation as the open-field test. The ani-

mal that has been manipulated in infancy shows no physiological stress response in this situation although it exhibits a vigorous and immediate endocrine response when challenged by the pain and threat of an electric shock.

In this connection we have made the interesting discovery that stimulation by handling and stress hastens the maturation of the stress response in the infant animal. Although the adrenal glands begin to function shortly after birth and the pituitary appears to contain ACTH early

in the course of development, the nerve mechanism that controls the release of ACTH does not seem to come into operation until the rat is about 16 days of age. When we exposed infant rats that had been handled from birth to severe cold stress, however, they showed a significant ACTH response as early as 12 days of age. This four days' difference represents a considerable acceleration of development in the rat, equivalent to several months in the growth of a human infant. The manipulated animals, more-

PAINFUL STIMULATION consisted of subjecting the infant rats to an electric shock lasting from several seconds to several minutes. The effects on the rat's behavior as an adult were indistinguishable from those produced by the routine shown on the opposite page.

146

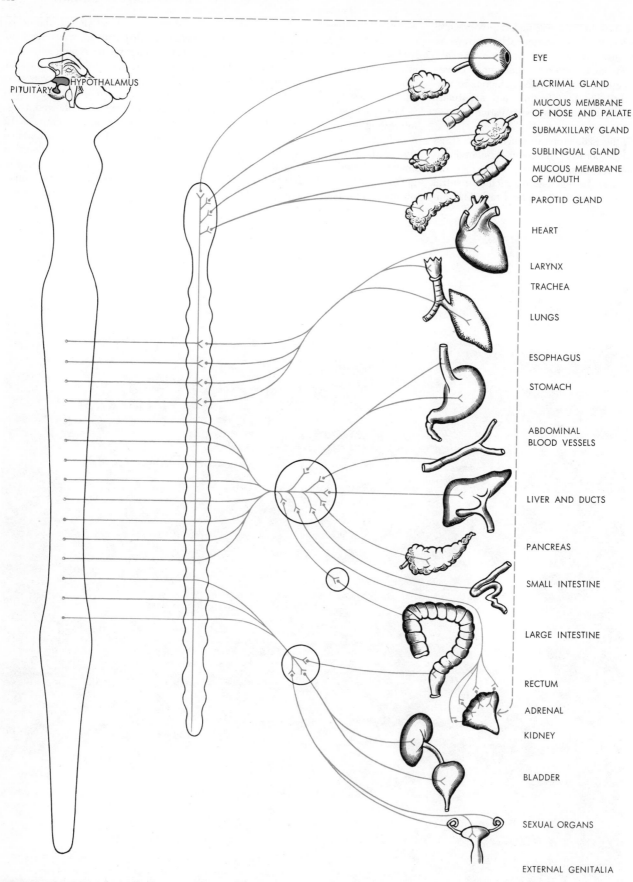

EYE

LACRIMAL GLAND

MUCOUS MEMBRANE
OF NOSE AND PALATE

SUBMAXILLARY GLAND

SUBLINGUAL GLAND

MUCOUS MEMBRANE
OF MOUTH

PAROTID GLAND

HEART

LARYNX

TRACHEA

LUNGS

ESOPHAGUS

STOMACH

ABDOMINAL
BLOOD VESSELS

LIVER AND DUCTS

PANCREAS

SMALL INTESTINE

LARGE INTESTINE

RECTUM

ADRENAL

KIDNEY

BLADDER

SEXUAL ORGANS

EXTERNAL GENITALIA

RESPONSES TO STRESS are partly controlled by the pathways shown in this diagram of the human sympathetic nervous system. Sympathetic fibers (*solid colored lines*) originating in the spinal cord (*far left*) innervate the internal organs via the chain ganglia (*left center*) and the ganglia of the celiac plexus (*right center*).

Extreme stress upsets the normal rhythm of this system, causing disturbances such as loss of bladder control and increased pulse rate. Stress also stimulates the hypothalamus and the pituitary to produce ACTH, which reaches the adrenals via the bloodstream (*broken line*) and stimulates them to produce steroid hormones.

over, reached an adult level of response considerably earlier than their untreated litter mates.

From the evidence it may be inferred that stimulation must have accelerated the maturation of the central nervous system in these animals. We have direct evidence that this is so from analysis of the brain tissue of our subjects. The brains of infant rats that have been handled from birth show a distinctly higher cholesterol content. Since the cholesterol content of the brain is related principally to the brain's white matter, this is evidence that in these animals the maturation of structure parallels the maturation of function.

In all respects, in fact, the manipulated infants exhibit a more rapid rate of development. They open their eyes earlier and achieve motor coordination sooner. Their body hair grows faster, and they tend to be significantly heavier at weaning. They continue to gain weight more rapidly than the nonstimulated animals even after the course of stimulation has been completed at three weeks of age. Their more vigorous growth does not seem to be related to food intake but to better utilization of the food consumed and probably to a higher output of the somatotrophic (growth) hormone from the pituitary. These animals may also possess a higher resistance to pathogenic agents; they survive an injection of leukemia cells for a considerably longer time.

Another contrast between the stimulated and unstimulated animals developed when we electrically destroyed the septal region of their brains, the region between and under the hemispheres of the midbrain. Such damage makes an animal hyperexcitable, vicious and flighty. It will attack a pencil extended to it, react with extreme startle to a tap on the back, is exceedingly difficult to capture and upon capture will bite wildly and squeal loudly. In systematic observation of these responses we found that manipulated animals are far tamer postoperatively than nonmanipulated ones. The latter rank as the most excitable and vicious rats we have ever observed in the laboratory; it was not unusual for one of these animals to pursue us around the room, squealing and attacking our shoes and pants legs.

At the very least our experiments yield an additional explanation for the variability among laboratory animals that so often confuses results in experimental biology. This has been attributed to genetic differences, unknown factors and sometimes to experimental error. It

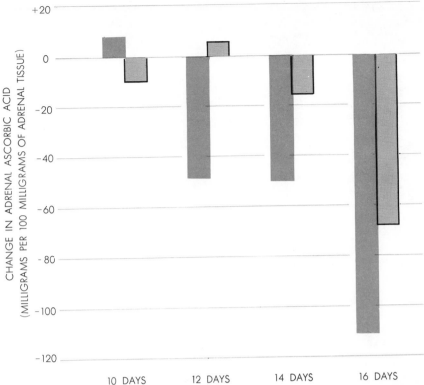

EXPOSURE TO COLD produced a marked drop in the ascorbic acid (vitamin C) concentration in the adrenal glands of stimulated rats more than 10 days old (*colored bars*), but produced no significant effect on the nonstimulated rats until they were 16 days old.

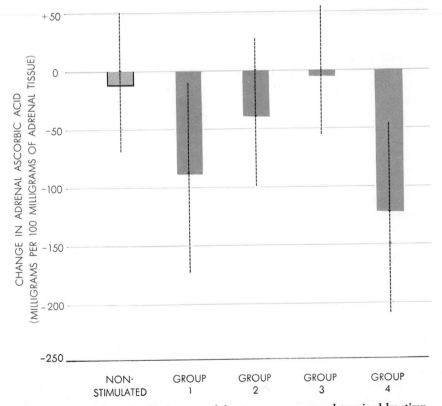

CRITICAL PERIOD in the development of the stress response was determined by stimulating infant rats at different stages of life. They were then exposed to cold and the drop in their adrenal ascorbic acid level was analyzed. Rats in Group 1 (stimulated from the second to the fifth days of life) and in Group 4 (stimulated from the second to the 13th days) responded better than both the nonstimulated rats and those in Groups 2 and 3 (stimulated from the sixth to the ninth and from the 10th to the 13th days, respectively). The bars show the average drop in the concentration and the broken lines the range.

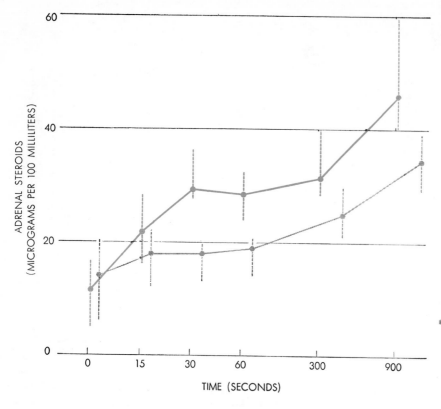

SLUGGISH RESPONSE to an electric shock is indicated by the slow rise in the concentration of circulating steroid hormones in previously nonstimulated rats (*gray curve*). In the stimulated animals (*colored curve*) the level increases rapidly for about 15 minutes. The points on the curve indicate the average level and the broken lines the range of values.

is apparent that the character of early infant experience is another important determinant of individual differences in animals.

The same consideration leads to the broader question of "nature v. nurture," that is, the contribution of genetic factors as opposed to the influence of the environment. Both sets of factors are essential and they interact to give rise to the individual organism. The basic patterns of development are most likely determined by heredity. But the genetic determinants do not find expression except in interaction with various aspects of the environment. In the normal course of events the environment provides the substance, energy and milieu for the unfolding of the organism's potentialities; in the extreme, environmental influences can determine whether the process of development will continue and produce an organism. In other words, organisms do not grow in a vacuum. This is true even of our nontreated animals. litter mates and the routine laboratory procedures furnish stimulation of all kinds. Such stimulation does not compare, however, with that provided by our experimental treatments. We have dealt with only a limited range of effects, and

have focused primarily on the physiological and behavioral responses to stress. But our results clearly indicate that stimulation of the infant organism has quite universal consequences upon the behavior and physiology of the adult.

One must be careful in attempting to bridge the gap between animal experimentation and human biology. The effects of early experience have proved to be significant, however, in many species of mammal, including the monkey, dog, cat, guinea pig and mouse, and in such nonmammals as fish and fowl. It cannot be said that the phenomenon is species-limited. A great deal of clinical evidence, moreover, clearly indicates that infant experience in humans has a profound effect in shaping the character and constitution of the adult. Investigators concerned with maternal deprivation report that children raised in foundling homes develop at a retarded rate and are more susceptible to disease. These observations are similar to those we have made in our animal experiments. It may be that the detrimental effects of the foundling home have less to do with maternal deprivation than with the simple lack of stimulation that is inevitable in most such environments. The character of early experience may thus also underlie

many problems in psychosomatic medicine and may explain in part why one individual develops ulcers, another migraine headaches and yet another shows little or no psychosomatic involvement under the same pressures of living.

One of the most encouraging aspects. of our research is that it has raised more questions than it has answered. We have not yet, for example, identified the critical element in our stimulation procedures that leads to such predictable and profound effects. Painful and extreme forms of stimulation seem to have effects indistinguishable from those produced by merely picking up an animal and placing it in another location for a brief period of time. Is picking up an infant organism as casual and insignificant a procedure as it appears? Or is the experience of the infant closer to that of Gulliver in Brobdingnag? Mere handling may, in fact, constitute a stimulation as compelling and severe as the more obviously traumatic forms of stimulation. It may be that some degree of stressful experience in infancy is necessary for successful adaptation of the organism to the environment it encounters in later life.

Another important question is whether there is a critical infantile period (or periods) during which stimulation is most effective. The evidence so far points to a period following immediately after birth. In one study we handled the animals in three separate groups for four days each, from the second through the fifth day, from the sixth through the ninth day and from the 10th through the 13th day. When we tested them for stress response on the 14th day, only the first group showed any evidence that they were capable of an endocrine response. Other investigators have had similar results. This should not be taken to mean, however, that stimulation has no effect after the critical period is past or that one critical period sets all responses.

Still other questions have not yet been satisfied by even partial answers. There is, for example, the question of therapy: Can the effects of lack of stimulation in the critical period be counteracted by stimulation of any sort after the critical period has passed? The most pressing question—the most "stimulating" question—is how stimulation causes change in the infant organism. The answer to this question should lead to a fuller understanding of the differences between individual constitutions and of the physiological mechanisms that are involved in behavior.

Deprivation Dwarfism

by Lytt I. Gardner

July 1972

*Children raised in an emotionally deprived environment
can become stunted. The reason may be that abnormal
patterns of sleep inhibit the secretion of pituitary
hormones, including the growth hormone*

It has long been known that infants will not thrive if their mothers are hostile to them or even merely indifferent. This knowledge is the grain of truth in a tale, which otherwise is surely apocryphal, told about Frederick II, the 13th-century ruler of Sicily. It is said that Frederick, himself the master of six languages, believed that all men were born with an innate language, and he wondered what particular ancient tongue—perhaps Hebrew—the language was. He sought the answer through an experiment. A group of foster-mothers was gathered and given charge of certain newborn infants. Frederick ordered the children raised in silence, so that they would not hear one spoken word. He reasoned that their first words, owing nothing to their upbringing, would reveal the natural language of man. "But he labored in vain," the chronicler declares, "because the children all died. For they could not live without the petting and the joyful faces and loving words of their foster mothers."

Similar emotional deprivation in infancy is probably the underlying cause of the spectacularly high mortality rates in 18th- and 19th-century foundling homes [see "Checks on Population Growth: 1750–1850," by William L. Langer; SCIENTIFIC AMERICAN Offprint 674]. This, at least, was the verdict of one Spanish churchman, who wrote in 1760: "In the foundling home the child becomes sad, and many of them die of sorrow." Disease and undernourishment certainly contributed to the foundlings' poor rate of survival, but as recently as 1915 James H. M. Knox, Jr., of the Johns Hopkins Hospital noted that, in spite of adequate physical care, 90 percent of the infants in Baltimore orphanages and foundling homes died within a year of admission. Only in the past 30 years or so have

the consequences of emotional deprivation in childhood been investigated in ways that give some hope of understanding the causative mechanisms. One pioneer in the field, Harry Bakwin of New York University, began in 1942 to record the physiological changes apparent in infants removed from the home environment for hospital care. These children, he noted, soon became listless, apathetic and depressed. Their bowel movements were more frequent, and even though their nutritional intake was adequate, they failed to gain weight at the normal rate. Respiratory infections and fevers of unknown origin persisted. All such abnormalities, however, quickly disappeared when the infants were returned to their home and mother.

Another early worker, Margaret A. Ribble, studied infants at three New York maternity hospitals over a period of eight years; in several instances she was able to follow the same child from birth to preadolescence. When normal contact between mother and infant was disrupted, she noted, diarrhea was more prevalent and muscle tone decreased. The infants would frequently spit up their food and then swallow it again. This action is called "rumination" by pediatricians. Largely as the result of studies by Renata and Eugenio Gaddini of the University of Rome, it is recognized today as a symptom of psychic disturbance. Ribble concluded that alarm over a lack of adequate "mothering" was not mere sentimentality. The absence of normal mother-infant interaction was "an actual privation which may result in biological, as well as psychological, damage to the infant."

Two other psychiatrically oriented investigators of this period, René Spitz of the New York Psychoanalytic Institute and his colleague Katherine Wolf, took

histories of 91 foundling-home infants in the eastern U.S. and Canada. They found that the infants consistently showed evidence of anxiety and sadness. Their physical development was retarded and they failed to gain weight normally or even lost weight. Periods of protracted insomnia alternated with periods of stupor. Of the 91, Spitz and Wolf reported, 34 died "in spite of good food and meticulous medical care." The period between the seventh and the 12th month of life was the time of the highest fatalities. Infants who managed to survive their first year uniformly showed severe physical retardation.

The comparative irrelevance of good diet, as contrasted with a hostile environment, was documented with startling clarity in Germany after World War II. In 1948 the British nutritionist Elsie M. Widdowson was stationed with an army medical unit in a town in the British Zone of Occupation where two small municipal orphanages were located. Each housed 50-odd boys and girls between four and 14 years of age; the children's average age was 8½. They had nothing except official rations to eat and were below normal in height and weight. The medical unit instituted a program of physical examinations of the orphans every two weeks and continued these observations for 12 months. During the first six months the orphanages continued to receive only the official rations. During the last six months the children in what I shall call Orphanage *A* received in addition unlimited amounts of bread, an extra ration of jam and a supply of concentrated orange juice.

The matron in charge of Orphanage *A* at the start of this study was a cheerful young woman who was fond of the children in her care. The woman in charge of Orphanage *B* was older, stern and a strict disciplinarian toward all the

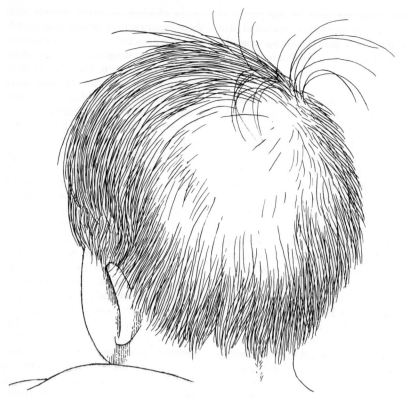

BALD SPOT had developed on the head of child shown on preceding page. This occurs frequently among emotionally deprived children, who will lie in one position for long periods.

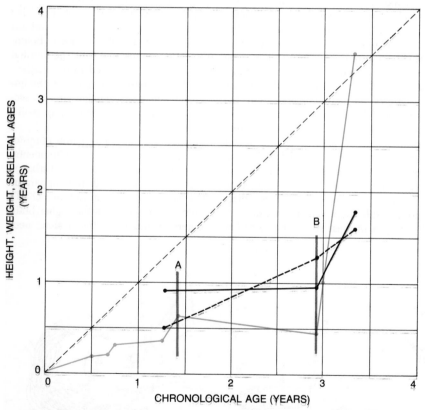

GROWTH RECORD of the boy shown on the preceding page traces his slow decline in weight (color) following his release from the hospital at about 18 months of age (A). By the age of three he weighed no more than the average six-month-old and was also severely retarded in height (black curve) and in bone maturation (broken curve). Reentering the hospital (B), the child gained weight at a dramatic rate. Placed in a foster home thereafter, he reached a weight above average for his age. Height, weight and skeletal ages of the vertical scale refer to the height, weight or skeletal development of an average child of that age.

ral pathways to the hypothalamus and thence, by neurohumoral mechanisms, exert influence on the pituitary gland. Research on "releasing factors" secreted by the hypothalamus, which in turn are responsible for the secretion of various trophic hormones by the anterior pituitary, has shown that hypothalamic centers exercise a major influence over this neighboring gland. Moreover, it is now known that virtually all the blood reaching the pituitary has first bathed the hypothalamic median eminence. Apparently the releasing factors are transported to the pituitary in the blood flowing from the median eminence through the pituitary portal veins.

Evidence of pituitary involvement in deprivation dwarfism is now becoming increasingly abundant. For example, one of the six children that Patton and I had studied came to the attention of my colleague Mary Voorhess in 1963. Following the child's earlier discharge from the hospital, he had spent two years in his disordered home environment and once again exhibited deprivation dwarfism. He was depressed and in an advanced state of malnutrition. An X-ray examination showed that in terms of maturation his bone age was three years less than his chronological age.

A determination was made of the boy's reserve of one important hormone secreted by the anterior pituitary: the adrenocorticotrophic hormone (ACTH). This hormone stimulates the secretion of the steroid hydrocortisone by the adrenal cortex. The steroid is an important regulator of carbohydrate metabolism; it promotes the conversion of protein into sugar, thereby raising the level of blood sugar. The examination showed that the child's reserve of ACTH was abnormally low.

The child entered the hospital, and following his recovery he was placed in a foster home with a favorable emotional environment. Eighteen months after the first measurement of his ACTH reserve the examination was repeated. The child's pituitary function was normal with respect to ACTH reserve. Furthermore, his formerly lagging bone age had almost caught up with his age in years. Both observations provide support for the working hypothesis that the chain of events leading to the clinical manifestations of deprivation dwarfism involves disturbances in pituitary trophic hormone secretion.

Another important pituitary hormone is somatotrophin, the growth hormone. Like most chemical messengers, the growth hormone is present in the bloodstream in very small quantities; the normal adult on waking will have less than three micrograms of growth hormone in each liter of blood plasma. Even with refined assay techniques measurements at this low level of concentration are difficult. As a result ways have been sought to briefly stimulate the production of growth hormone. Such stimulation not only makes measurement easier but also, if there is little or no response to the stimulus, provides direct evidence for a pituitary deficiency.

Bernardo A. Houssay of Argentina was the first investigator to point to the metabolic antagonism between insulin and the growth hormone. Subsequent research showed that an injection of insulin lowers the recipient's blood-sugar level, and the reaction is normally followed by an increased output of growth hormone. Jesse Roth and his colleagues at the Veterans Administration Hospital in the Bronx perfected an insulin-stimulation test for growth-hormone concentration in the blood in 1963.

Several quantitative assessments of the relation between emotional deprivation and abnormalities in growth-hormone concentration have been made in recent years. For example, in 1967 George Powell and his colleagues at the Johns Hopkins Hospital worked with a group of children over three years old suffering from deprivation dwarfism. They found that the children's growth-hormone response to insulin stimulation was subnormal. After the children had been transferred to an adequate emotional environment insulin stimulation was followed by the normal increase in growth-hormone concentration. At the Children's Hospital of Michigan patients over the age of three with deprivation dwarfism showed a similar picture. Working there in 1971, Ingeborg Krieger and Raymond Mellinger found that the response to insulin stimulation in these children was subnormal. When they repeated the test with deprived children below the age of three, however, they were surprised to find that the concentration of growth hormone under fasting conditions was abnormally high, and that after insulin stimulation the concentration was normal. It is likely that the difference in the responses of the two age groups is related to maturation, perhaps the maturation of the cerebral cortex. In any event both the Johns Hopkins and the Michigan studies are further evidence of the connection between deprivation dwarfism and impaired pituitary function.

Loss of appetite is a well-known com-

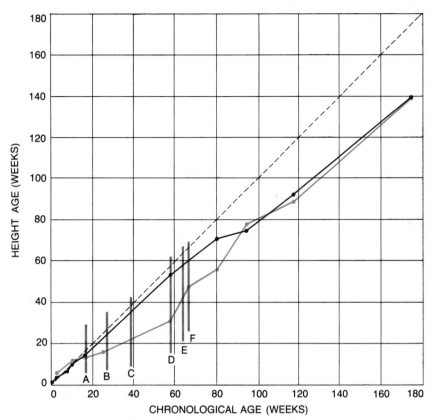

TWINS' INCREASE IN HEIGHT over a 3½-year period is plotted on this graph in terms of normal children's average growth. (The diagonal line marks the 50th percentile.) The boy who developed deprivation dwarfism also exhibited "rumination," that is, he spat up his food and swallowed it again. Nevertheless, his increase in height (*color*) was equal to or better than his sister's until his mother became pregnant (*A*). From that time on he fell steadily behind his sister as his father first lost his job (*B*) and then left home (*C*). Recovery of lost growth, begun when the boy was admitted to the hospital (*D*), continued on his return home (*F*), an event that followed his father's return (*E*). Before his second birthday the boy once more equaled his sister in height; both, however, were below normal.

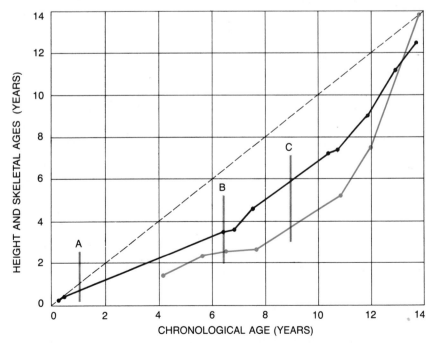

THIRTEEN YEARS of subnormal growth are shown in a child with a disordered home environment. Deserted by her husband (*A*), the mother worked full time. Six years later (*B*) the parents were divorced; then (*C*) the mother remarried and resumed keeping house. The child's bone development (*color*) but not his stature (*black*) finally became average.

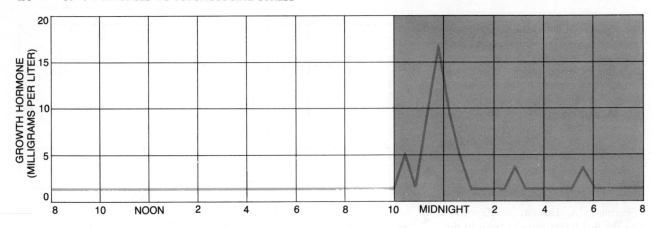

RELEASE OF GROWTH HORMONE increases during early hours of sleep (*shaded area*). This record of the growth hormone in the blood of a preadolescent child was collected by Jordan W. Finkelstein and his associates at Montefiore Hospital in the Bronx.

plaint in adolescents; in a few cases it is so extreme that it is termed anorexia nervosa. The disorder is usually attributable to adverse interpersonal relations between parent and child, particularly between mother and daughter. The clinical similarities between adolescent anorexia nervosa and infant and child deprivation dwarfism have long interested investigators. For example, adolescent girls with anorexia nervosa may stop menstruating; the adverse emotional climate evidently halts secretion of the pituitary hormones that mediate ovarian function. It now appears that some patients with anorexia nervosa also respond to insulin stimulation with a subnormal release of growth hormone.

Because it can be difficult to distinguish between anorexia nervosa and organic disorders of the pituitary, John Landon and his associates at St. Mary's Hospital in London recently tested the pituitary function of five patients with anorexia nervosa. All five showed the reaction typical of individuals with impaired pituitary function: after injections of insulin their level of blood sugar was excessively low. Two of the five also showed the reaction that typifies many instances of deprivation dwarfism: the concentration of growth hormone in the blood did not increase. The investigators also noted that before the injection of insulin the concentration of growth hormone in their patients' blood was significantly higher than normal. This paradoxical finding is not unlike the reaction observed among the younger deprivation-dwarfism patients in Michigan.

The significance of the St. Mary's findings is not yet clear, but they could prove to be important clues in unraveling the relations among the higher brain centers, the hypothalamus and the pituitary. We may be observing here a series of differing, age-mediated physiological responses to what are essentially identical psychosocial stimuli.

An important new direction for future investigation is suggested by the recent discovery of a connection between the release of growth hormone and individual modes of sleep. This connection was uncovered in 1965 by W. M. Hunter and W. M. Rigal of the University of Edinburgh, who observed that the total amount of growth hormone secreted during the night by older children was many times greater than the amount secreted during the day. Soon thereafter Hans-Jürgen Quabbe and his colleagues at the Free University of Berlin measured the amount of growth hormone secreted by adult volunteers during a 24-hour fast and detected a sharp rise that coincided with the volunteers' period of sleep. Pursuing this coincidence, Yasuro Takahashi and his co-workers at Washington University and Yutaka Honda and his colleagues at the University of Tokyo found that the rise in growth-hormone concentration occurs in adults during the first two hours of sleep, and that it equals the increase produced by insulin stimulation. If the subject remains awake, the growth hormone is not secreted. Honda and his associates propose that activation of the cerebral cortex somehow inhibits the secretion of growth hormone, whereas sleep—particularly the sleep that is accompanied by a high-voltage slow-wave pattern in encephalograph readings—induces the secretion of a growth-hormone-releasing factor in the hypothalamus.

Collecting such data calls for the frequent drawing of blood samples from sleeping subjects through an indwelling catheter, with the result that studies of growth-hormone concentration in infants and young children have been relatively few. What findings there are, however, fall into a pattern. For example, there appears to be no correlation between the concentration of growth hormone in the blood of normal newborn infants and the infants' cycle of alternate sleep and wakefulness. Such a correlation does not appear until after the third month of life. In normal children between the ages of five and 15 the maximum growth-hormone concentration is found about an hour after sleep begins. Maturation therefore appears to be a factor in establishing the correlation. Whether there is any link between these maturation-dependent responses and similar responses in instances of deprivation dwarfism is not clear.

The existence of a cause-and-effect relation between deprivation dwarfism and abnormal patterns of sleep had been suggested by Joseph Schutt-Aine and his associates at the Children's Hospital in Pittsburgh. They and others found that in children with deprivation dwarfism there were spontaneous and transient decreases in reserves of ACTH, the hormone that stimulates the secretion of the adrenal-cortex steroid that raises the level of blood sugar in the bloodstream. They also found that the decrease in ACTH reserve was accompanied by a temporary lowering of the child's blood-sugar level. Taking a position much like Honda's with respect to the growth hormone, Schutt-Aine and his colleagues suggest that any preponderance of inhibitory cerebral-cortex influences on the hypothalamus would tend to interfere with the normal release of ACTH and other pituitary hormones. An abnormal sleep pattern, they conclude, could lead to such a preponderance.

Is it in fact possible to attribute the retarded growth of psychosocially deprived children to sleep patterns that in-

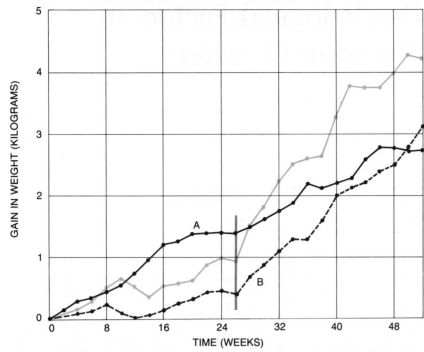

QUALITY OF CARE proved more important than quality of food in two postwar German orphanages studied by Elsie M. Widdowson. For six months during 1948 the 50-odd war orphans in each home received nothing but basic rations, yet the children in Orphanage *A*, supervised by a kindly matron, gained more weight than most of those in Orphanage *B*, whose matron was a stern disciplinarian. An exception was a group of favorites of the stern matron at *B* (*color*); they did better than their companions. After six months the matron at *B* was transferred to *A* and brought her favorites with her. Simultaneously the children at *A* were given extra rations, whereas the children at *B* remained on the same basic diet. (The transition is indicated by the vertical gray line.) Relieved of the stern matron's discipline, the children at *B* began to show a sharp increase in weight; those at *A* showed a weight gain that averaged somewhat less than it had during the preceding six months in spite of the larger ration. Again matron's favorites were an exception: their gain was greatest of any.

Georg Wolff and John W. Money of the Johns Hopkins Hospital have attempted a quantitative assessment of this question by studying the sleep patterns of a group of children with deprivation dwarfism. Analysis of their data suggests that the sleep pattern of children with deprivation dwarfism is disturbed in periods of subnormal growth and undisturbed in periods of normal growth. Thus far, however, the data are insufficient to establish whether or not there is an abnormal pattern in the secretion of growth hormone by sleeping children with deprivation dwarfism.

It has been proposed that the human infant is born with an innate, species-specific repertory of responses that includes clinging, sucking, "following" with the eyes, crying and smiling. These are the responses that John Bowlby of the Tavistock Institute of Human Relations in London identifies as having survival value for the infant; he believes their existence is a product of natural selection. In Bowlby's view the primary function of the mother is to integrate these responses into "attachment behavior," a more mature and more complicated pattern. In addition there evidently are "sensitive" periods in the course of human development, such as those familiar from animal experimentation. Exactly when these periods occur in human infancy, however, and just what conditions and experiences are necessary if the child is to develop normally remain uncertain. One conclusion nevertheless seems clear. Deprivation dwarfism is a concrete example—an "experiment of nature," so to speak—that demonstrates the delicacy, complexity and crucial importance of infant-parent interaction.

hibit the secretion of growth hormone? Certainly there is evidence that deprivation dwarfism and sleep abnormalities, both of commission and omission, often go together. The infants observed by Spitz and Wolf alternated between insomnia and stupor. One of the six thin dwarfs that Patton and I studied slept as much as 18 hours a day; another spent the hours while his family slept roaming the dark house. The tube-fed 15-month-old in Rochester, when admitted to the hospital, appeared to use sleep as a means of withdrawing from the world.

16

Psychological Factors in Stress and Disease

by Jay M. Weiss
June 1972

*A new technique separates the psychological
and physical factors in stressful conditions. In studies
with rats the psychological factors were the
main cause of stomach ulcers and other disorders*

One of the most intriguing ideas in medicine is that psychological processes affect disease. This concept is not new; it dates back to antiquity, and it has always been controversial. To counter those skeptics who believed that "no state of mind ever affected the humors of the blood," Daniel Hack Tuke, a noted 19th-century London physician, compiled an exhaustive volume, *Illustrations of the Influence of the Mind on the Body*. He concluded:

"We have seen that the influence of the mind upon the body is no transient power; that *in health* it may exalt the sensory functions, or suspend them altogether; excite the nervous system so as to cause the various forms of convulsive action of the voluntary muscles, or to depress it so as to render them powerless; may stimulate or paralyze the muscles of organic life, and the processes of Nutrition and Secretion—causing even death; that *in disease* it may restore the functions which it takes away in health, reinnervating the sensory and motor nerves, exciting healthy vascularity and nervous power, and assisting the *vis medicatrix Naturae* to throw off disease action or absorb morbid deposits." Through the years many other individuals have voiced their belief in the importance of psychological factors in disease, and they have carved out the field known as psychosomatic medicine. It is a field filled with questions, and we are still seeking better evidence on the role of the psychological factors.

Our ability to determine the influence of psychological factors on disease entered a new phase recently with the application of experimental techniques. Formerly the evidence that psychological factors influence disease came from the observations of astute clinicians who noted that certain psychological conditions seemed to be associated with particular organic disorders or with the increased severity of disorders. But such evidence, no matter how compelling, is correlational in nature. Although a psychological characteristic or event may coincide with the onset or advance of a disorder, one cannot be certain that it actually has any effect on the disease process; the psychological event may simply occur together with the disease or may even be caused by the disease. In addition it is always possible that the apparent correlation between the psychological variable and the disease is spurious; that among the myriad of other factors in the physical makeup of the patient and in his life situation lies a different critical element the observer has failed to detect. Such considerations can be ruled out only by accounting for every possible element in the disease process, which obviously is impractical.

The development of experimental techniques for inducing disease states in animals has made the task of determining whether or not psychological factors affect disease much easier. When an investigator can establish conditions that will cause a pathology to develop in experimental animals in the laboratory, he does not have to wait for the disease to arise and then attempt to determine if a particular factor had been important; instead he can introduce that factor directly into his conditions and see if it does indeed affect the development of the disease. Moreover, the use of experimental procedures enables the investigator to deal with the numerous other variables that might influence the disease. Even so, the investigator certainly cannot regulate or even be aware of all the variables that affect a disease. Such variables will, however, be distributed randomly throughout the entire population of experimental subjects. When the experimenter applies some treatment to one randomly selected group of subjects and not to another, he knows that any consistent difference between the treatment group and the control group will have been caused by the experimental treatment and not by the other variables, since those variables are distributed randomly throughout both groups.

Recently I have been studying the influence of psychological factors on certain experimentally induced disorders, particularly the development of gastric lesions or stomach ulcers. As the prob-

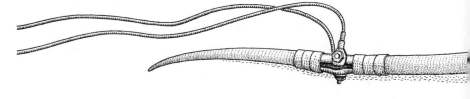

TRANSPARENT-PLASTIC CHAMBER is one type of apparatus used in stress ulcer experiments with rats. The rat is housed in the chamber for the entire 48-hour stress session.

lem of gastrointestinal ulcers has become more widespread (the disease currently causes more than 10,000 deaths each year in the U.S. and afflicts one out of every 20 persons at some point in their life) pathologists have continued to refine techniques of studying ulceration experimentally. Within the past 15 years, beginning with the pioneering work of Serge Bonfils and his associates at the Institut d'Hygiène in Paris, investigators have discovered that experimental animals can develop gastric lesions under stressful environmental conditions. The finding that lesions can be induced by manipulating an animal's external environment opened the way for experimental study of the psychological conditions that are brought about by stressful environmental events.

The experimental techniques had to be refined still further, however, for studying the influence of psychological variables. The reason is that when experimental animals are exposed to an environmental stressor (a stressor is a stress-inducing agent), the effects of psy-

chological variables may be confounded with the effects of the physical stressors. For example, suppose rats are made to swim for an hour in a tank from which they cannot escape, and these animals then develop organic pathology whereas control animals (those not exposed to the swim stressor) show no pathology at all. Although it is evident that the pathology was induced by the swim stressor, how can we assess the role of any psychological factor in producing this pathology? Certainly the pathology might have been affected by the fear the animal experienced, by its inability to escape, by the constant threat of drowning. But what about the extraordinary muscular exertion the stressful situation required, with its attendant debilitation and exhaustion of tissue resources? Clearly in such an experiment it is not possible to determine if psychological variables influenced the development of pathology, since the pathology might have been due simply to the direct impact of the swim stressor itself. Thus in order to study the role of psychological factors one must

devise a means of assessing the importance of psychological variables apart from the impact of the physical stressor on the organism.

The method I have used is to expose two or more animals simultaneously to exactly the same physical stressor, with each animal in a different psychological condition. I then look to see if a consistent difference results from these conditions; such a difference must be due to psychological variables since all the animals received the same physical stressor. To illustrate this technique, let us consider an experiment on the effects of the predictability of an electric shock on ulceration.

Two rats received electric shocks simultaneously through electrodes placed on their tail, while a third rat served as a control and received no shocks. Of the two rats receiving shocks, one heard a beeping tone that began 10 seconds before each shock. The other rat also heard the tone, but the tone sounded randomly with respect to the

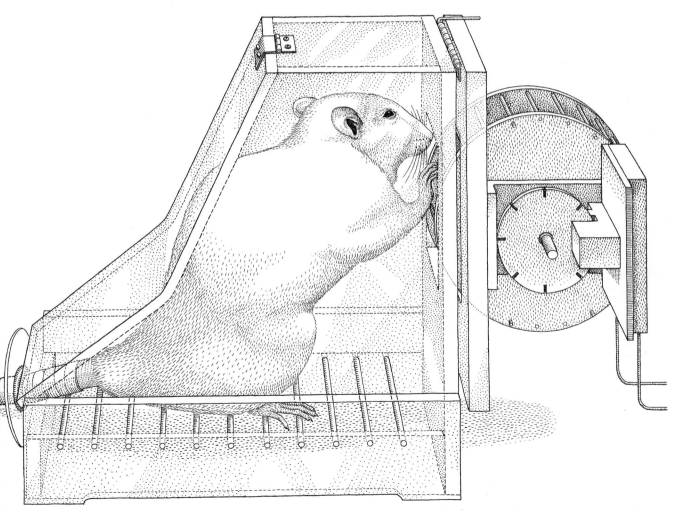

Shocks are delivered through electrodes attached to the tail. A disk secured to the tail by a piece of tubing prevents the rat from pulling its tail into the chamber and biting off the electrodes. The rat receives no food during the experiment, but water is available.

shock. Thus both these animals received the same shocks, but one could predict when the shocks would occur whereas the other could not. Since the physical stressor was the same for the two animals, any consistent difference between them in the amount of ulceration would be the result of the difference in the predictability of the stressor, the psychological variable being studied.

This raises a most important point: If such experiments are to be valid, one must be sure that the physical stressor, in this case the electric shock, is the same for all animals in the test. When these studies were begun, the standard way of administering electric shock to experimental rats was to place them on a grid floor whose bars were electrically charged. That method of delivering an electric shock was clearly inadequate for the present experiments, since rats can lessen the shock on a grid floor by changing their posture or can even terminate the shock completely by jumping. The experimenter is faced with a serious problem if one group of rats is able to perform such maneuvers more effectively than another group. In the predictability experiment, for example, the rats that are able to predict shock would surely have been able to prepare for such postural changes more effectively than the rats that were unable to predict shock, so that the groups would have differed with respect not only to the predictability of the stressor but also to the amount of shock received. That would essentially invalidate the experiment. The tail electrode, which is used in all the experiments discussed here, was developed specifically to avoid the possibility of unequal shock. Because the electrode is fixed to the tail, the rat cannot reduce or avoid the shock by moving about. In addition, with the fixed electrode it is possible to wire the electrodes of matched animals in series, so that both animals are part of the same circuit. Thus all the shocks received by the matched subjects are equal in duration and have an identical current intensity, which appears to be the critical element in determining the discomfort of shock.

As was to be expected, the control rats that received no shock developed very little gastric ulceration or none. A

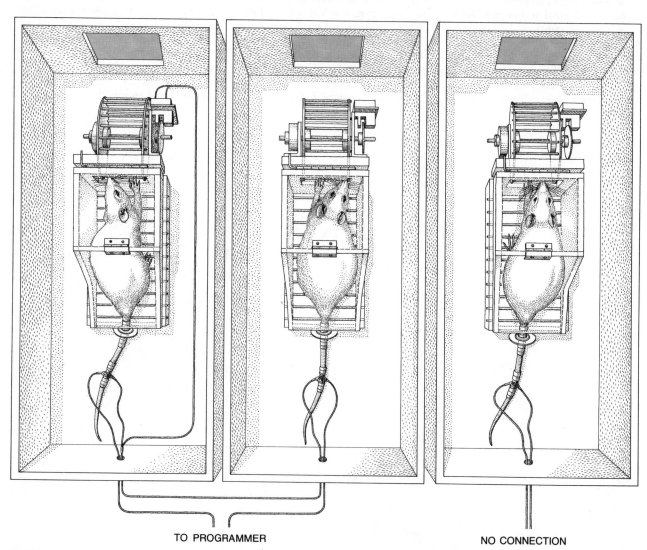

TO PROGRAMMER   NO CONNECTION

BASIC TRIPLET PARADIGM for the ulceration experiments consists in placing three rats, matched for weight and age, in individual soundproof compartments with one-way-mirror windows. Each rat is prepared in exactly the same way and then randomly assigned to one of the three experimental conditions: avoidance-escape, yoked and control. In this illustration the rat on the left is the avoidance-escape subject. It can terminate the programmed shock by turning the wheel. Moreover, turning the wheel between shocks will postpone the shock. The rat in the center is electrically wired in series to the first rat, so that when the first rat receives a shock, the yoked rat simultaneously receives a shock of the same intensity and duration. The actions of the yoked rat do not affect the shock sequence. The electrodes on the tail of the control rat on the right are not connected, and this rat does not receive shocks at any time. At the end of the experimental session the rats are sacrificed and the length of their gastric lesions is measured.

striking result of the experiment was that rats able to predict when the shocks would occur also showed relatively little ulceration, whereas those that received the same shocks unpredictably showed a considerable amount of ulceration [*see top illustration at right*]. In short, the results demonstrated clearly that the psychological variable of predictability, rather than the shock itself, was the main determinant of ulcer severity.

Even though some rats in the foregoing experiment could predict the shock, they were helpless in that they could not avoid the shock. How will stress reactions, such as gastric ulcers, be altered if an animal has control over a stressor instead of being helpless? A recent series of experiments conducted in our laboratory at Rockefeller University has yielded a considerable amount of new information on effects of coping behavior, and also has given us some insight into why these effects arise.

To study the effects of coping behavior, three rats underwent experimental treatment simultaneously, just as in the predictability experiment. Two of the rats again received exactly the same shocks through fixed tail electrodes wired in series, while the third rat served as a nonshock control. In the coping experiments, however, the difference between the two rats receiving shock was not based on predictability (all matched shocked rats in these experiments received the same signals) but rather was based on the fact that one rat could avoid or escape the shock whereas the other could not do anything to escape it.

Since albino rats tend to lose weight when stressed, at first I simply measured the effects of avoidance-escape versus helplessness in terms of changes in the rate of weight gain. In this experimental arrangement one rat of each shocked pair could avoid the shock by jumping onto a platform at the rear of its enclosure during a warning signal, thereby preventing the shock from being given to itself or its partner. If the avoidance-escape rat failed to jump in time, so that shock occurred, it could still jump onto the platform to terminate the shock for itself and its partner. Thus the avoidance-escape rat could affect the occurrence and duration of shock by its responses, whereas its partner, called the yoked subject, received exactly the same shocks but was helpless: its responses had no effect at all on shock. The control rat, which received no shock, was simply allowed to explore its apparatus during the experiment. It is important to note that in all these experiments the three rats were assigned to their respec-

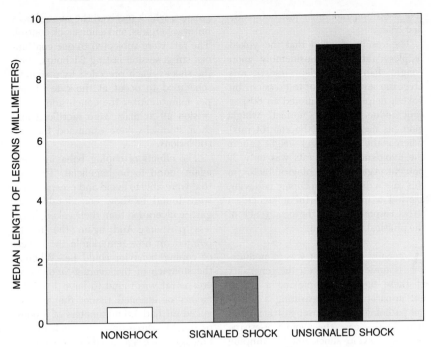

PREDICTABILITY EXPERIMENT showed that if a rat could predict when shocks would occur, it developed less gastric ulceration than a rat that received the same shocks unpredictably. One rat heard a "beep" before each shock, whereas for the other rat the "beep" occurred randomly with respect to the shock. A third rat, a control subject, heard the sound but received no shocks. Thus the only difference between the rats receiving shocks was psychological, and this psychological variable strongly affected the degree of ulceration.

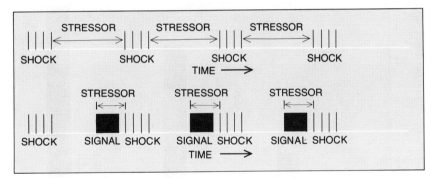

STRESSFUL SITUATION (called "stressor") is created by the electric shock and the stimuli associated with the shock. When there is no warning-signal stimulus, the stressor condition extends through the entire time between shocks. When a warning of the shock is given, however, the stressor condition tends to be restricted to the warning signal.

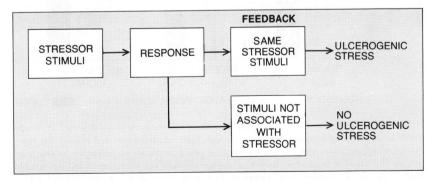

COPING ATTEMPTS OR RESPONSES made by the rat in trying to escape from the stressor are related to ulcerogenic stress only if the responses fail to produce stimuli that are not associated with the stressor. If coping responses produce stimuli not associated with the stressor, the rat receives relevant feedback and ulcerogenic stress does not occur.

tive condition randomly just before the first trial.

The results showed that the yoked, helpless rats lost considerably more weight than the avoidance-escape rats. Over the several days of test sessions the yoked, helpless rats suffered an 80 percent reduction from "normal" weight gain (as measured in the control rats), whereas the reduction of weight gain in the avoidance-escape rats was only 30 percent. Again a psychological factor, in this case a difference in coping behavior, exerted a more powerful influence on stress responses than the occurrence of the physical stressor did.

A second experiment was immediately undertaken to test the generality of these findings, employing a different apparatus and measuring a different pathological response: gastric ulceration. In this case the avoidance-escape rat could avoid shock not by jumping onto a platform but by reaching through a hole in a small restraint cage to touch a panel mounted just outside. The avoidance-escape animal again had a yoked

partner, which received the same shocks but was helpless, and a nonshock control. The rats were subjected to one continuous stress session lasting 21 hours, with the shocks—each preceded by a signal—scheduled to occur at the rate of one per minute. After the conclusion of the session all animals were sacrificed and their stomachs were examined for gastric lesions.

The effects of coping behavior were again found to be beneficial. The rats that were able to avoid and escape shock were found to show considerably less gastric ulceration than their yoked, helpless partners. And again the results pointed up how remarkable the effects of coping behavior could be. Whereas the stomach of the average avoidance-escape rat was found to have 1.6 millimeters of lesioned tissue, the average yoked rat had 4.5 millimeters of lesions, or roughly three times as much ulceration as the average avoidance-escape rat.

In both avoidance-escape experiments the shock was always preceded by a warning, so that the rats could predict when a shock was going to occur. Hence

the avoidance-escape rat always had a signal to inform it when to respond. What would happen if there was no signal before the shock? Would the avoidance-escape rat again show less ulceration than a yoked subject?

To find out a large experiment was conducted in which three different warning-signal conditions were set up: no warning signal, a single uniform signal preceding the shock as in the earlier experiments, and a series of different signals that acted like a clock and therefore gave more information about when a shock would occur than the single uniform signal did. For these studies each rat was placed in a chamber with a large wheel [see illustration on pages 156 and 157]. If the avoidance-escape rat turned the wheel at the front of the apparatus, the shock was postponed for 200 seconds, or if shock had begun, it was immediately terminated and the next shock did not occur for 200 seconds. Thus the avoidance-escape conditions were exactly the same except for the difference in the warning signals. Each avoidance-escape rat had a yoked, helpless partner and both received exactly the same signals and shocks. A rat that never received a shock also was included in every case as a control subject.

The results showed that regardless of the warning-signal condition avoidance-escape rats developed less gastric ulceration than yoked, helpless rats [see illustration at left]. Although the presence of a warning signal did reduce ulceration in both avoidance-escape and yoked groups, the avoidance-escape rats always developed less ulceration.

All the experimental findings on coping behavior that I have described up to this point have been opposite to the result found by Joseph V. Brady, Robert Porter and their colleagues in an experiment with monkeys [see "Ulcers in 'Executive' Monkeys," by Joseph V. Brady; SCIENTIFIC AMERICAN Offprint 425]. They reported that in four pairs of monkeys the animals that could avoid shock by pressing a lever developed severe gastrointestinal ulcers and died, whereas yoked animals that received the same shocks but could not perform the avoidance response survived with no apparent ill effects. Why were these results so markedly different?

With careful study of the data from the 180 rats that were used in the coping-behavior and warning-signal experiment, it was possible to develop a theory that may explain how coping behavior affects gastric ulceration. This theory can account for the results I have

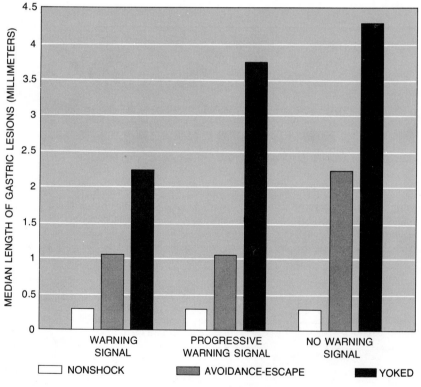

"HELPLESS" RATS develop more ulcers than their counterparts that can avoid or escape shock by performing a simple task, even though both rats have received exactly the same shocks. In all situations the avoidance-escape rat could terminate or postpone the shock by turning the wheel in front of it; its yoked partner received the same shocks but was unable to affect the shock sequence by its behavior. The control rat was never shocked. Regardless of the warning-signal conditions, rats that could do something to stop the shock developed less ulceration than their yoked, helpless mates. Ulceration was more extensive in both groups when there was no warning signal. The yoked, helpless rats unexpectedly developed almost as much ulceration with the progressive warning signal as with no warning signal.

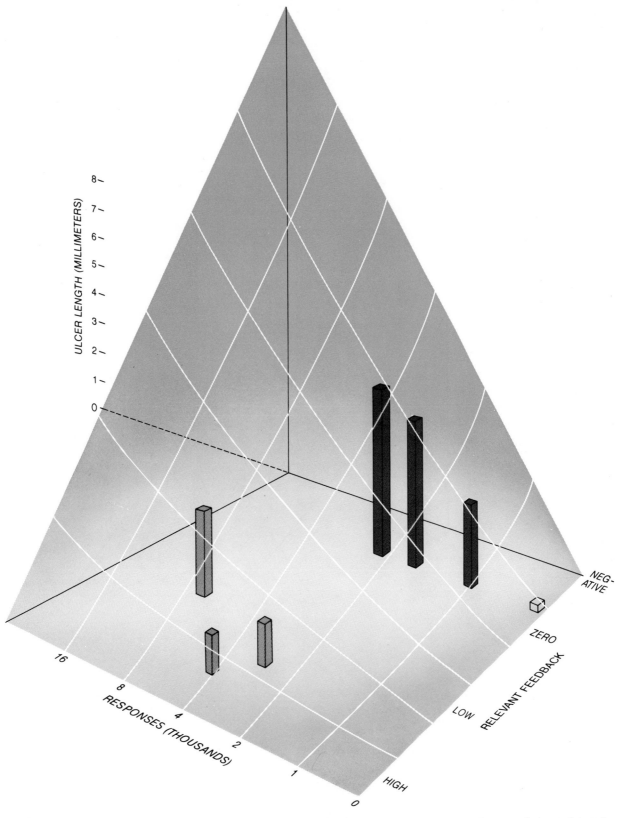

RESPONSE RATE was found to be related to the amount of ulceration: the greater the number of responses, the more the ulceration. Moreover, increasing the amount of relevant feedback decreases ulceration. Combining these variables produces this three-dimensional graph. Here data from the illustration on the opposite page are replotted. The yoked, helpless rats given the progressive warning signal (*middle black bar*) made more responses than helpless rats given only a brief warning signal (*lowest black bar*). Helpless rats shocked without a warning signal made more responses than the other helpless rats and had the highest ulceration (*tallest black bar*). The single white bar represents the control rats that received no shocks in all three conditions, since their response rate and ulceration was nearly the same in all cases. (They developed some ulceration because they too were in a mildly stressful condition.) The avoidance-escape rats that received no warning signal made the greatest number of coping responses and had a high amount of ulceration (*tallest gray bar*). The avoidance-escape rats given the progressive warning signal (high feedback) made more responses (*middle gray bar*) than avoidance-escape rats that heard the brief warning signal before the shock (*right gray bar*).

162

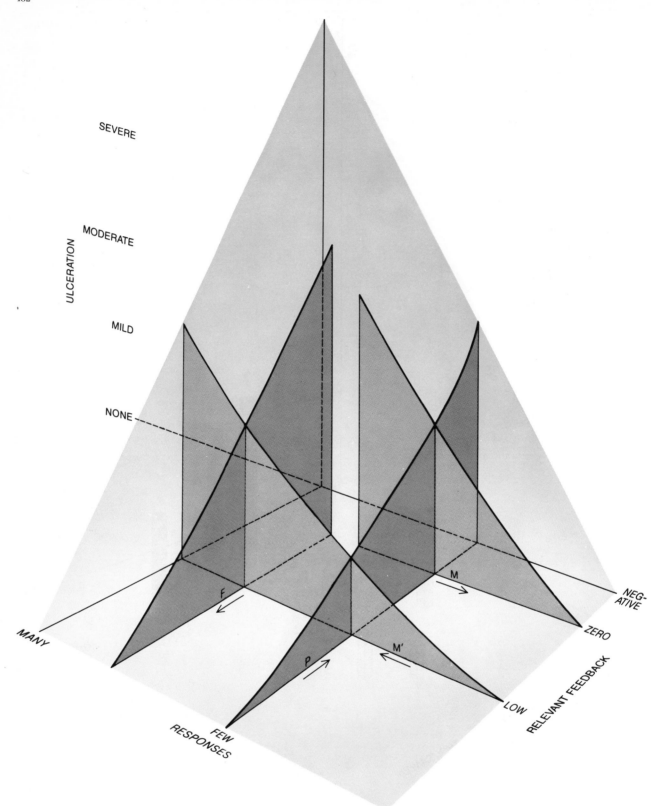

SEVERE

MODERATE

MILD

NONE

ULCERATION

MANY

FEW
RESPONSES

F

P

M'

M

LOW

ZERO

NEG-
ATIVE

RELEVANT FEEDBACK

PREDICTIVE MODEL of the relation between responses, feedback and ulceration is shown here. It predicts that the highest rate of ulceration will occur when the number of coping responses is very high and the feedback is very low or negative. The model can explain why the "executive" monkeys developed more ulcers than their "helpless" counterparts. The executive monkeys could postpone a shock by pressing a lever. Their response rate was quite high, but the amount of relevant feedback they received was low. Thus increased response rate along the low-feedback plane (*M' in illustration*) led to increased ulceration in the executive monkeys.

For the "helpless" monkeys, even though they received zero feedback (*Plane M*), their low response rate put them in the low-ulceration area. Similarly, if a high-relevant-feedback situation is changed into a negative-relevant-feedback situation (for example, a previously correct response suddenly begins to produce punishment), the model predicts a rapid increase in ulceration (*Plane P*) even at low levels of response. Finally, increasing feedback to a very high level should reduce ulceration to a very low amount (*Plane F*) even if the response rate is high. Results of experiments conducted to test these predictions are shown on page 164.

obtained and also can reconcile the seemingly contradictory results of the "executive monkey" experiment.

The theory states that stress ulceration is a function of two variables: the number of coping attempts or responses an animal makes, and the amount of appropriate, or "relevant," feedback these coping attempts produce. When an animal is presented with a stressor stimulus, the animal will make coping attempts or responses. The first proposition is simply that the more responses one observes, the greater is the ulcerogenic (ulcer-producing) stress. (Note that this does not say that the behavioral responses themselves cause ulceration, only that the amount of coping behavior and the amount of ulcerogenic stress tend to rise together.) If the responses, however, immediately produce appropriate feedback —that is, if the responses bring about stimuli that have no connection with the stressor—ulcerogenic stress will not occur. On the other hand, if the responses fail to produce such stimuli, then ulcerogenic stress will occur [see bottom illustration on page 159].

Perhaps the most important concept in this theory is that of feedback. The appropriate feedback is called relevant feedback. It consists of stimuli that are not associated with the stressful situation. Relevant feedback occurs when a response produces stimuli that differ from the stressor. The amount of relevant feedback produced depends on how different the stimulus situation becomes and how far removed these new stimuli are from any association with the stressor.

We can now specify how the two variables, responding and relevant feedback, are related to ulceration. Ulceration increases as the number of responses increases, and ulceration decreases as the amount of relevant feedback increases. Combining these two produces a function that forms a three-dimensional plane [see illustration on page 161]. From this model one can predict the amount of ulceration that is expected to occur in any stressful situation by specifying the number of coping attempts or responses the animal makes and the amount of relevant feedback these responses produce.

The model explains why animals able to perform effective coping responses usually develop fewer ulcers than helpless animals. Whenever a helpless animal makes a coping attempt, the response necessarily produces no relevant feedback because it has no effect on the stimuli of the animal's environment. Thus if

helpless animals make an appreciable number of coping attempts, which many of them do, they will develop ulcers because of the lack of relevant feedback. Animals that have control in a stressful situation, however, do receive relevant feedback when they respond. In my experiments, for example, the avoidance-escape rats could terminate warning signals and shocks (thereby producing silence and the absence of shocks), so that their responses produced stimuli that were dissociated from the stressor. Hence animals in control of a stressor can usually make many responses and not develop ulcers because they normally receive a substantial amount of relevant feedback for responding.

It is evident that, according to the theory, the effectiveness of coping behavior in preventing ulceration depends on the relevant feedback that coping responses produce; simply to have control over the stressor is in and of itself not beneficial. This means that conditions certainly can exist wherein an animal that has control will ulcerate severely. Specifically, in cases of low relevant feedback ulceration will be severe if the number of responses made is high [see illustration on opposite page].

I believe the foregoing statement tells us precisely why the executive monkeys died of severe gastrointestinal ulceration while performing an avoidance response. First of all, the responding of the monkeys was maintained at a very high rate in that experiment because they had to respond once every 20 seconds to avoid shock. In addition, the executives were actually selected for their high rate of responding. On the basis of a test before the experiment began, the monkey in each pair that responded at the higher rate was made the avoidance animal while its slower partner was assigned to the yoked position. Thus on the basis of their response rate the executive monkeys were more ulcer-prone from the beginning than their yoked partners. With regard to the relevant feedback for responding, the feedback for avoidance responding was quite low. There were no warning signals, and so the executives' rapid-fire responses could not turn off any external signals and therefore did not change the external-stimulus environment at all. As a result the relevant feedback came entirely from internal cues. Evidently this feedback was not sufficient to counteract the extremely high response condition, so that the executive animals developed ulcers and died. At the same time the yoked animals probably made very few responses or coping attempts because

the shocks were few and far between, thanks to the high responding of the executives. It is no wonder, then, that the yoked animals in this case survived with no apparent ill effects.

Further evidence in support of this model has emerged, both in an analysis of earlier experiments and in new direct tests with rats. Reviewing those experiments in which hyperactive avoidance-escape rats happened to be paired with low-responding yoked rats under conditions similar to those of the executive-monkey experiment, I found that high-responding avoidance-escape animals developed more ulceration than their helpless partners, which replicated the results of the monkey pairs [see top illustration on next page]. I then went on to test the model directly by examining the effects of very poor relevant feedback and excellent relevant feedback.

The first experiment examined the effects of very poor relevant feedback, which should, of course, produce severe ulceration. In this case avoidance-escape rats, having spent 24 hours in a normal avoidance situation with a warning signal, were given a brief pulse of shock every time they performed the correct response. Although the avoidance-escape rats had control over the stressor, their responses now produced the wrong kind of feedback: the stressor stimulus itself. The feedback in this condition is even worse than it is in the zero-relevant-feedback, or helplessness, condition. The results showed that even though these rats had control over the stressor, they developed severe gastric ulceration, in fact more ulceration than helpless animals receiving the same shocks [see middle illustration on next page].

Having found that very poor feedback would cause severe ulceration, I conducted an experiment to determine if excellent feedback could reduce and possibly eliminate ulceration in a stressful situation. Initially the shock was administered without a warning signal, and under this condition the avoidance-escape rats will normally develop a considerable amount of gastric ulceration. Presumably ulceration occurs because the relevant feedback for responding is low, as in the case of the executive monkeys. Then a brief tone was added to the experiment. When the rat now performed its coping response, it not only postponed shock but also sounded the tone. Because the tone immediately followed the response, and the response postponed the shock, the tone was not associated with the shock. Thus the tone produced a change in the stimulus situa-

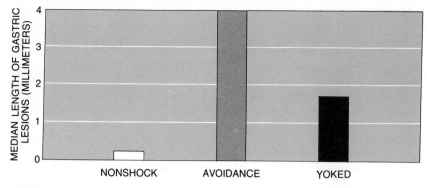

REPLICATION of the executive-monkey situation with data from experiments in which rats received unsignaled shocks offers support for the theoretical model of ulcerogenic stress. Matched pairs of high-responding avoidance rats and low-responding yoked rats were statistically selected and their ulcers measured. As the model predicts, the avoidance rats showed higher ulceration and the low-responding yoked rats had less ulceration.

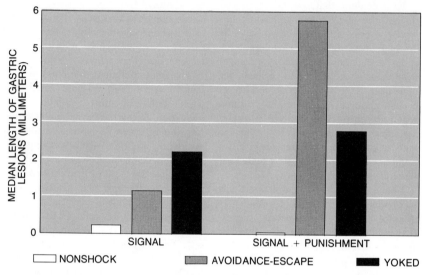

NEGATIVE RELEVANT FEEDBACK produces severe ulceration even when the animal has control over the shock. In the warning-signal condition avoidance-escape rats learned to perform a response to avoid a shock whenever a tone sounded. In the punishment situation during the last half of the experiment the rats received a shock every time they performed the previously learned correct response. With this negative relevant feedback the avoidance-escape rats developed more ulcers than their yoked, helpless mates did.

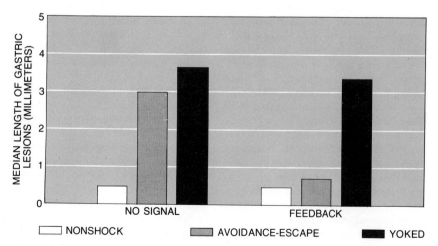

EXCELLENT RELEVANT FEEDBACK following a correct response drastically reduces the amount of ulceration in rats. In the no-signal condition the avoidance-escape rat could postpone the shock by turning a wheel but there was no feedback other than the absence of shock. In the feedback condition a signal followed immediately after every correct response. The number of responses by avoidance-escape rats in both conditions was the same.

tion that constituted excellent relevant feedback. The result was striking. Although the rats in this situation received about as many shocks as the counterparts that were not given the tone feedback, they developed a small amount of ulceration; in fact, they developed only slightly more ulceration than controls receiving no shocks at all did [*see bottom illustration at left*].

Hence by manipulating the feedback consequences of responding, rats could be made to develop extensive gastric ulceration in an otherwise nonulcerogenic condition or could be protected almost completely from developing ulcers in a condition that was normally quite ulcerogenic. The fact that these results are consistent with the proposed model means that we are beginning to develop some idea of why the remarkable effects of psychological variables in stress situations occur.

It appears that the principles discovered in these animal experiments may operate in human situations as well. For example, Ronald Champion of the University of Sydney and James Geer and his associates at the State University of New York at Stony Brook have shown that if people are given inescapable shocks, the individuals who think they can terminate these shocks by clenching a fist or pressing a button show less emotional arousal as measured by electrical skin resistance than individuals who receive the same shocks and are also asked to clench a fist or press the button but are told that the shocks are inescapable. These findings can be explained using the model derived from the animal experiments, again emphasizing the role of relevant feedback. The people who thought they had control over the shock perceived their responses as producing the shock-free condition, that is, they saw their responses as producing relevant feedback. In contrast, the people who thought they were helpless necessarily perceived their responses as producing no relevant feedback. Thus for humans, as for rats, the same variables seem important in describing the effects of behavior in stress situations. On the other hand, the experiments with humans alert us to how important higher cognitive processes are in people, showing that verbal instructions and self-evaluation can determine feedback from behavior, which will subsequently affect bodily stress reactions.

Other stress responses have been measured in addition to gastric ulceration, for example the level of plasma corticosterone in the blood and the amount of body weight the animals lost during the stress session. In many instances the results reflect those found with gastric ulcers, showing that ulceration may be only one manifestation of a more general systemic stress response. The correlation between measures is by no means perfect, so that it is evident that all physiological systems participating in stress reactions are not affected in the same way by a given stress condition. Certain systems in the body may be severely taxed by one set of conditions whereas other systems may be hardly affected at all, or actually may be benefited. For example, I have observed that heart weight tends to decrease in certain conditions, and this effect is seen more often in animals that are able to avoid and escape shock. We do not yet even know what this change indicates, but it suggests that certain conditions protecting one organ system, such as the gastrointestinal tract, might adversely affect another, such as the cardiovascular system.

Perhaps the most exciting biochemical system we have begun to study involves the catecholamines of the central nervous system. Eric A. Stone, Nell Harrell and I have studied changes in the level of norepinephrine in the brain. This substance is a suspected neurotransmitter that is thought to play a major role in mediating active, assertive responses, and several investigators have suggested that depletion of norepinephrine is instrumental in bringing about depression in humans. We found that animals able to avoid and escape shock showed an increase in the level of brain norepinephrine, whereas helpless animals, which received the same shocks, showed a decrease in norepinephrine. At the same time Martin Seligman, Steven Maier and Richard L. Solomon of the University of Pennsylvania have found that dogs given inescapable shocks will subsequently show signs of behavioral depression, but that dogs that are able to avoid and escape shocks do not show such depression. It may well be that the causal sequence leading from "helplessness" to behavioral depression depends on biochemical changes in the central nervous system, such as changes in brain norepinephrine. This would indicate that depressed behavior often can be perpetuated in a vicious circle: the inability to cope alters neural biochemistry, which further accentuates depression, increasing the inability to cope, which further alters neural biochemistry, and so on. We need to know more about this cycle and how to break it.

166

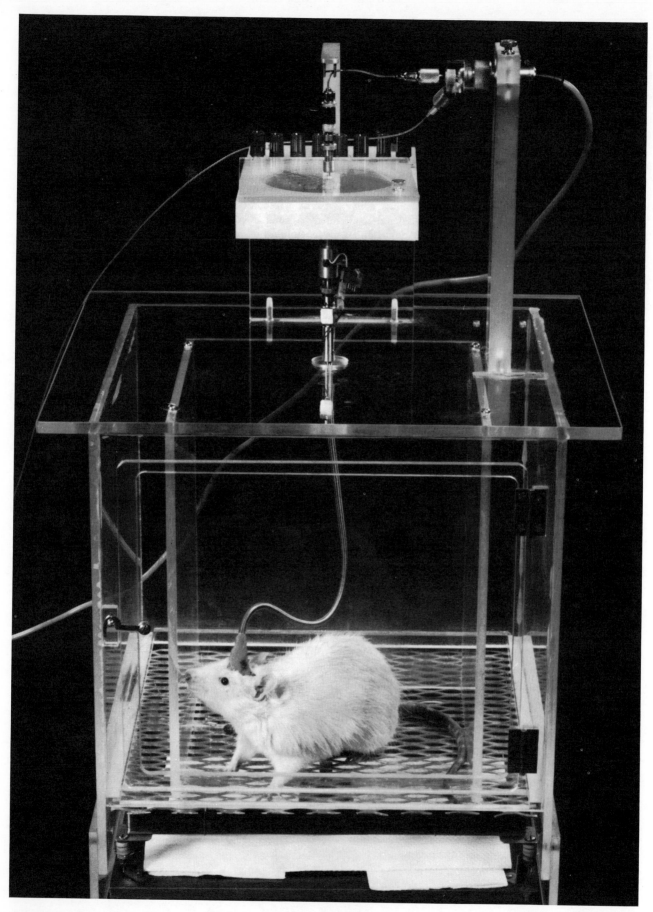

VISCERAL RESPONSES of active rats were measured with this experimental setup. The necessary tubes and wires (in this case including a plastic catheter to sense blood pressure in an abdom-inal artery) are led from the rat's skull through a protective steel spring to a mercury connector, a fluid swivel that permits free movement. The platform is wired to record the rat's activity.

17

Learning in the Autonomic Nervous System

by Leo V. DiCara
January 1970

It has long been assumed that body functions such as heartbeat and intestinal contraction were involuntary and could not be influenced by learning. Recent experiments indicate that this is not the case

The heart beats and the stomach digests food without any obvious training, effort or even attention. That may be the basis of a curious prejudice against the visceral responses—the responses of glands, of cardiac muscle and of the smooth muscle of the alimentary canal and blood vessels—and against the autonomic nervous system, which controls them. Such responses are assumed to be quite different from, and somehow inferior to, the highly coordinated voluntary responses of skeletal muscles and the cerebrospinal nervous system that controls them. A corollary of this attitude has been the assumption that visceral responses can be "conditioned" but cannot be learned in the same way as skeletal responses. It turns out that these long-standing assumptions are not valid. There is apparently only one kind of learning; supposedly involuntary responses can be genuinely learned. These findings, which have profound significance for theories of learning and the biological basis of learning, should lead to better understanding of the cause and cure of psychosomatic disorders and of the mechanisms whereby the body maintains homeostasis, or a stable internal environment.

Learning theorists distinguish between two types of learning. One type, which is thought to be involuntary and therefore inferior, is classical, or Pavlovian, conditioning. In this process a conditioned stimulus (a signal of some kind) is presented along with an innate unconditioned stimulus (such as food) that normally elicits a certain innate unconditioned response (such as salivation); after a time the conditioned stimulus elicits the same response. The other type of learning—clearly subject to voluntary control and therefore considered superior—is instrumental, or trial-and-error, learning, also called operant conditioning. In this process a reinforcement, or reward, is given whenever the desired conditioned response is elicited by a conditioned stimulus (such as a certain signal). The possibilities of learning are limited in classical conditioning, because the stimulus and response must have a natural relationship to begin with. In instrumental learning, on the other hand, the reinforcement strengthens any immediately preceding response; a given response can be reinforced by a variety of rewards and a given reward can reinforce a variety of responses.

Differences in the conditions under which learning occurs through classical

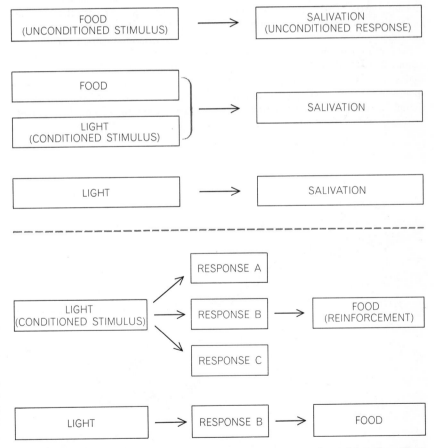

TWO TYPES OF LEARNING are classical conditioning and instrumental learning. Classical conditioning (*top*) begins with an unconditioned stimulus. The conditioned stimulus that is paired with it comes to substitute for it in producing the unconditioned response. In instrumental learning (*bottom*) a conditioned stimulus is presented along with an opportunity to respond in various ways. The correct response is reinforced, or rewarded. After several reinforcements the stimulus serves as a signal to perform the learned response.

conditioning and through instrumental learning have been cited to show that the two processes are two distinct phenomena that operate through different neurophysiological mechanisms. The traditional belief has been that the involuntary and inferior visceral responses can be modified only by the correspondingly inferior type of learning—classical conditioning—and not by the superior and voluntary instrumental learning, which has been thought to modify only voluntary, skeletal responses.

Not all learning theorists accepted this distinction. For many years Neal E. Miller of Rockefeller University has held that classical conditioning and instrumental learning are not two basically different phenomena but rather two manifestations of the same phenomenon under different conditions—that there is, in fact, only one kind of learning. To support such a position he had to show that instrumental training procedures can produce learning of any visceral responses that can be acquired through classical conditioning, and the demonstration had to be very clear and convincing in the face of the ingrained belief that such learning is simply not possible.

Research on the instrumental modification of visceral responses comes up against a basic problem: most such responses can be affected by voluntary activities such as the tensing of muscles or

changes in the rate or pattern of breathing. It is therefore hard to rule out completely the possibility that the experimental subject has not directly learned to control a visceral response through the autonomic system but rather has learned to execute some subtle and undetectable skeletal response that in turn modifies the visceral behavior. (A skilled disciple of yoga, for example, can stop his heart sounds by controlling his rib cage and diaphragm muscles so that pressure within the chest is increased to the point where the venous return of blood to the heart is considerably retarded.)

To guard against the contamination of experimental results by such "cheating," careful controls and detailed statistical analysis of data are required. The primary control Miller and I apply in our animal experiments is paralysis of the subject's skeletal muscles. This is accomplished by administering a drug of the curare family (such as *d*-tubocurarine) that blocks acetylcholine, the chemical transmitter by which cerebrospinal nerve impulses are delivered to skeletal muscles, but does not interfere with consciousness or with the transmitters that mediate autonomic responses. A curarized animal cannot breathe and must therefore be maintained on a mechanical respirator. Moreover, it cannot eat or drink, and so the possibilities of rewarding it are limited. We rely on two methods of reinforcement. One is elec-

trical stimulation of a "pleasure center" in the brain, the medial forebrain bundle in the hypothalamus, and the other is the avoidance of or escape from a mildly unpleasant electric shock.

Utilizing these techniques, we have shown that animals can learn visceral responses in the same way that they learn skeletal responses. Specifically, we have produced, through instrumental training, increases and decreases in heart rate, blood pressure, intestinal contractions, control of blood-vessel diameter and rate of formation of urine. Other investigators have demonstrated significant instrumental learning of heart-rate and blood-pressure control by human beings and have begun to apply the powerful techniques developed in animal experiments to the actual treatment of human cardiovascular disorders.

After Miller and his colleagues Jay Trowill and Alfredo Carmona had achieved promising preliminary results (including the instrumental learning of salivation in dogs, the classical response of classical conditioning), he and I undertook in 1965 to show that there are no real differences between the two kinds of learning: that the laws of learning observed in the instrumental training of skeletal responses all apply also to the instrumental training of visceral responses. We worked with curarized rats, which we trained to increase or to decrease their heart rate in order to obtain pleasurable brain stimulation. First we rewarded small changes in the desired direction that occurred during "time in" periods, that is, during the presentation of light and tone signals that indicated when the reward was available. Then we set the criterion (the level required to obtain a reward) at progressively higher levels and thus "shaped" the rats to learn increases or decreases in heart rate of about 20 percent in the course of a 90-minute training period [*see illustration on page 170*].

These changes were largely overall increases or decreases in the "base line" heart rate. We were anxious to demonstrate something more: that heart rate, like skeletal responses, could be brought under the control of a discriminative stimulus, which is to say that the rats could learn to respond specifically to the light and tone stimuli that indicated when a reward was available and not to respond during "time out" periods when they would not be rewarded. To this end we trained rats for another 45 minutes at the highest criterion level. When we began discrimination training, it took the

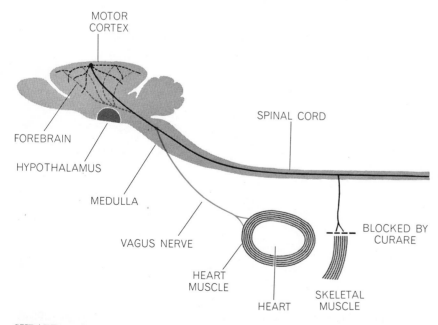

CURARE paralyzes skeletal muscles. It ensures, for example, that a change in heart rate has been controlled by autonomic impulses from the vagus nerve and not by cerebrospinal impulses to skeletal muscles. The two nervous systems are not completely separate: visceral responses have representation at higher brain centers in the cortex (**broken lines**).

rats some time after the beginning of each stimulus period to meet the criterion and get the reward; by the end of the training they were changing their heart rate in the rewarded direction almost immediately after the time-in period began [see illustration on page 171].

In skeletal instrumental training discrimination is also learned between a positive stimulus, response to which is rewarded, and a negative stimulus, response to which is not rewarded. Our animals learned to respond with the proper visceral behavior to one stimulus (such as a light) and not to respond to another (such as a tone). Moreover, once an animal has learned to discriminate between positive and negative cues for a given skeletal response, it is easier for it to respond similarly with a different response for the same reward. We found that this phenomenon of transfer also appeared in visceral training: rats that showed the best discrimination between a positive and a negative stimulus for a skeletal response (pressing a bar) also showed the best discrimination when the same stimuli were used for increased or decreased heart rate.

Two other properties of instrumental training are retention and extinction. To test for retention we gave rats a single training session and then returned them to their home cages for three months. When they were again curarized and tested, without being reinforced, rats in both the increase group and the decrease group showed good retention by exhibiting reliable changes in the direction for which they had been rewarded three months earlier. Although learned skeletal responses are remembered well, they can be progressively weakened, or experimentally extinguished, by prolonged trials without reward. We have observed this phenomenon of extinction in visceral learning also. To sum up, all the phenomena of instrumental training that we have tested to date have turned out to be characteristic of visceral as well as skeletal responses.

The experiments I have described relied on electrical stimulation of the brain as a reinforcement. In order to be sure that there was nothing unique about brain stimulation as a reward for visceral learning, Miller and I did an experiment with electric-shock avoidance, the other of the two commonly used rewards that can conveniently be administered to paralyzed rats. A shock signal was presented to the curarized rats. After it had been on for five seconds it was accompanied by brief pulses of mild

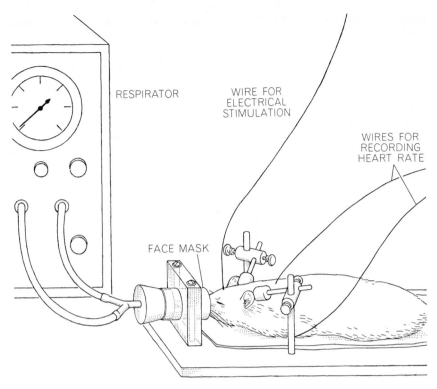

CURARIZED RATS cannot breathe and must be fitted with a face mask connected to a respirator. Such usual instrumental-learning rewards as food and water cannot be used.

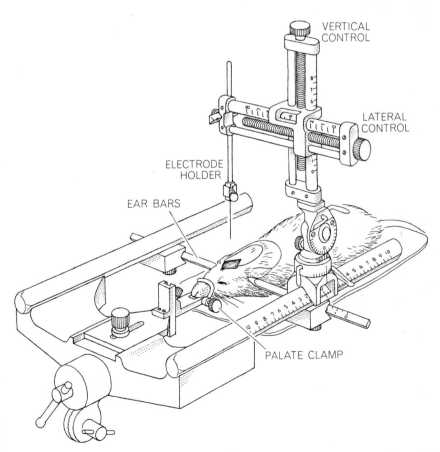

REWARD for visceral learning is either electrical stimulation of the brain or avoidance of electric shock. For brain stimulation an electrode implanted in the brain of an anesthetized rat is guided to a "pleasure center" in the hypothalamus with the aid of a stereotactic device.

shock delivered to the rat's tail. During the first five seconds the animal could turn off the shock signal and avoid the shock by making the correct heart-rate response; failing that, it could escape the shock by making the correct response and thus turning off both the signal and the shock.

In the course of a training session we mixed shock trials with "safe" trials and "blank" trials at random. During a safe trial we presented a different signal and did not administer a shock; during a blank trial there was no signal or shock. For half of the rats the shock signal was a tone and the safe signal a flashing light; for the other half the stimuli were reversed. The rats that were rewarded for increasing their heart rate learned to increase it and those that were rewarded for decreasing the rate learned to decrease it. In part the learning represented a general change in base line, as indicated by the trend of the heart rate during blank trials. Beyond this, however, the rats clearly learned to discriminate. As their training progressed, the shock signal began to elicit a greater change in the rewarded direction than the blank trials did. Conversely, the safe signal elicited a trend in the opposite direction—toward the base line represented by the data for the blank trials [see illustration on page 172].

At this point we had shown that instrumental learning of visceral responses follows the laws of skeletal instrumental training and that it is not limited to a particular kind of reward. We also showed that the response itself is not limited: we trained rats to raise and lower their systolic blood pressure in much the same way. These results were all obtained, however, with animals that were paralyzed. Would normal, active animals also learn a visceral response? If so, could that response be shown to be independent of skeletal activity? We designed a special experimental cage and the necessary equipment to make possible the recording of various responses of active rats [see illustration on page 17], and we established that heart-rate and blood-pressure changes could be learned by noncurarized animals. The heart-rate learning persisted in subsequent tests during which the same animals were paralyzed by curare, indicating that it had not been due to the indirect effects of overt skeletal responses. This conclusion was strengthened when, on being retrained without curare, the two groups of animals displayed increasing differences in heart rate, whereas any differences in respiration and general level of activity continued to decrease.

We noted with interest that initial learning in the noncurarized state was slower and less effective than it had been in the previous experiments under curare. Moreover, a single training session under curare facilitated later learning in the noncurarized state. It seems likely that paralysis eliminated "noise" (the confusing effects of changes in heart action and blood-vessel tone caused by skeletal activity) and perhaps also made it possible for the animal to concentrate on and sense the small changes accomplished directly by the autonomic system.

In all these studies the fact that the same reward could produce changes in opposite directions ruled out the possibility that the visceral learning was caused by some innate, unconditioned effect of the reward. Furthermore, the fact that the curarized rats were completely paralyzed, which was confirmed by electromyographic traces that would have recorded any activity of the skeletal muscles, ruled out any obvious effect of the voluntary responses. It was still possible, however, that we were somehow inducing a general pattern of arousal or were training the animals to initiate impulses from the higher brain centers that would have produced skeletal movements were it not for the curare, and that it was the innate effect of these central commands to struggle and relax that were in turn changing the heart rate. Such possibilities made it desirable to discover whether or not changes in heart rate could be learned independently of changes in other autonomic responses that would occur as natural concomitants of arousal.

To this end Miller and Ali Banuazizi compared the instrumental learning of heart rate with that of intestinal contraction in curarized rats. They chose these two responses because the vagus nerve innervates both the heart and the gut, and the effect of vagal activation on both organs is well established. In order to record intestinal motility they inserted a water-filled balloon in the large intestine. Movement of the intestine wall caused fluctuations in the water pressure that were changed into electric voltages by a pressure transducer attached to the balloon.

The results were clear-cut. The rats rewarded (by brain stimulation) for increases in intestinal contraction learned an increase and those rewarded for decreases learned a decrease, but neither

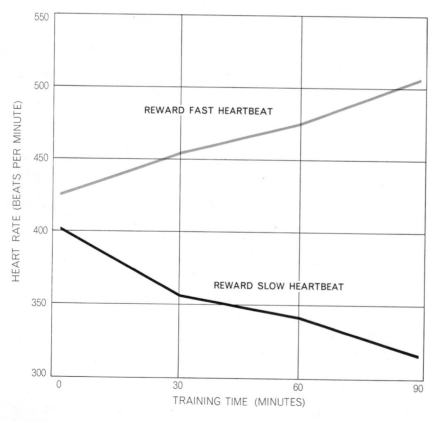

HEART-RATE CHANGES are shown for rats rewarded for increasing the rate (*color*) and for decreasing it (*black*). Animals were curarized and rewarded with brain stimulation.

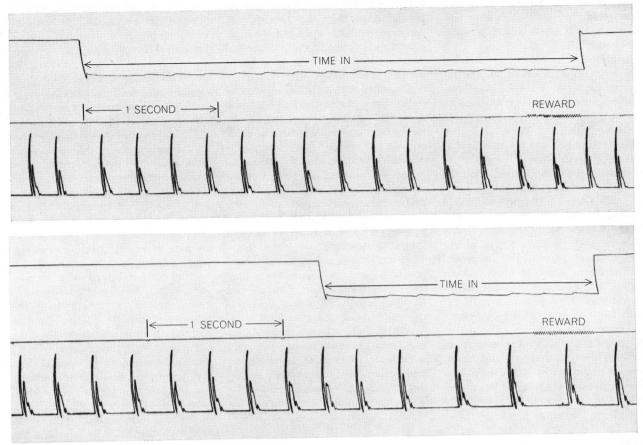

ELECTROCARDIOGRAMS made at the beginning and at the end of an extra period of training demonstrate discrimination. At first the rat takes some time after the onset of stimulus ("time in") to respond (by slowing its heartbeat) and earn a reward (*top*). After 45 minutes of discrimination training the rat responds more directly: it slows its heartbeat soon after time-in period begins (*bottom*).

group showed an appreciable change in heart rate. The group that was rewarded for increases in heart rate learned an increase and the group rewarded for decreases learned a decrease, but neither heart-rate group showed an increase in intestinal contraction [*see illustrations on page 175*]. Moreover, the heart-rate and intestinal learning were negatively correlated: the better the response being rewarded was learned, the less change there was in the unrewarded response. These results showed that the instrumental learning of two visceral responses can occur independently of each other and that what is learned is specifically the rewarded response. They ruled out the possibility that the learning was mediated by a general reaction such as arousal.

There was still a remote possibility to be eliminated: The central impulses I mentioned might be initiated selectively toward muscles that affect the intestines when intestinal changes are rewarded and toward muscles that affect heart rate when heart-rate changes are rewarded. Miller and I therefore trained curarized rats to increase or decrease their heart rate and then tested them in the non-curarized state for transfer of learning. We reasoned that if heart-rate changes were not directly learned under curare but rather were mediated by the learning of central impulses to skeletal muscles, movement of such muscles would betray the fact if the learning was transferred to the noncurarized state. We found that learned increases and decreases of about 10 percent did transfer independently of muscle movement: the differences between the two groups in heart rate were too large to be accounted for by the differences between them in respiration or general level of activity.

The strongest argument against attempts to explain visceral learning as a response to skeletal movement or central motor impulses is this kind of specificity. As more and more different visceral responses are recorded and the learning of them is shown to be specific, it becomes harder to think of enough different voluntary responses to account for them all. We have shown, for example, that curarized rats can learn to make changes in the dilation and constriction of blood vessels in the skin and to make these vasomotor changes independently of changes in heart rate and blood pressure. Indeed, the rats can be trained to make these changes specific to a single structure: they can dilate the blood vessels in one ear more than those in the other ear! This could not be the result of heart-rate or blood-pressure changes, which would affect both ears equally. We also obtained instrumental learning in the rate of urine formation by the kidneys, independent of blood pressure or heart rate. The increases and decreases in the amount of urine produced were achieved by specific changes in the arteries of the kidneys that resulted in an increase or decrease in the blood flow through the kidneys.

In addition to buttressing the case for instrumental learning of visceral responses, these striking results suggest that vasomotor responses, which are mediated by the sympathetic division of the autonomic nervous system, are capable of much greater specificity than was believed possible. This specificity is compatible with an increasing body of evidence that various visceral responses have specific representation at the cerebral cortex, that is, that they have neural connections of some kind to higher brain centers.

Some recent experiments indicate that not only visceral behavior but also the electrical activity of these higher brain centers themselves can be modified by direct reinforcement of changes in brain activity. Miller and Carmona trained noncurarized cats and curarized rats to change the character of their electroencephalogram, raising or lowering the voltage of the brain waves. A. H. Black of McMaster University in Canada trained dogs to alter the activity of one kind of brain wave, the theta wave. More recently Stephen S. Fox of the University of Iowa used instrumental techniques to modify, both in animals and in human subjects, the amplitude of an electrical event in the cortex that is ordinarily evoked as a visual response.

We are now trying to apply similar techniques to modify the electrical activity of the vagus nerve at its nucleus in the lowermost portion of the brain. Preliminary results suggest that this is possible. The next step will be to investigate the visceral consequences of such modi-fication. This kind of work may open up possibilities for modifying the activity of specific parts of the brain and the functions they control and thereby learning more about the functions of different parts of the brain.

Controlled manipulation of visceral responses by instrumental training also makes it possible to investigate the mechanisms that underlie visceral learning. We have made a beginning in this direction by considering the biochemical consequences of heart-rate training and specifically the role of the catecholamines, substances such as epinephrine and norepinephrine that are synthesized in the brain and in sympathetic-nerve tissues. Norepinephrine serves as a nerve-impulse transmitter in the central nervous system. Both substances play roles in the coordination of neural and glandular activity, influencing the blood vessels, the heart and several other organs. Alterations in heart rate produced by increased sympathetic-nerve activity in the heart, for example, are accompanied by changes in the synthesis, uptake and utilization of catecholamines in the heart, suggesting that it may be possible to influence cardiac catecholamine me-tabolism through instrumental learning of heart-rate responses. This would be important in view of the possible role of norepinephrine in essential hypertension (high blood pressure) and congestive heart failure; it might also help to establish the role of learning and experience in the development of certain psychosomatic disorders.

Eric Stone and I found that the level of catecholamines in the heart varies with heart-rate training. After three hours of training under curare, rats trained to increase their heart rate have a significantly higher concentration of cardiac catecholamines than rats trained to decrease their heart rate. Experiments are now under way to determine how long such biochemical differences between the two groups persist after training and whether the heart-rate conditioning has long-range effects on the heart and on the excitability of the sympathetic nerves. When we examined the brains of rats in the two groups we found a similar biochemical difference: the animals trained to increase their heart rate had a significantly higher level of norepinephrine in the brain stem than rats trained to decrease heart rate. Brain norepinephrine helps to determine the excitability of the central nervous system and is involved in emotional behavior. We have therefore started experiments to see whether or not changes in sympathetic excitability obtained by cardiovascular instrumental training are related to changes in the metabolism of norepinephrine and, if so, in which areas of the brain these metabolic changes are most apparent.

Is the capacity for instrumental learning of autonomic responses just a useless by-product of the capacity for cerebrospinal, skeletal-muscle learning? Or does it have a significant adaptive function in helping to maintain homeostasis, a stable internal environment? Skeletal responses operate on the external environment; there is obvious survival value in the ability to learn a response that brings a reward such as food, water or escape from pain. The responses mediated by the autonomic system, on the other hand, do not have such direct effects on the external environment. That was one of the reasons for the persistent belief that they are not subject to instrumental learning. Yet the experiments I have described demonstrate that visceral responses are indeed subject to instrumental training. This forces us to think of the internal behavior of the visceral organs in the same way we think of the external,

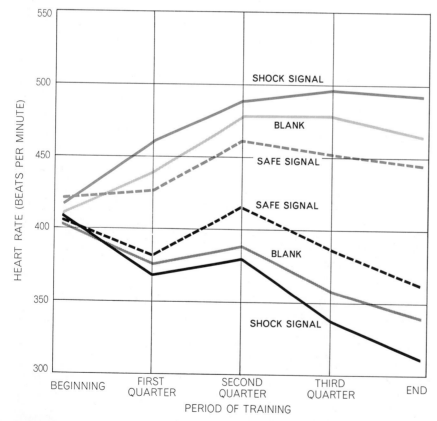

DISCRIMINATION is demonstrated by these curves for rats that were trained to increase (color) and decrease (black) heart rate and were rewarded by avoidance of shock. The results for blank trials (no signal or shock) show "base line" learning. The results for shock-signal and safe-signal trials show discriminating responses to more specific stimuli.

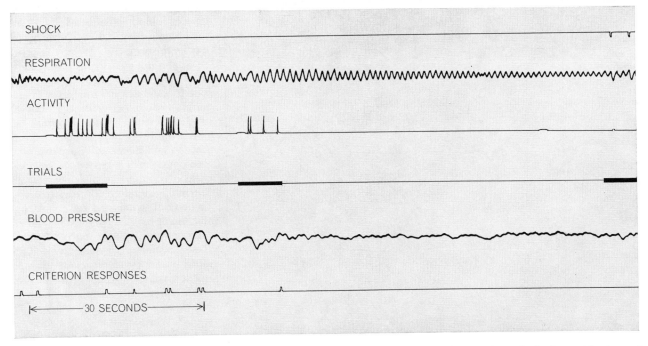

SHOCK

RESPIRATION

ACTIVITY

TRIALS

BLOOD PRESSURE

CRITERION RESPONSES

|←————30 SECONDS————→|

POLYGRAPH RECORD, a small portion of which is reproduced, records a free-moving rat's respiration, activity and systolic blood pressure. It also shows when trials took place, whether the blood pressure increase met the criterion and whether, not having met criterion, the animal received an electric shock. This record was made by an animal being tested as in the illustration on page 166.

observable behavior of the skeletal muscles, and therefore to consider its adaptive value to homeostasis.

In a recent experiment George Wolf, Miller and I found that the correction of a deviation from homeostasis by an internal, glandular response (rather than by an external response such as eating or drinking) can serve as a reward to reinforce learning. We injected albino rats with an antidiuretic hormone (ADH) if they chose one arm of a *T*-shaped maze and with a control solution (a minute amount of isotonic saline solution) if they chose the other arm. Before running the maze each rat had been given an excess of water through a tube placed in the stomach, so that the antidiuretic hormone was maladaptive: it interfered with the kidney response that was necessary to get rid of the excess water and restore homeostasis, whereas the control solution did not interfere. The rats learned to select the side of the maze that ensured an injection of saline solution, so that their own glandular response to the excess water could restore homeostasis. Then we did the same experiment with rats that suffered from diabetes insipidus, a disorder in which too much urine is passed and it is insufficiently concentrated. These rats had been tube-fed an excess of a highly concentrated salt solution. Now the homeostatic effects of the two injections were reversed: the ADH was adaptive, tend-

ing to concentrate the urine and thereby get rid of the excess salt, whereas the control solution had no such effect. This time the rats selected the ADH side of the maze. As a control we tested normal rats that were given neither water nor concentrated saline solution, and we found they did not learn to choose either side of the maze in order to obtain or avoid the antidiuretic hormone.

In many experiments a deficit in water or in salt has been shown to serve as a drive to motivate learning; the external response of drinking water or saline solution—thus correcting the deficit—functions as a reward to reinforce learning. What our experiment showed was that the return to a normal balance can be effected by action that achieves an internal, glandular response rather than by the external response of drinking.

Consider this result along with those demonstrating that glandular and visceral responses can be instrumentally learned. Taken together, they suggest that an animal can learn glandular and visceral responses that promptly restore a deviation from homeostasis to the proper level. Whether such theoretically possible learning actually takes place depends on whether innate homeostatic mechanisms control the internal environment so closely and effectively that deviations large enough to serve as a drive are not allowed. It may be that innate controls are ordinarily accurate

enough to do just that, but that if abnormal circumstances such as disease interfere with innate control, visceral learning reinforced by a return to homeostasis may be available as an emergency replacement.

Are human beings capable of instrumental learning of visceral responses? One would think so. People are smarter than rats, and so anything rats can do people should be able to do better. Whether they can, however, is still not completely clear. The reason is largely that it is difficult to subject human beings to the rigorous controls that can be applied to animals (including deep paralysis by means of curare) and thus to be sure that changes in visceral responses represent true instrumental learning of such responses.

One recent experiment conducted by David Shapiro and his colleagues at the Harvard Medical School indicated that human subjects can be trained through feedback and reinforcement to modify their blood pressure. Each success (a rise in pressure for some volunteers and a decrease for others) was indicated by a flashing light. The reward, after 20 flashes, was a glimpse of a nude pinup picture. (The volunteers were of course male.) Most subjects said later they were not aware of having any control over the flashing light and did not in fact know what physiological function was being

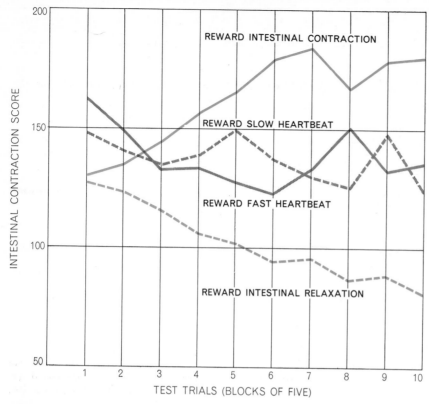

INTESTINAL CONTRACTION is learned independently of heart-rate changes. Contractions are increased by rats rewarded for increases (*colored line*) and decreased by rats rewarded for decreases (*broken colored line*). The intestinal-contraction score does not change appreciably, however, in rats rewarded for increasing or decreasing heart rate (*gray*).

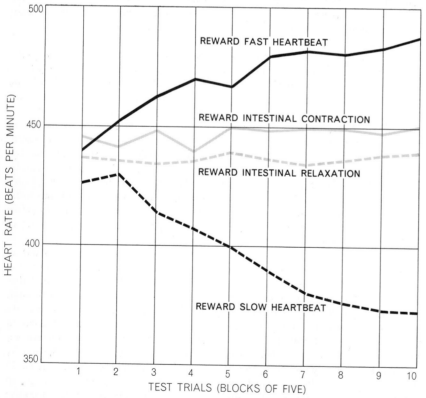

SPECIFICITY of learning is shown by this graph and the one at the top of the page. Here the results for heart rate rather than intestinal contraction are shown for the same animals. Rats rewarded for changing their heart rate change it in the appropriate direction (*black lines*). Rats rewarded for intestinal changes do not change heart rate (*light colored lines*).

measured, and so they presumably had not exerted any voluntary effort (at least not consciously and deliberately) to modify the response.

Whatever is actually being learned by such subjects, the extent of learning is clearly less than can be achieved in animals. In one of our experiments the average difference in blood pressure between the two groups of curarized rats was 58 millimeters of mercury. Shapiro's two human groups, in contrast, yielded a comparable difference of about four millimeters. Clearly curarized rats do better than noncurarized people, but that is not really surprising. The difference between the noncurarized rats and the noncurarized human subjects is much smaller [*see bottom illustration on opposite page*].

The curare effect here is in line with what is seen in experiments with a single species. What does it mean? I mentioned above that initial training under curare facilitated further training in the noncurarized state. Perhaps the curare keeps the animal from being confused (as it may be in the noncurarized state) when a small change in the correct direction that is produced by direct control of the visceral response is obscured by a larger change in the opposite direction that is accomplished through skeletal activity and is therefore not rewarded. It is also possible that the curare helps to eliminate variability in the stimulus and to shift the animal's attention from distracting skeletal activity to the relevant visceral activity. It may be possible to facilitate visceral learning in humans by training people (perhaps through hypnosis) to breathe regularly, to relax and to concentrate in an attempt to mimic the conditions produced by curarization.

The evidence for instrumental learning of visceral responses suggests that psychosomatic symptoms may be learned. John I. Lacey of the Fels Research Institute has shown that there is a tendency for each individual to respond to stress with his own rather consistent sequence of such visceral responses as headache, queasy stomach, palpitation or faintness. Instrumental learning might produce such a hierarchy. It is theoretically possible that such learning could be carried far enough to create an actual psychosomatic symptom. Presumably genetic and constitutional differences among individuals would affect the susceptibility of the various organ systems. So would the extent to which reinforcement is available. (Does a child's mother keep him home from school when he complains of head-ache? When he looks pale?) So also would the extent to which visceral learning is effective in the various organ systems.

We are now trying to see just how far we can push the learning of visceral responses—whether it can be carried far enough in noncurarized animals to produce physical damage. We also want to see if there is a critical period in the animal's infancy during which visceral learning has particularly intense and long-lasting effects. Some earlier experiments bear on such questions. For example, during training under curare seven rats in a group of 43 being rewarded for slowing their heart rate died, whereas none of 41 being rewarded for an increase in heart rate died. This statistically reliable difference might mean one of two things. Either training to speed the heart rate helps a rat to resist the stress of curare or the reward for slowing the heart rate is strong enough to overcome innate regulatory mechanisms and induce cardiac arrest.

If visceral responses can be modified by instrumental learning, it may be possible in effect to "train" people with certain disorders to get well. Such therapeutic learning should be worth trying on any symptom that is under neural control, that can be continuously monitored and for which a certain direction of change is clearly advisable from a medical point of view. For several years Bernard Engel and his colleagues at the Gerontology Research Center in Baltimore have been treating cardiac arrhythmias (disorders of heartbeat rhythm) through instrumental training. Heart function has been significantly improved in several of their patients. Miller and his colleagues at the Cornell University Medical College treated a patient with long-standing tachycardia (rapid heartbeat). For two weeks the patient made almost no progress, but in the third week his learning improved; since then he has been able to practice on his own and maintain his slower heart rate for several months. Clark T. Randt and his colleagues at the New York University School of Medicine have had some success in training epileptic patients to suppress paroxysmal spikes, an abnormal brain wave.

It is far too early to promise any cures. There is no doubt, however, that the exciting possibility of applying these powerful new techniques to therapeutic education should be investigated vigorously at the clinical as well as the experimental level.

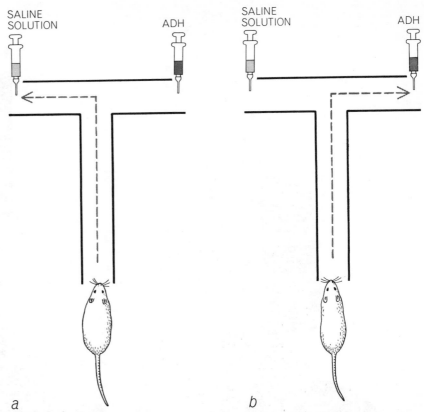

VISCERAL RESPONSE that adjusts the internal environment can serve as a reward to reinforce learning. Rats "loaded" with water (*a*) learned to choose the side of a *T*-maze that resulted in an injection of a control solution rather than one of antidiuretic hormone (ADH), which would interfere with water excretion. (The arms associated with each reward were changed at random.) Rats loaded with salt (*b*), on the other hand, for whom the hormone would induce the proper kidney response, learned to pick the ADH-associated arm.

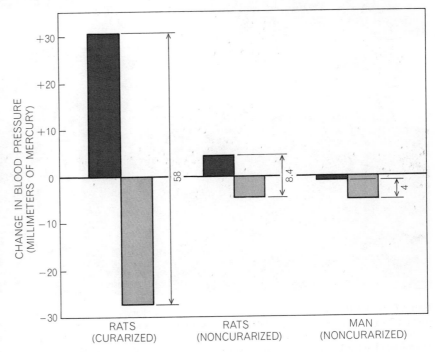

COMPARISON of blood-pressure learning in rats and humans rewarded for increasing (*dark gray bars*) and for decreasing (*light gray*) blood pressure shows that the difference between curarized and noncurarized subjects is greater than the difference between species.

2176

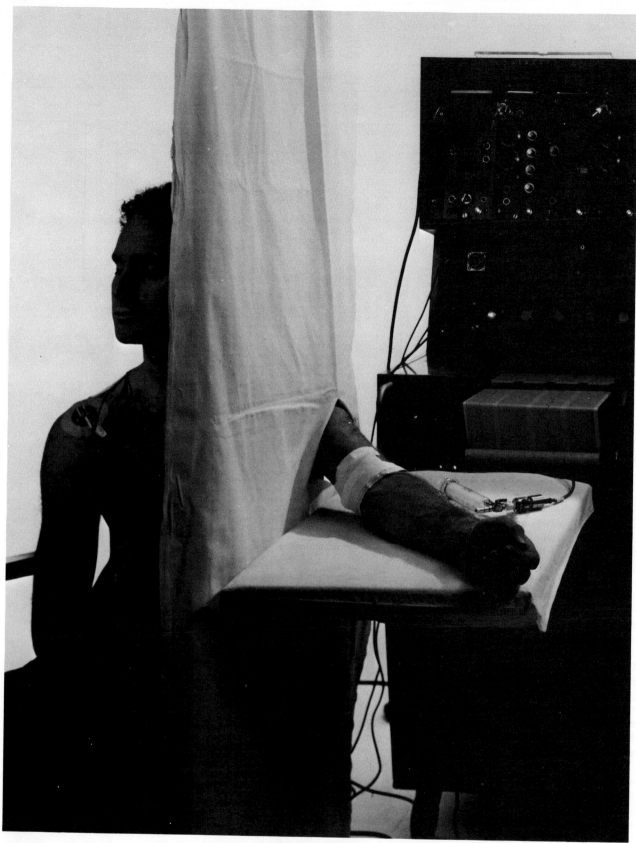

ISOLATED SUBJECT is connected with an instrument array that continuously records such physiological variables as heart rate and blood pressure. A catheter in the subject's left arm draws samples of arterial blood at 10-minute intervals; these samples are analyzed for oxygen and carbon dioxide content and for blood acidity and blood-lactate level. The subject's arm is screened from his view to minimize the psychological effects of blood withdrawal. Each sub-ject first sat quietly for an interval and then was invited to medi-tate for a 30-minute period. At the end of the period the subject was asked to stop meditating but to continue sitting quietly during a further recording interval. Thirty-six qualified "transcendental" meditators from 17 to 41 years old volunteered as subjects for the study, which was conducted both at the Harvard Medical Unit of the Boston City Hospital and at the University of California at Irvine.

18

The Physiology of Meditation

by Robert Keith Wallace and Herbert Benson
February 1972

*Is the meditative state that is achieved by yogis
and other Far Eastern mystics accompanied
by distinct physiological changes? A study of
volunteer subjects in the U.S. indicates that it is*

How capable is the human organism of adjusting to psychologically disturbing changes in the environment? Our technological age is probably testing this capacity more severely than it was ever tested in the past. The impact of the rapid changes—unprecedented in scale, complexity and novelty—that technology is bringing about in our world seems to be having a deleterious effect on the mental and physical health of modern man. Some of the common disorders of our age, notably "nervous stomach" and high blood pressure, may well be attributable in part to the uncertainties that are burgeoning in our environment and daily lives. Since the environment is not likely to grow less complex or more predictable, it seems only prudent to devote some investigative attention to the human body's resources for coping with the vicissitudes of the environment.

There are in fact several ways in which an individual can control his physiological reactions to psychological events. Among the claims for such control the most notable have come from practitioners of meditation systems of the East: yoga and Zen Buddhism. This article will review and discuss recent studies of the effects of meditation that have been made by ourselves and by other investigators.

Yogis in India have long been reputed to perform phenomenal feats such as voluntarily stopping the heartbeat or surviving for extended periods in an "airtight" pit or in extreme cold without food or in a distorted physical posture. One of the first investigators to look into these claims in an objective way was a French cardiologist, Thérèse Brosse, who went to India in 1935 equipped with a portable electrocardiograph so that she could monitor the activity of the heart. Brosse concluded from her tests that one

of her subjects actually was able to stop his heart. In 1957 two American physiologists, M. A. Wenger of the University of California at Los Angeles and B. K. Bagchi of the University of Michigan Medical School, conducted a more extensive investigation in collaboration with B. K. Anand of the All-India Institute of Medical Sciences in New Delhi. None of the yogis they studied, with more elaborate equipment than Brosse had used, showed a capability for stopping the heart. Wenger and Bagchi concluded that the disappearance of the signal of heart activity in Brosse's electrocardiogram was probably an artifact, since the heart impulse is sometimes obscured by electrical signals from contracting muscles of the thorax. (In attempting to stop the heart the yogis usually performed what is called the Valsalva maneuver, which increases the pressure within the chest; it can be done by holding one's breath and straining downward.) Wenger, Bagchi and Anand did find, however, that some of the yogis could slow both heartbeat and respiration rate.

Reports of a number of other investigations by researchers in the 1950's and 1960's indicated that meditation as practiced by yoga or Zen meditators could produce a variety of physiological effects. One of the demonstrated effects was reduction of the rate of metabolism. Examining Zen monks in Japan who had had many years of experience in the practice of deep meditation, Y. Sugi and K. Akutsu found that during meditation the subjects decreased their consumption of oxygen by about 20 percent and reduced their output of carbon dioxide. These signs of course constitute evidence of a slowing of metabolism. In New Delhi, Anand and two collaborators, G. S. Chhina and Baldeu Singh, made a similar finding in examination of a yoga

practitioner; confined in a sealed metal box, the meditating yogi markedly reduced his oxygen consumption and carbon dioxide elimination.

These tests strongly indicated that meditation produced the effects through control of an "involuntary" mechanism in the body, presumably the autonomic nervous system. The reduction of carbon dioxide elimination might have been accounted for by a recognizably voluntary action of the subject—slowing the breathing—but such action should not markedly affect the uptake of oxygen by the body tissues. Consequently it was a reasonable supposition that the drop in oxygen consumption, reflecting a decrease in the need for inhaled oxygen, must be due to modification of a process not subject to manipulation in the usual sense.

Explorations with the electroencephalograph showed further that meditation produced changes in the electrical activity of the brain. In studies of Zen monks A. Kasamatsu and T. Hirai of the University of Tokyo found that during meditation with their eyes half-open the monks developed a predominance of alpha waves—the waves that ordinarily become prominent when a person is thoroughly relaxed with his eyes closed. In the meditating monks the alpha waves increased in amplitude and regularity, particularly in the frontal and central regions of the brain. Subjects with a great deal of experience in meditation showed other changes: the alpha waves slowed from the usual frequency of nine to 12 cycles per second to seven or eight cycles per second, and rhythmical theta waves at six to seven cycles per second appeared. Anand and other investigators in India found that yogis, like the Zen monks, also showed a heightening of alpha activity during meditation. N. N. Das and H. Gastaut, in an electroencephalographic examination of seven yogis, observed that as the meditation progressed the alpha waves gave way to fast-wave activity at the rate of 40 to 45 cycles per second and these waves in turn subsided with a return of the slow alpha and theta waves.

Another physiological response tested by the early investigators was the resistance of the skin to an electric current. This measure is thought by some to reflect the level of "anxiety": a decrease in skin resistance representing greater anxiety; a rise in resistance, greater relaxation. It turns out that meditation increases the skin resistance in yogis and somewhat stabilizes the resistance in Zen meditators.

W e decided to undertake a systematic study of the physiological "effects," or, as we prefer to say, the physiological correlates, of meditation. In our review of the literature we had found a bewildering range of variation in the cases and the results of the different studies. The subjects varied greatly in their meditation techniques, their expertise and their performance. This was not so true of the Zen practitioners, all of whom employ the same technique, but it was quite characteristic of the practice of yoga, which has many more adherents. The

state called yoga (meaning "union") has a generally agreed definition: a "higher" consciousness achieved through a fully rested and relaxed body and a fully awake and relaxed mind. In the endeavor to arrive at this state, however, the practitioners in India use a variety of approaches. Some seek the goal through strenuous physical exercise; others concentrate on controlling a particular overt function, such as the respiratory rate; others focus on purely mental processes, based on some device for concentration or contemplation. The difference in technique may produce a dichotomy of physiological effects; for instance, whereas those who use contemplation show a decrease in oxygen consumption, those who use physical exercise to achieve yoga show an oxygen-consumption increase. Moreover, since most of the techniques require rigorous discipline and long training, the range in abilities is wide, and it is difficult to know who is an "expert" or how expert he may be. Obviously all these complications made the problem of selecting suitable subjects for our systematic study a formidable one.

Fortunately one widely practiced yoga technique is so well standardized that it enabled us to carry out large-scale studies under reasonably uniform con-

ditions. This technique, called "transcendental meditation," was developed by Maharishi Mahesh Yogi and is taught by an organization of instructors whom he personally qualifies. The technique does not require intense concentration or any form of rigorous mental or physical control, and it is easily learned, so that all subjects who have been through a relatively short period of training are "experts." The training does not involve devotion to any specific beliefs or life-style. It consists simply in two daily sessions of practice, each for 15 to 20 minutes.

The practitioner sits in a comfortable position with eyes closed. By a systematic method that he has been taught, he perceives a "suitable" sound or thought. Without attempting to concentrate specifically on this cue, he allows his mind to experience it freely, and his thinking, as the practitioners themselves report, rises to a "finer and more creative level in an easy and natural manner." More than 90,000 men and women in the U.S. are said to have received instruction in transcendental meditation by the organization teaching it. Hence large numbers of uniformly trained subjects were available for our studies.

What follows is a report of the detailed measurements made on a group of 36 subjects. Some were observed at the Thorndike Memorial Laboratory, a part of the Harvard Medical Unit at the Boston City Hospital. The others were observed at the University of California at Irvine. Twenty-eight were males and eight were females; they ranged in age from 17 to 41. Their experience in meditation ranged from less than a month to nine years, with the majority having had two to three years of experience.

During each test the subject served as his own control, spending part of the session in meditation and part in a normal, nonmeditative state. Devices for continuous measurement of blood pressure, heart rate, rectal temperature, skin resistance and electroencephalographic events were attached to the subject, and during the period of measurement samples were taken at 10-minute intervals for analysis of oxygen consumption, carbon dioxide elimination and other parameters. The subject sat in a chair. After a 30-minute period of habituation, measurements were started and continued for three periods: 20 to 30 minutes of a quiet, premeditative state, then 20 to 30 minutes of meditation, and finally 20 to 30 minutes after the subject was asked to stop meditating.

The measurements of oxygen consumption and carbon dioxide elimination confirmed in precise detail what had

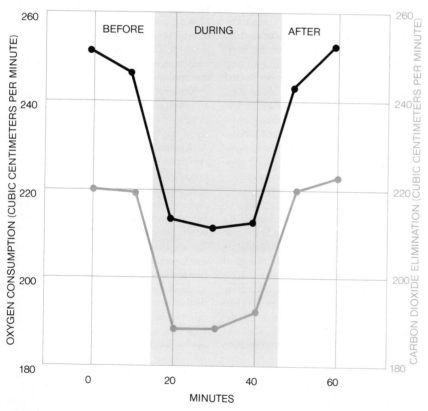

EFFECT OF MEDITATION on the subjects' oxygen consumption (*black*) and carbon dioxide elimination (*color*) was recorded in 20 and 15 cases respectively. After the subjects were invited to meditate both rates decreased markedly (*colored area*). Consumption and elimination returned to the premeditation level soon after the subjects stopped meditating.

been reported earlier. Oxygen consumption fell sharply from 251 cubic centimeters per minute in the premeditation period to 211 cubic centimeters during meditation, and in the postmeditation period it rose gradually to 242 cubic centimeters. Similarly, carbon dioxide elimination decreased, from 219 centimeters per minute beforehand to 187 cubic centimeters during meditation, and then returned to about the premeditation level afterward. The ratio of carbon dioxide elimination to oxygen consumption (in volume) remained essentially unchanged throughout the three periods, which indicates that the controlling factor for both was the rate of metabolism. The reduction in metabolic rate (and hence in the need for oxygen) during meditation was reflected in a decrease, essentially involuntary, in the rate of respiration (off two breaths per minute) and in the volume of air breathed (one liter less per minute).

For the measurement of arterial blood pressure and the taking of blood samples we used a catheter, which was inserted in the brachial artery and hidden with a curtain so that the subject would not be exposed to possible psychological trauma from witnessing the drawing of blood. Since local anesthesia was used at the site of the catheter insertion in the forearm, the subject felt no sensation when blood samples were taken. The blood pressure was measured continuously by means of a measuring device connected to the catheter.

We found that the subjects' arterial blood pressure remained at a rather low level throughout the examination; it fell to this level during the quiet premeditation period and did not change significantly during meditation or afterward. On the average the systolic pressure was equal to 106 millimeters of mercury, the diastolic pressure to 57 and the mean pressure to 75. The partial pressures of carbon dioxide and oxygen in the arterial blood also remained essentially unchanged during meditation. There was a slight increase in the acidity of the blood, indicating a slight metabolic acidosis, during meditation, but the acidity was within the normal range of variation.

Measurements of the lactate concentration in the blood (an indication of anaerobic metabolism, or metabolism in the absence of free oxygen) showed that during meditation the subjects' lactate level declined precipitously. During the first 10 minutes of meditation the lactate level in the subjects' arterial blood decreased at the rate of 10.26 milligrams per 100 cubic centimeters per hour, nearly four times faster than the rate of

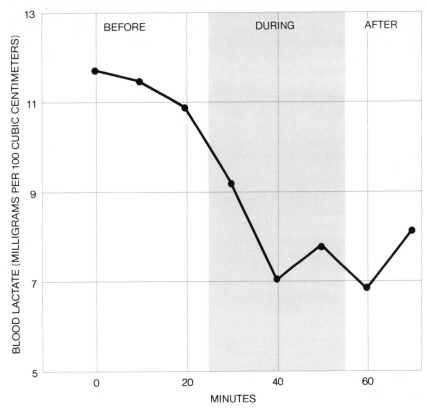

RAPID DECLINE in the concentration of blood lactate is apparent following the invitation to start meditating (*colored area*). Lactate is produced by anaerobic metabolism, mainly in muscle tissue. Its concentration normally falls in a subject at rest, but the rate of decline during meditation proved to be more than three times faster than the normal rate.

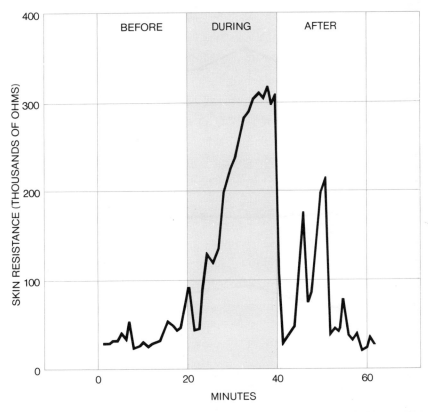

RAPID RISE in the electrical resistance of the skin accompanied meditation (*colored area*) in a representative subject. The 15 subjects tested showed a rise of about 140,000 ohms in 20 minutes. In sleep skin resistance normally rises but not so much or at such a rate.

decrease in people normally resting in a supine position or in the subjects themselves during their premeditation period. After the subjects ceased meditating the lactate level continued to fall for a few minutes and then began to rise, but at the end of the postmeditation period it was still considerably below the premeditation level. The mean level during the premeditation period was 11.4 milligrams per 100 cubic centimeters, during meditation 8.0 milligrams and during postmeditation 7.3 milligrams.

How could one account for the fact that lactate production, which reflects anaerobic metabolism, was reduced so much during meditation? New experiments furnished a possible answer. These had to do with the rate of blood flow in meditating subjects; the explanation they suggest appears significant with respect to the psychological benefits that can be obtained from meditation.

In studies H. Rieckert conducted at the University of Tübingen, he reported that during transcendental meditation his subjects showed a 300 percent increase in the flow of blood in the forearm. In similar measurements on our subjects we found the increase in forearm blood flow to be much less: 32 percent. Still, this increase was interesting, and it offered an explanation of the relatively

large decrease in blood-lactate concentration. The main site of lactate production in the body is the skeletal muscle tissue. Presumably the observed acceleration of blood flow to the forearm muscles during meditation speeds up the delivery of oxygen to the muscles. The resulting gain in oxidative metabolism may substitute for anaerobic metabolism, and this would explain the sharp drop in the production of lactate that accompanies meditation.

The intriguing consequence of this view is that it brings the autonomic nervous system further into the picture. In a situation of constant blood pressure (which is the case during meditation) the rate of blood flow is controlled basically by dilation or constriction of the blood vessels. The autonomic nervous system, in turn, controls this blood-vessel behavior. One element in this system, a part of the sympathetic nerve network, sometimes gives rise to the secretion of acetylcholine through special fibers and thereby stimulates the blood vessels to dilate. Conversely, the major part of the sympathetic nerve network stimulates the secretion of norepinephrine and thus causes constriction of the blood vessels. Rieckert's finding of a large increase in blood flow during meditation suggested that meditation increased the activity of

the sympathetic nerve network that secretes the dilating substance. Our own finding of a much more modest enhancement of blood flow indicated a different view: that meditation reduces the activity of the major part of the sympathetic nerve network, so that its constriction of the blood vessels is absent. This interpretation also helps to account for the great decrease in the production of lactate during meditation; norepinephrine is known to stimulate lactate production, and a reduction in the secretion of norepinephrine, through inhibition of the major sympathetic network, should be expected to diminish the output of lactate.

Whatever the explanation of the fall in the blood-lactate level, it is clear that this could have a beneficial psychological effect. Patients with anxiety neurosis show a large rise in blood lactate when they are placed under stress [see "The Biochemistry of Anxiety," by Ferris N. Pitts, Jr.; SCIENTIFIC AMERICAN Offprint 521]. Indeed, Pitts and J. N. McClure, Jr., a co-worker of Pitts's at the Washington University School of Medicine, showed experimentally that an infusion of lactate could bring on attacks of anxiety in such patients and could even produce anxiety symptoms in normal subjects. Furthermore, it is significant that patients with hypertension (essential and renal) show higher blood-lactate levels in a resting state than patients without hypertension, whereas in contrast the low lactate level in transcendental meditators is associated with low blood pressure. All in all, it is reasonable to hypothesize that the low level of lactate found in subjects during and after transcendental meditation may be responsible in part for the meditators' thoroughly relaxed state.

Other measurements on the meditators confirmed the picture of a highly relaxed, although wakeful, condition. During meditation their skin resistance to an electric current increased markedly, in some cases more than fourfold. Their heart rate slowed by about three beats per minute on the average. Electroencephalographic recordings disclosed a marked intensification of alpha waves in all the subjects. We recorded the waves from seven main areas of the brain on magnetic tape and then analyzed the patterns with a computer. Typically there was an increase in intensity of slow alpha waves at eight or nine cycles per second in the frontal and central regions of the brain during meditation. In several subjects this change was also accompanied by prominent theta waves in the frontal area.

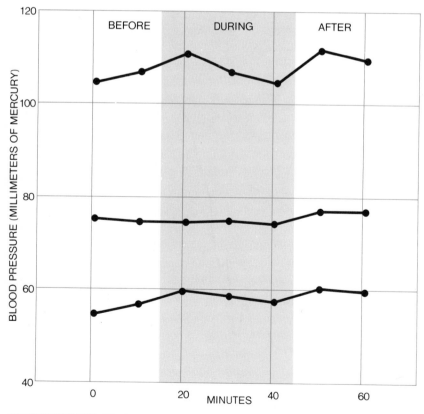

NO SIGNIFICANT CHANGE was observed in nine subjects whose arterial blood pressure was recorded before, during and after meditation. Systolic pressure (*top*), mean pressure (*middle*) and diastolic pressure (*bottom*), however, stayed relatively low throughout.

To sum up, our subjects during the practice of transcendental meditation manifested the physiological signs of what we describe as a "wakeful, hypometabolic" state: reductions in oxygen consumption, carbon dioxide elimination and the rate and volume of respiration; a slight increase in the acidity of the arterial blood; a marked decrease in the blood-lactate level; a slowing of the heartbeat; a considerable increase in skin resistance, and an electroencephalogram pattern of intensification of slow alpha waves with occasional theta-wave activity. These physiological modifications, in people who were practicing the easily learned technique of transcendental meditation, were very similar to those that have been observed in highly trained experts in yoga and in Zen monks who have had 15 to 20 years of experience in meditation.

How do the physiological changes during meditation compare with those in other relaxed states, such as sleep and hypnosis? There is little resemblance. Whereas oxygen consumption drops rapidly within the first five or 10 minutes of transcendental meditation, hypnosis produces no noticeable change in this metabolic index, and during sleep the consumption of oxygen decreases appreciably only after several hours. During sleep the concentration of carbon dioxide in the blood increases significantly, indicating a reduction in respiration. There is a slight increase in the acidity of the blood; this is clearly due to the decrease in ventilation and not to a change in metabolism such as occurs during meditation. Skin resistance commonly increases during sleep, but the rate and amount of this increase are on a much smaller scale than they are in transcendental meditation. The electroencephalogram patterns characteristic of sleep are different; they consist predominantly of high-voltage (strong) activity of slow waves at 12 to 14 cycles per second and a mixture of weaker waves at various frequencies—a pattern that does not occur during transcendental meditation. The patterns during hypnosis have no relation to those of the meditative state; in a hypnotized subject the brain-wave activity takes the form characteristic of the mental state that has been suggested to the subject. The same is true of changes in heart rate, blood pressure, skin resistance and respiration; all these visceral adjustments in a hypnotized person merely reflect the suggested state.

It is interesting to compare the effects obtained through meditation with those that can be established by means of operant conditioning. By such conditioning

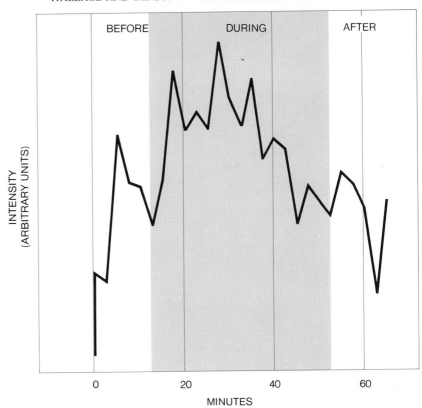

INCREASE IN INTENSITY of "slow" alpha waves, at eight to nine cycles per second, was evident during meditation (*colored area*) in electroencephalograph readings of the subjects' frontal and central brain regions. This is a representative subject's frontal reading. Before meditation most subjects' frontal readings showed alpha waves of lower intensity.

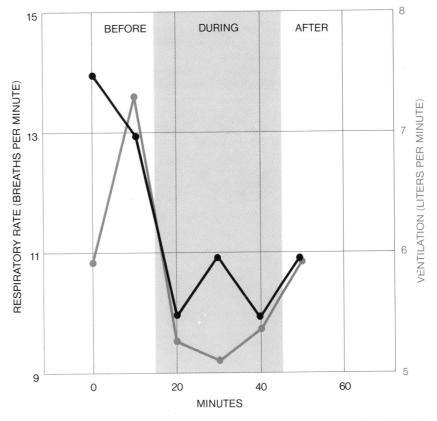

DECREASES OCCURRED in respiratory rate (*black*) and in volume of air breathed (*color*) during meditation. The ratio between carbon dioxide expired and oxygen consumed, however, continued unchanged and in the normal range during the entire test period.

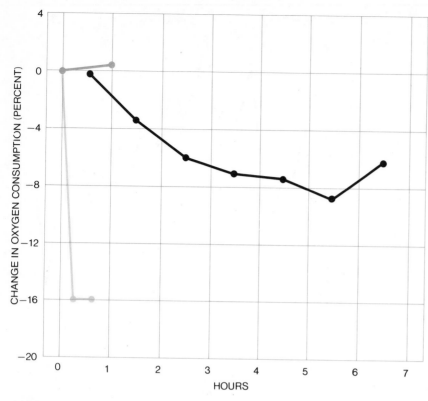

CONSUMPTION OF OXYGEN is compared in three different circumstances: during hypnosis (*color*), sleep (*black*) and meditation (*light color*). No significant change occurs under hypnosis. One study shows that oxygen consumption is reduced by about 8 percent after five hours' sleep. Meditation brings twice the reduction in a fraction of the time.

animals and people have been trained to increase or decrease their heart rate, blood pressure, urine formation and certain other autonomic functions [see the article "Learning in the Autonomic Nervous System," by Leo V. DiCara, beginning on page 166]. Through the use of rewards that act as reinforcers a subject is taught to make a specific visceral response to a given stimulus. This procedure and the result are quite different, however, from what occurs in transcendental meditation. Whereas operant conditioning is limited to producing specific responses and depends on a stimulus and feedback of a reinforcer, meditation is independent of such assistance and produces not a single specific response but a complex of responses that marks a highly relaxed state.

The pattern of changes suggests that meditation generates an integrated response, or reflex, that is mediated by the central nervous system. A well-known reflex of such a nature was described many years ago by the noted Harvard physiologist Walter B. Cannon; it is called the "fight or flight" or "defense alarm" reaction. The aroused sympathetic nervous system mobilizes a set of physiological responses marked by increases in the blood pressure, heart rate, blood flow to the muscles and oxygen consumption. The hypometabolic state produced by meditation is of course opposite to this in almost all respects. It looks very much like a counterpart of the fight-or-flight reaction.

During man's early history the defense-alarm reaction may well have had high survival value and thus have become strongly established in his genetic makeup. It continues to be aroused in all its visceral aspects when the individual feels threatened. Yet in the environment of our time the reaction is often an anachronism. Although the defense-alarm reaction is generally no longer appropriate, the visceral response is evoked with considerable frequency by the rapid and unsettling changes that are buffeting modern society. There is good reason to believe the changing environment's incessant stimulations of the sympathetic nervous system are largely responsible for the high incidence of hypertension and similar serious diseases that are prevalent in our society.

In these circumstances the hypometabolic state, representing quiescence rather than hyperactivation of the sympathetic nervous system, may indicate a guidepost to better health. It should be well worthwhile to investigate the possibilities for clinical application of this state of wakeful rest and relaxation.

V

IMMUNE DEFENSES
OF THE BODY

IMMUNE DEFENSES OF THE BODY

V

INTRODUCTION

The immune defenses are the complex physiological mechanisms which allow the body to recognize materials as foreign to itself and to neutralize or eliminate them. They operate not only against microbes — viruses, bacteria, and other parasites — but against cancer cells and transplanted cells as well.

Immune responses can be classified into two general categories: (1) nonspecific mechanisms and (2) specific mechanisms. The nonspecific immune responses nonselectively protect against foreign materials without having to recognize their actual identities. In contrast, specific immune responses require prior exposure to the specific foreign material; then, upon subsequent exposure, the material is recognized as being foreign. Thus, the specific immune responses can be viewed as a form of acclimatization in that they are adaptive physiological responses which develop as a result of exposure to a particular stressor (in this case a foreign invader), the net result being a greatly improved reaction against that stressor.

The basic nonspecific response to invasion by a microbe or other foreign substance is inflammation, which is described in "Wound Healing" by Ross. The local inflammatory response brings phagocytic cells into the damaged area so that they can destroy or, at least, neutralize the foreign invaders and set the stage for tissue repair. As a result of chemicals released or generated locally, white blood cells (mostly neutrophils) are attracted to the area, migrate through the capillary walls (themselves made leaky by these same chemicals), engulf the foreign substance, and digest it. The identities of the chemical mediators of inflammation and the exact nature of their interactions with the neutrophils and other phagocytes have been the subject of much physiological study for many years. Only more recently has the remarkable process of wound healing been given the attention it deserves, and this is the main theme of Ross' article.

As is so often the case, the implications of the basic research in a particular area extend into many other areas of biology and medicine. Thus, the phenomenon of "contact inhibition," the process by which cells cease dividing when they contact their neighbors, is of central importance not only to wound healing but to the problem of cancer formation, as well; that is, when cells are cancerous, contact inhibition between them fails. "Wound Healing" also well illustrates how adaptive responses may become maladaptive if they are excessive. For example, Ross points out how inflammation may cause damage to normal tissue (as in rheumatoid arthritis) and how the excessive formation of scar tissue during wound healing can cause disease (as in rheumatic heart disease).

With "The Immune System " by Jerne, we move into the realm of specific immune responses. It must be emphasized that the specific responses do not eliminate the need for the nonspecific defenses; rather the former amplify the basic inflammatory process so as to ensure accomplishment of the ultimate goal of phagocytosis and death (or at least neutralization) of the foreign particle. Again, the analogy to any other physiological adaptation should be apparent since specific immune responses are really "improvements" upon the normally occurring homeostatic response (inflammation).

Specific immune responses are mediated by two populations of lymphocytes, B and T cells. In general, the B cells, through the antibodies they secrete, confer specific immune resistance against bacteria, whereas T cells are active primarily against fungi, viruses, parasites, cancer cells, and solid-tissue transplants. However, as Jerne rightly emphasizes, this separation of function is not completely justified, since the B and T cells manifest important interactions, some inhibitory and some facilitatory. The physiology of the specific immune responses is incredibly complex, involving not only T and B cells, antigens and antibodies, but a large number of other mediator chemicals. However, the basic events of the specific immune response are straightforward: the presence of a foreign substance is detected and the substance is specifically identified by the protein markers it carries; the population of lymphocytes capable of launching an attack against that substance is mobilized: the attack itself is mediated by antibodies released from the B cells or by a different set of chemicals released from the T cells; the net result is that the invader is neutralized, directly killed, or phagocytized. As Jerne describes, it seems likely that any given B lymphocyte is capable of making only one type of antibody. This means there must be millions of different types of lymphocytes, each differentiated in a highly specific manner, and the same may also be true for the T cells. How this comes about is still an unsettled question.

Central to the entire concept of specific immune responses is the problem of "recognition." This is introduced in Jerne's article and is the major subject of "Markers of Biological Individuality" by Reisfeld and Kahan. The latter article describes the protein markers which exist on the surfaces of all nucleated cells, the polymorphic genetic system which codes for them, and the mechanism which allows the host to recognize them as foreign and destroy the cells which bear them. These "transplant markers" (so named because of their role in triggering rejection of transplants) are simple proteins and their solubilization and isolation have made possible exciting new approaches to pretransplant tissue-typing (analogous to the blood-typing done before a transfusion) and to methods for suppression of the rejection phenomenon.

Given the complexity of the nonspecific and specific immune defenses, it should not be surprising that a large number of factors may importantly influence the body's resistance to infection or cancer, for these diseases are multicausal in origin—that is, the presence of the microbe or newly arisen cancer cell is the necessary but frequently not sufficient cause of the disease. The nutritional background of the person is important, as are the degree of psychosocial stress he is subjected to, the presence of drugs, environmental pollutants, or concurrent illness. Most of the mechanisms by which these factors alter resistance are not yet fully understood, but it is thought that they all act by altering one or more components of the normal defense mechanisms. For example, the hormone cortisol, when present in very large amounts, inhibits virtually every step of the nonspecific inflammatory response. Indeed, this is why many patients with rheumatoid arthritis are given this hormone: it is an attempt to reduce the excessive inflammation which contributes to the disease; of course, the price to be paid for this beneficial effect is a decreased resistance to infection (as well as other pharmacological effects of cortisol).

Excessive nonspecific wound-healing is not the only type of maladaptive immune response. Indeed, of late it has become clear that the body's specific immune defenses are frequently, themselves, the cause of serious disease. The article "How the Immune Response to a Virus Can Cause Disease," by Notkins and Koprowski, describes how the common denominator of many diseases is the destruction of body cells and tissues not by viruses themselves, but rather by a maladaptive specific immune response to the relatively harmless viruses. The phenomena they describe also explain a good deal of what has been called "autoimmune disease," the breakdown in self-recognition which turns the body's immune mechanisms against its own tissues. The number of serious diseases suspected to be autoimmune in nature is rapidly increasing as is investigation into their underlying pathophysiology (viral infection is certainly not the only trigger). One can only ponder why such deleterious effects are so often the product of a system which obviously evolved to protect us from disease.

19

Wound Healing

by Russell Ross
June 1969

*In man and other higher animals an intricate,
three-step process involving a variety of cells
renews injured tissue. Studies of this process
could yield ways of controlling several diseases*

When a salamander loses a leg, it can grow a new one, but man has not retained this ability to duplicate injured tissue. The liver and the surface layer of the skin are among the few mammalian tissues that can regenerate themselves; otherwise man must rely on wound healing. This is an intricate physiological process in which several different kinds of cells appear at successive intervals in order to absorb foreign matter, destroy bacteria and repair the injury.

Clearly wound repair is essential to health and comfort. It also represents a biological adaptation without which complex multicellular organisms could neither survive nor evolve. Because of its medical importance the general character of the process has been known for years. Recently, however, the advance of biochemistry and the development of the electron microscope and other research tools have enabled investigators to observe and understand in greater detail the molecular events that manifest themselves on the macroscopic level as inflammation, scar formation, restoration of the skin's surface and the other stages of wound repair. Furthermore, because the study of wound repair involves examining some of the most basic features of living tissue, progress in this field of inquiry could yield information that might lead to new understanding of such illnesses as cancer, rheumatoid arthritis and rheumatic heart disease.

The process of wound repair differs little from one kind of tissue to another and is generally independent of the form of injury. Wound repair can therefore be somewhat arbitrarily divided into three overlapping and probably related stages, each characterized by the activities of particular populations of cells.

These stages are clearly seen in a deep skin cut. Initially blood flows into the gap created by the cutting instrument, fills the space and clots, uniting the edges of the wound. Within several hours the clot loses fluid and the surface becomes dehydrated, so that it forms the hard scab that serves to protect the wound. Inflammation begins as fluids enter the wound around the clot. The injury at this stage may become swollen and painful. Then, about six hours after wounding, various kinds of white blood cells start to migrate into the wound and begin removing and breaking down cellular debris, bacteria and other foreign material. Subsequently in the dermis, or subsurface layer, the cells called fibroblasts enter the wound and build scar tissue by manufacturing collagen fibers and other proteins. Meanwhile the epidermis, or surface layer, creates a new surface similar to the old one. When this layer is almost completely formed, the scab sloughs off.

At the University of Washington School of Medicine we have been investigating the repair of both the surface and subsurface layers. Our tools are the light microscope, the electron microscope, autoradiography and substances labeled with radioactive isotopes. Medical-student volunteers provide some of the material for our study. To collect specimens we make one-centimeter incisions on the inside surface of the upper arm of each student and remove the wounded areas at successive intervals with a rotating-punch biopsy tool. This painless procedure enables us to study each stage of wound repair in detail, just as biology students observe the development of a chicken embryo by incubating eggs and opening them at different intervals.

Healing in the dermal layer begins when a clot forms from the blood that flows into the wound immediately after injury. As the blood coagulates, fibrinogen molecules from the blood quickly link up into interconnected strands of fibrin. These strands provide a network throughout the wound defect that somewhat tenuously unites its edges. Meanwhile at the surface fibrin and other proteins in the blood serum dehydrate and form the scab.

After the clot has formed, the dying and broken tissue produces substances that cause blood vessels in the nearby uninjured tissue to leak, perhaps by altering the intercellular structure of the vessel walls so that the cells do not fit together as tightly as they did before. The resulting flow of serum contains a number of proteins such as globulin, albumin and antibodies. If the wound contains viruses (or is a lesion caused by viruses), the globulin and antibodies may attack them, but normally this fluid merely provides a sustaining environment for the white cells that begin to follow it into the wound about six hours later.

The first of these cells are the neutrophils. To get into the wound the neutrophil and the other white cells that follow it must migrate through the walls of the blood vessels in the nearby tissue. In order to get through the wall of a blood vessel a white cell inserts a bit of its cytoplasm between the cells lining the vessel and forces them apart. The blood cell can then slip through from the bloodstream into the adjacent tissue [*see illustration on page 191*].

Once inside the wound, a neutrophil can provide two kinds of defense. First, the cell can ingest organisms, by the process termed phagocytosis. In many cases the neutrophils kill the bacteria and digest most of their remains. In a

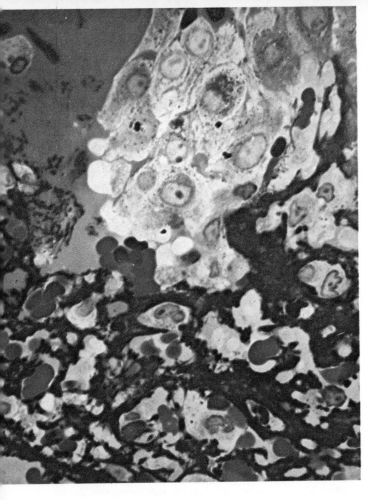
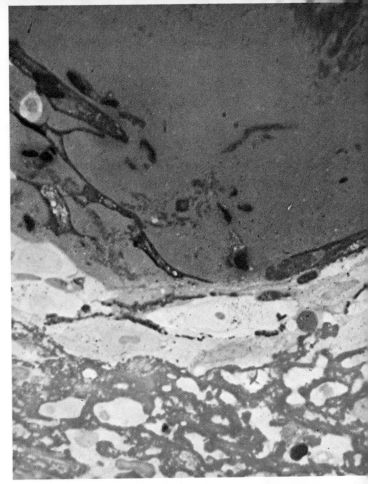

STAGES OF WOUND HEALING are shown in these photomicrographs. At top left, 24 hours after wounding, a wine red scab has already formed and purple fibrin strands unite the edges of the gap. Surface cells have intruded into the wound and begun to ingest fibrin and red and white blood cells. At top right, 24 hours later, surface cells have formed a layer under the scab. Inflammatory cells are lodged among the fibrin strands. At bottom left, 14 days after wounding, surface layers are complete and white connective-tissue cells have restored the bluish subsurface layer. At bottom right, six days later, high magnification reveals a subsurface layer of parallel collagen bundles and a few cells. Micrographs were made by James D. Huber of University of Washington.

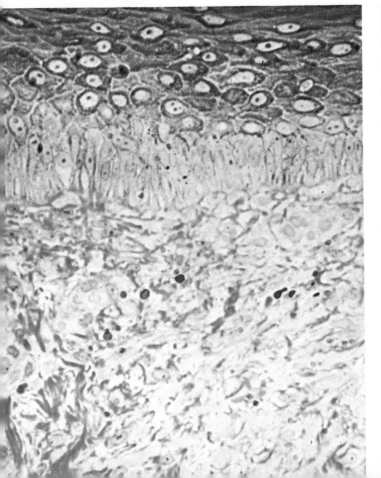
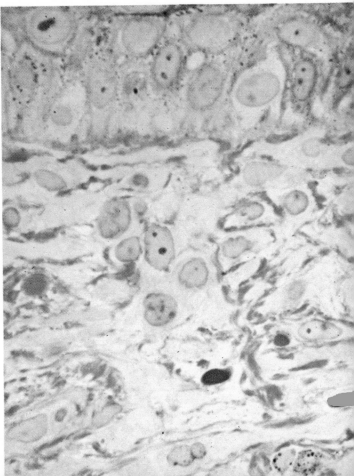

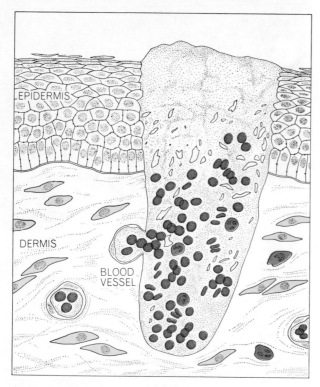

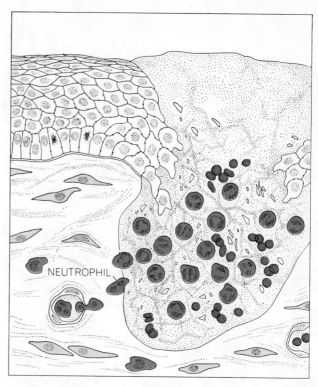

AT THE TIME OF WOUNDING gap fills with blood from broken vessels and cellular debris from the surface (epidermal) and subsurface (dermal) layers. Blood contains red cells and fibrinogen molecules, which form into strands that unite the wound's edges. Yellow cells of the dermis called fibroblasts surround the wound.

A DAY LATER bluish neutrophils enter the wound from surrounding blood vessels and begin to ingest bacteria and debris. The epidermal cells, held together by desmosomes, have also intruded. Other epidermal cells begin to reproduce (*note dividing nuclei*) so that those that migrate into the wound will be replaced.

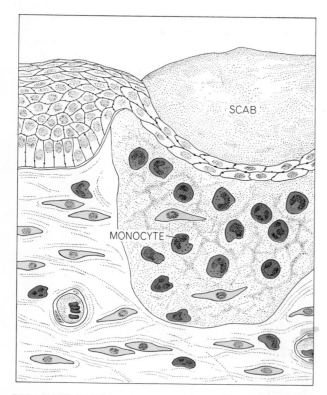

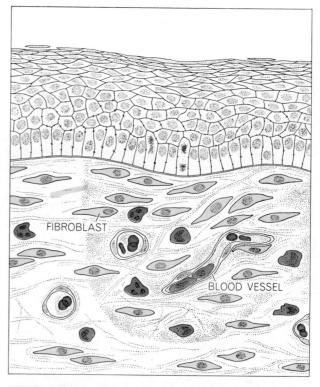

TWO DAYS LATER wedges of epidermis have met under the scab, and the next generation of inflammatory cells, called monocytes, have entered the wound to complete the scavenging process. Meanwhile the fibroblasts that make the collagen and other connective-tissue proteins of the dermis have migrated into the defect.

SEVEN DAYS LATER the scab has sloughed from the restored epidermis. A few monocytes and neutrophils remain in the wound. At this stage the fibroblasts actively make collagen and other dermal-tissue proteins. The new connective tissue is denser than the original, unwounded dermis and is permeated by blood vessels.

sterile wound such as the one made by a surgeon a neutrophil has no bacteria to ingest. Under these circumstances the cell appears to degenerate and die. At this point the neutrophil's outer membrane ruptures, pouring enzyme-containing granules of different sizes into the wound. As the enzymes are released from the granules, they may attack the extracellular debris at the site of injury; such material can then be more easily removed by the cells that subsequently appear in the wound. Earl P. Benditt and I first observed this dissolution of neutrophils and the release of their granules when we studied healing skin wounds in guinea pigs in 1960. George F. Odland and I have since made identical observations in man. The mature neutrophil seems particularly suited for its specific task. Its cytoplasm contains only a few of the organelles found in most other cells. It has a life-span of only a few days and can neither synthesize proteins nor divide to replicate itself.

Within the first 12 hours after injury a second kind of white blood cell, the monocyte, begins to migrate into the wound. On entering the wound a monocyte becomes a macrophage: a phagocytic cell that removes most of the debris from the injured area by ingesting and then partially digesting it. Unlike the neutrophil, the monocyte has a fairly long life-span, it can synthesize proteins (particularly the enzymes it uses in phagocytosis) and it remains extremely active in the latter part of the inflammatory phase of wound repair. There is no evidence thus far that the monocyte can divide.

Electron-microscope studies of wounds in both guinea pigs and humans show that the monocyte goes through a series of changes before it begins to envelop and digest bacteria and debris. At first the cell contains comparatively few of the organelles responsible for the synthesis of protein within its cytoplasm. After the cell appears in the wound, however, it forms organelles. Subsequently small granules appear within the cytoplasm. This organelle system is called the rough endoplasmic reticulum, and it consists of membrane-lined channels with small, dense particles attached to the outer surface of the membrane. These particles are ribosomes; they synthesize the enzymes that degrade the debris ingested by the monocyte. It is remarkable to see how a mature monocyte becomes engorged with large vacuoles containing various forms of material, including whorls and membrane-like structures. It is believed that these structures

are the remnants of the debris the cell is unable to digest completely [see illustration at bottom right on page 193].

Macrophages were once thought to originate in the connective and lymphoid tissues near the wound. Alvin Volkman and James L. Gowans of the University of Oxford recently demonstrated in a definitive series of experiments, however, that they stem from a small pool of rapidly dividing precursor cells in the bone marrow. Then they travel to the wound in the bloodstream.

Although neutrophils and monocytes play a vital part in wound repair, there are times when they can be destructive. This is particularly true when exaggerated inflammation persists for abnormally long periods of time, as it does in some diseases such as rheumatoid arthritis. In this debilitating disease there is often marked destruction of the synovial membranes of the joints, which secrete the fluids that lubricate the joints, and in later stages of the disease the cartilage and bones of the joints may be partially destroyed and may subsequently adhere to each other. Repetitive inflammation, with its overabundance of inflammatory cells and release of their enzymes, may be partially responsible for much of the damage. It is for this reason that cortisone and other drugs that suppress inflammation have been used to treat arthritis. A better understanding of the role played by neutrophils and monocytes could lead to further improvement in the methods for controlling the disease.

Toward the end of the inflammatory response another kind of cell, the fibroblast, appears and begins to repair the injury by secreting the collagen and the protein polysaccharides that form scar tissue. Collagen endows the scar, which eventually serves to replace the wound defect, with great tensile strength. The most abundant protein in the animal kingdom, it is the major fibrous constituent of skin, tendon, ligament, cartilage, bone and the stroma (a form of connective tissue that supports most glands and organs).

Scar tissue begins to form when the collagen molecules aggregate into fibrils that in the electron microscope exhibit a regular pattern of bands. The fibrils in turn are bound by the protein polysaccharides into fibers, which can be seen

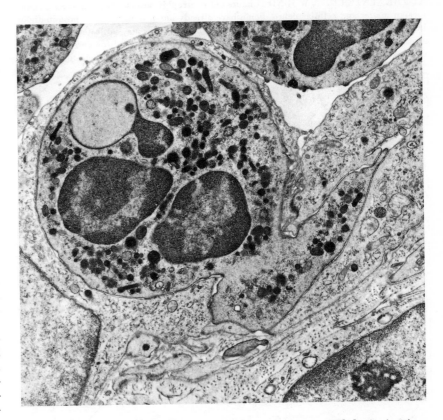

NEUTROPHIL, enlarged by 14,000 diameters in this electron micrograph, begins its trip to the wound from inside a blood vessel embedded in connective tissue near the wound. To escape from the vessel into the connective tissue the granule-filled neutrophil squeezes a process, or extension of itself (below center), between the cells of the vessel wall and forces them apart. At upper left sections of two more neutrophils can be seen in the vessel's hollow.

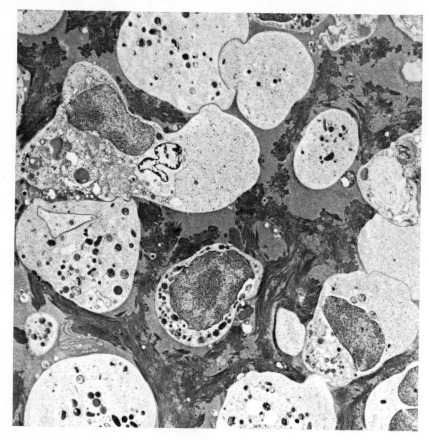

AT THE CENTER OF A WOUND 24 hours old neutrophils are cleaning up fibrin strands, foreign matter and cellular debris. Some have already burst, pouring their protein-digesting granules into the gap. Fuzzy "halos" around some granules may indicate that they have already started to break down the plasma protein. A monocyte has also entered the wound and has begun to ingest material. Map below identifies cells, granules and other features.

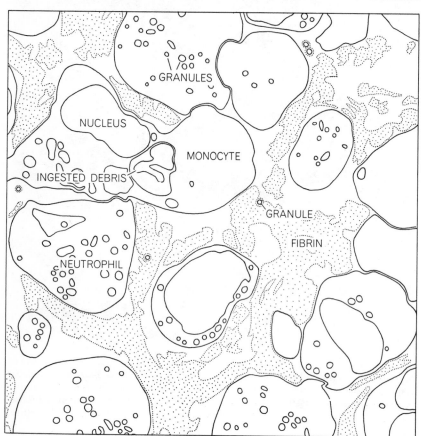

with the light microscope. After about two weeks (in the case of a linear incision) the synthesis of connective-tissue proteins begins to decrease and a process of remodeling takes place; it is this process that gives scar tissue its great strength. During remodeling many of the randomly oriented smaller collagen fibers that had formed earlier are broken down and reassembled into thick bundles. As the bundles form they orient themselves along the lines of stress in the healing wound. Eventually they gather into large fibers that look much like those in unwounded tissue. In fact, the difference between the original dermis and the scar tissue is one of degree rather than kind. Scar tissue is denser, its strands tend to lie more parallel to one another, and it contains fewer cells and blood vessels.

Just as the direction of stress determines how collagen bundles orient themselves in a wound, so the amount of scar tissue that forms depends in part on how much stress the wounded area normally receives as the body moves. More scar tissue may be needed to strengthen an injury on the hand or the leg than is needed for one on a comparatively quiet surface such as the chest or the abdomen. Heredity also plays a role. Negroes and Orientals, for instance, are more likely than whites to produce the raised scars called keloids.

The rough endoplasmic reticulum of the fibroblast synthesizes the collagen and other proteins that form scar tissue. In the fibroblast this organelle is highly developed, and the ribosomes attached to its membranes have a characteristic spiral or curved configuration [*see upper illustration on page 194*]. The rough endoplasmic reticulum of the monocyte is much less developed, which probably reflects the fact that the small amounts of protein synthesized by this cell are enzymes that work only within the cell to digest material in the wound. In contrast, the fibroblast synthesizes proteins for the construction of scar tissue and secretes them into the extracellular space.

In a series of studies that confirmed this finding Benditt and I traced the manufacture of collagen through the organelles of individual fibroblasts. In order to follow the process we had to label some amino acid constituent of the protein with a radioactive atom. We chose the amino acid proline because it has the advantage of being the source of both proline and hydroxyproline, which together account for some 25 percent of the collagen molecule. We then followed labeled proline by making a series of electron-microscope autoradiographs.

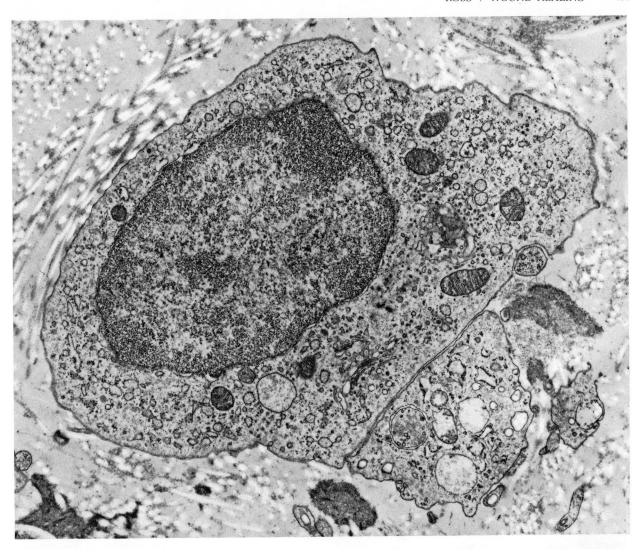

MONOCYTE does not develop the capacity to ingest and digest foreign material until it has entered the wound. This cell in the dermis outside a wound displays features typical of the young monocyte: a large, round nucleus and a vesicle-filled cytoplasm.

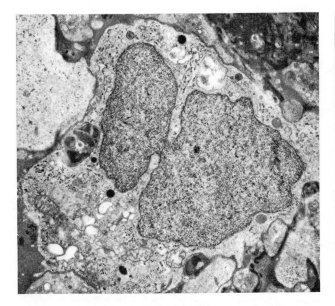

MONOCYTE BECOMES PHAGOCYTIC soon after it enters wound. The monocyte ingests foreign material by enclosing it in a vacuole, such as the one near the cell's nucleus. This particular vacuole contains strands of fibrin and serum protein from the clot.

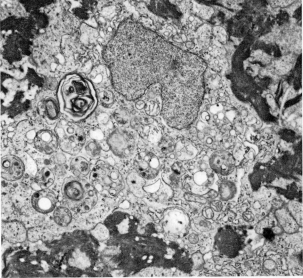

MATURE MONOCYTE, surrounded by dense fibrin strands, has a distended cytoplasm containing many vacuoles that are filled with ingested material, and whorl-like structures that may consist of un-digested lipoprotein. This cell has been enlarged 11,000 diameters.

An autoradiograph is made by placing a layer of photosensitive emulsion on a section of tissue in which cells are synthesizing protein from amino acids, in this case proline. Since the silver grains of the emulsion lie close to the radioactively labeled proline they are exposed just as the grains on an ordinary photographic negative are. Under the high resolution of the electron microscope the exposed silver grains thus reveal which organelles in an individual fibroblast use proline. By quantitatively analyzing a large number of these autoradiographs made in sequence we have traced the pathways followed by the amino acid and the substances into which it became incorporated through the rough endoplasmic reticulum and the other organelles of the fibroblast [*see bottom illustration on opposite page*].

The fibroblast manufactures proteins as other cells do, but it appears that it may secrete one of the proteins, collagen, differently. For example, George E. Palade and his colleagues at Rockefeller University have demonstrated that the digestive enzymes synthesized by the exocrine cells of the pancreas are sequestered in the rough endoplasmic reticulum. Then the proteins are passed along to the region of the cell called the Golgi complex, which consists of a series of vesicles (cavities) and sacs that contain no ribosomes. Here the proteins are packaged into structures called condensing vacuoles. Subsequently these vacuoles form zymogen granules, in which the characteristic secretions of the exocrine cell are stored before they are released from the cell.

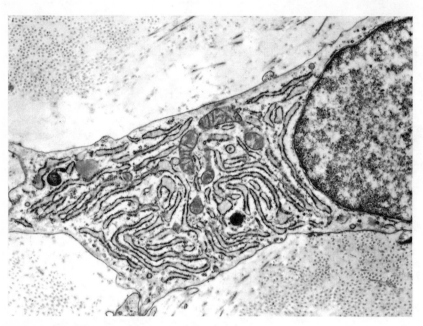

FIBROBLAST, enlarged 15,000 diameters, manufactures collagen molecules and other proteins in its mazelike "rough endoplasmic reticulum." Proteins are synthesized at the dark sites called ribosomes and are then transferred into the elongated channels. Many collagen fibrils, from which scar tissue forms, surround the cell. Part of nucleus can be seen at right.

Although fibroblasts also have a Golgi complex, the process of secretion in these cells may be much more complex than it is in the exocrine cells. Benditt and I have reasoned that if a fibroblast stores its proteins after manufacture and transports them to special secretory mechanisms and sites as the exocrine pancreatic cell does, considerable time would elapse between the moment a fibroblast picked up an amino acid and the moment the amino acid, now part of a protein molecule, would be secreted back into the intracellular space. In order to measure the time a fibroblast needs to incorporate proline and other amino acids into a finished protein molecule we made a series of light-microscope autoradiographs of tissue in which fibroblasts were utilizing labeled proline and secreting collagen into the extracellular space. We observed that within an hour after the labeled amino acid had been injected into the abdominal cavity of a guinea pig it was localized predominantly within the fibroblasts, and that after an hour it had already begun to appear in the space adjoining the cells [*see top illustrations on opposite page*]. If the fibroblast has an appreciable storage phase in its manufacturing and secretion cycle, we would have expected the labeled proteins to remain for a longer period of time in some compartment within the cell, as they do in the exocrine pancreas.

Fibroblasts, therefore, may secrete collagen directly from the rough endo-

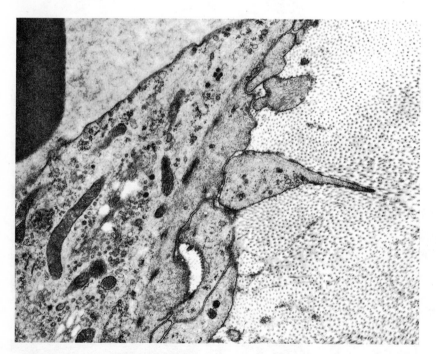

BLOOD VESSEL REGENERATION begins, as shown in this electron micrograph, when a sprout branches from the vessel wall into the connective tissue. The sprout can be seen at center jutting out to the right, away from the grayish wall of the vessel. Sprouts eventually divide and form new series of vessels until a network extends throughout the wound. Part of a red blood cell can be seen in the old vessel's interior, in the upper left-hand corner.

plasmic reticulum, probably by forming small vesicles that pinch off from the sacs of the reticulum and migrate a comparatively short distance to the outer membrane of the cell. The vesicles would then fuse with the membrane and release their contents into the extracellular space [*see illustration on next page*]. There may also be a second secretion channel for collagen. We have observed sites that could be potential openings between the cavities of the rough endoplasmic reticulum and the extracellular space. Such a site may evolve into a passage through which material sequestered within the cavities could be released directly to the exterior.

Since autoradiography is a somewhat imprecise technique, this view of protein synthesis in the fibroblast cannot be stated as a definite fact. The difficulty is this: fibroblasts make several proteins simultaneously, including collagen, and proline is a constituent of all of them. Therefore in an autoradiograph it is not possible to tell whether collagen or some other labeled protein is represented by a particular silver grain. We will be able to remove some of the uncertainty, however, by tracing the passage of proteins through the fibroblast, and using biochemical information to determine which labeled proteins are present in each cell compartment at a given time. Current evidence of this kind seems to support the view that the rough endoplasmic reticulum synthesizes both collagen and the proteins that will later be joined with polysaccharides. These latter proteins then travel in vesicles to the Golgi complex, where polysaccharides are added, whereas the collagen may be secreted directly from the rough endoplasmic reticulum into the space outside the cell.

It will be interesting to determine definitely which intracellular pathway or pathways the fibroblast can use. Such information would help in the interpretation of pathological changes and suggest ways of studying the regulation of synthesis and secretion by fibroblasts. If clinicians could control the synthesis and secretion of collagen, they might be able to ward off some of the disabling effects of rheumatic heart disease. Patients with rheumatic fever become disabled because the heart continuously forms a new connective-tissue scar in order to repair a valve that is being repeatedly injured by infection. Eventually much of the valve is replaced by the dense scar tissue, and the thin, pliable valve becomes a thick, rigid structure that can-

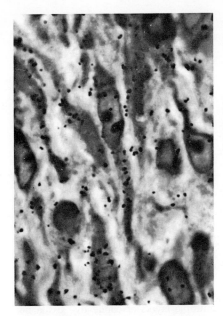
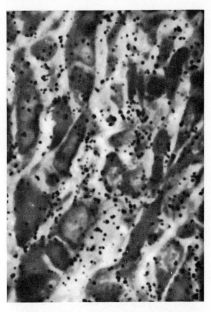

ONE HOUR after radioactively labeled proline has been administered to a guinea pig all this amino acid has been taken up by the dark fibroblasts, an event indicated by the localization of the exposed silver grains that appear as black dots around the dark fibroblasts in this light autoradiograph.

FOUR HOURS LATER most of the labeled proline has returned to the light gray space between the cells as part of the collagen they manufacture. This experiment shows that the fibroblasts secrete collagen immediately without storing it for long periods internally in organelles such as the Golgi complex.

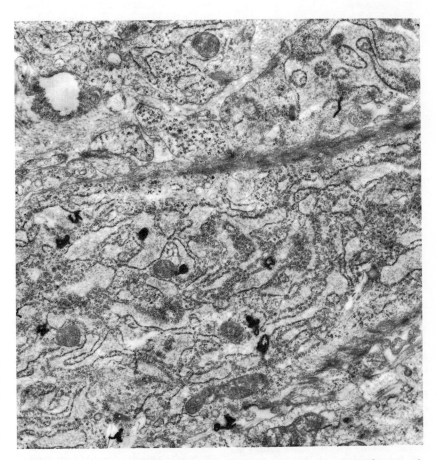

ROUGH ENDOPLASMIC RETICULUM of a fibroblast is enlarged by 14,000 diameters in this electron-microscope autoradiograph. Exposed silver grains mark sites in this part of the cell where ribosomes are synthesizing collagen and other proteins. This autoradiograph is one of a series taken so that the pathways of protein synthesis in the cell could be traced.

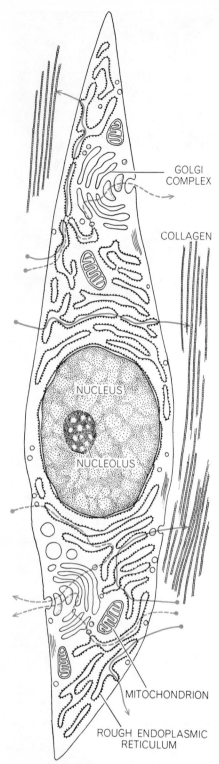

GOLGI
COMPLEX

COLLAGEN

NUCLEUS

NUCLEOLUS

MITOCHONDRION

ROUGH ENDOPLASMIC
RETICULUM

PROTEIN PATHWAYS are shown in this idealized fibroblast. The evidence suggests that collagen (*solid arrows*) is secreted in two ways: a vesicle may detach from the rough endoplasmic reticulum and empty the collagen through the cell's membrane wall, or a channel may open that connects the reticulum directly with the space outside the cell. Other proteins (*broken arrows*) are synthesized in the reticulum, transferred to smooth-chambered Golgi complexes where polysaccharides are added, and secreted through the cell wall by speckled vesicles.

not close in the normal fashion. If the production of scar tissue could be regulated at the fibroblast level, hearts affected by rheumatic fever might be saved. There are other diseases in which too little scar tissue forms. They might be controlled if fibroblasts could be induced to manufacture more collagen.

Many early workers postulated that wound fibroblasts, like monocytes, arise from white blood cells. Together with Jack W. Lillywhite we have conducted several experiments using white cells isolated from the blood of experimental animals to test this hypothesis, and we have found it wanting. We obtained the white blood cells by two different methods. In one method the blood was withdrawn by puncturing either the heart or a vein, so that the needle had to pass through multiple layers of connective tissue before reaching the blood. In the other method the blood was secured by carefully isolating an artery and inserting a plastic tube into it. Here only the layers of the vessel wall were penetrated.

White blood cells obtained by these two methods were then allowed to grow in special chambers, which were planted under the skin or in the abdominal cavity of an experimental animal. The cells taken from the heart or a vein divided actively, formed large amounts of collagen and looked somewhat like fibroblasts or modified smooth-muscle cells. The white cells obtained from an artery did not divide, and no collagen formed in the chamber. Special filters with a mesh too fine to pass cells eliminated the possibility of fibroblast precursors entering the chambers once they were sealed. Therefore the connective-tissue cells found within those chambers in which collagen had formed must have appeared as a result of contamination of the blood by a few connective-tissue cells that were picked up when the needle punctured the heart or a vein, a result suggesting that the fibroblasts in a wound originate in connective tissue nearby.

As the fibroblasts synthesize the collagen and the protein polysaccharides of the healing wound's matrix of connective tissue, large numbers of small blood vessels form throughout the wound. According to the classic observations of Eliot R. Clark and Eleanor L. Clark of the University of Pennsylvania, these capillaries originate as budlike structures on nearby vessels, penetrate the wound and grow into loops. The loops then ramify throughout the wound by the division of their cells. As the capillaries from different sites migrate through the

wound, they meet and form an interconnecting network of vessels.

In the early stages of wound repair this network of capillaries provides comparatively large quantities of oxygen for the cells that are actively synthesizing protein in the wound. John P. Remensnyder and Guido Majno of the Harvard Medical School have demonstrated that before the new capillary network forms there is a marked gradient of oxygen within the wound, the center of the wound being the most deficient in oxygen. This gradient, they say, may be partially responsible for the branching of new vessels into the region. Once the continuity of the connective tissue has been reestablished many of the new capillaries regress. Thus the wound changes from a tissue that is rich in blood vessels and actively dividing cells into one that has a much simpler cellular structure.

As scar tissue renews the dermis the epidermal cells begin to close the surface of the wound. In uninjured skin these ordered cells normally form a series of layers, each of which arises from the layer next to the dermis. The disruption of the epidermal layers causes some of the cells nearest the wound to degenerate, and most of the others lose the orderly, oriented appearance they had in their undisturbed state. Within 12 hours after wounding these cells become amorphous and develop ruffled borders and blisters, much like those seen in the actively moving cells grown in tissue culture.

The cells at the outer surface of the epidermis, the epithelial cells, are also highly mobile, and they travel quickly into the wound. The migration of the epithelial cell seems to be well organized. Normally these cells are attached to one another at sites called desmosomes, each of which is somewhat like a spot weld. In the course of migration the cells probably break up these attachments and re-form them. There are two reasons for believing that this happens; first, we have observed migrating cells attached to one another by desmosomes and, second, it would be difficult to explain the remarkable mobility of these if they were not able to rearrange their sites of attachment as they move into the wound.

The scaffolding of fibrin derived from the clotted blood that forms the first provisional "patch" in the wound also guides the migrating epidermal cells as they slide glacially from all sides under the scab and into the wound. As the cells migrate through this milieu they exhibit a feature quite unusual for epidermis.

They actively participate in the ingestion and digestion of the strands of serum protein and fibrin lying in their path [*see upper illustration on next page*]. This behavior is similar to the phagocytosis accomplished by the white blood cells found deeper in the wound. In both instances such activity serves to clear the wound of debris and of the fibrin scaffolding that united the wound margins. When the leading edges of the sheets of epidermal cells meet and form a continuous layer beneath the scab, each cell regains its normal identity, and as the thickness of this layer of cells is restored the rate of division among the epidermal cells returns to what it was before the injury.

Once the epidermal cells have closed the wound, how do they "know" that the time has come to resume their normal shape and growth rate? It has been suggested by many investigations, in particular those conducted in the laboratory of Michael Abercrombie at University College London, that one of the factors in the control of cell division is "contact inhibition." Abercrombie noted that cells in tissue culture continue to divide until they establish contact, at which point they stop dividing. Perhaps the same process occurs in man and other animals, but there is no evidence that this is so. Moreover, no one knows how contact inhibition works. Some investigators have suggested that as the cells touch each other the distribution of electric charge on their surface changes, and that the change serves as a signal that halts growth, but here again we have little evidence. The problem is an interesting and important one. If the process through which normal cell growth stops and starts can be learned, investigators might be able to figure out what has happened in tumor cells that continue to divide even though they repeatedly come in contact with one another.

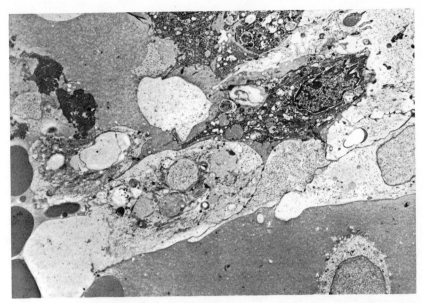

SURFACE HEALING BEGINS as the leading cell in a wedge of epidermis migrates into the wound. Fibrin strands and serum protein surround the cells. The cell's blunt, boxing-glove-like edge abuts a red blood cell. Desmosomes, which attach the epidermal cells to one another (*see map below*), can be seen at the cell's trailing edge, near the center of the picture. The cells and other bodies in this electron micrograph are enlarged by 4,500 diameters.

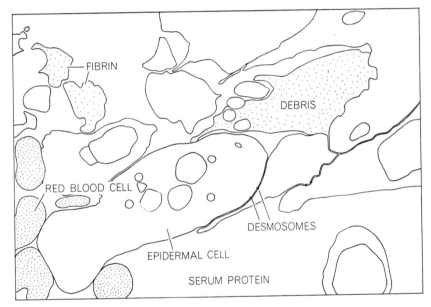

Since epidermis and dermis heal more or less simultaneously, it seems reasonable to ask if the two processes interact and are in any way dependent on each other. Hermes C. Grillo, who works in the laboratory of Jerome Gross at the Harvard Medical School, has found that the answer may well be affirmative. The connective tissue, it will be recalled, undergoes a process of remodeling in which bundles of collagen fiber form along the lines of stress running through the wound. This remodeling probably occurs as a result of the repeated dissolution and re-formation of collagen fibrils. What agent dissolves the fibrils?

According to one hypothesis, the agent is the enzyme collagenase, which comes primarily from the epidermis, perhaps in response to chemical signals from the dermis. Grillo's experiments provide evidence supporting this idea. He found that when he added epidermal cells to dermal cells in a tissue culture, they greatly enhanced the production of collagenase. As in the case of contact inhibition, however, someone must demonstrate that epidermis and dermis behave this way together in the living animal.

If the chemical events in simultaneous processes in the dermis and epidermis affect one another, it also seems possible that each stage in wound repair is an essential antecedent of the one following it. There is some evidence to support this hypothesis. As early as 1920 Alexis Carrel of the Rockefeller Institute succeeded in inhibiting the inflammatory response so that relatively few neutrophils appeared in the wound, and he found that the subsequent formation by fibroblasts of connective-tissue proteins was much delayed. Even earlier it was observed that the interruption of the formation of connective tissue can prevent wounds from healing. In his book *A Voyage around the World*, published in 1748, Richard Walter, chaplain to

George Anson, commander of the squadron that made the voyage, described the plight of seamen suffering from scurvy whose wounds failed to heal. Their scurvy was due to a lack of fresh fruit or vegetables in their diet, and it was later discovered that without such foods scurvy sufferers cannot form connective tissue because they have no source of vitamin C.

Burton Wolbach of the Harvard Medical School confirmed the importance of vitamin C to collagen formation in 1926. In the microscope he observed that fibroblasts from guinea pigs with scurvy produced a gelatinous, collagen-like substance with none of collagen's structural strength. It was only recently that the action of vitamin C in collagen formation has been clarified. Normally vitamin C allows hydroxyl groups (OH) to be added to the amino acids proline and lysine through the action of an enzyme, converting them into hydroxyproline and hydroxylysine. Hydroxylation, as the process is called, takes place after the amino acids have been synthesized into almost complete collagen molecules in the rough endoplasmic reticulum of the fibroblast. Without vitamin C the hydroxyl groups cannot be added, the collagen molecules remain incomplete and may not be secreted by the cell. Therefore collagen fibrils cannot form.

Our electron-microscope studies reveal that this chemical flaw is accompanied by a breakdown of the endoplasmic reticulum of the fibroblast. The walls of the cavities in the reticulum appear to be broken down, so that large, vacuole-like structures form. In addition, the ribosomes attached to the membrane of the endoplasmic reticulum are no longer arranged in the characteristic configuration seen in normal cells. We do not yet know how the morphologic changes observed in the fibroblasts are related to the biochemical changes that have been described in scurvy. The morphologic changes can be readily reversed, however, by feeding the experimental animal vitamin C; the cells resume their normal appearance within 24 hours, and recognizable collagen fibrils begin to appear in the extracellular space within 12 hours. It will be interesting to determine whether an increased rate of synthesis, or simply hydroxylation of a stored intracellular precursor of collagen, causes this rapid appearance of collagen.

Studies of wound repair have yielded much information about this important defense mechanism of the body. Like all research efforts, however, this one has generated new questions as well as new answers. For example, we should like to determine if there are important relationships between each of the different kinds of cells that migrate into the wound during the healing process, and to identify the factors that cause cells to start or stop multiplying. It may also be possible to modify with drugs the various protein-synthesizing activities of the cells in wounds in order to provide an optimal amount of connective tissue. These and other important problems remain to be solved, perhaps by further study of the healing wound.

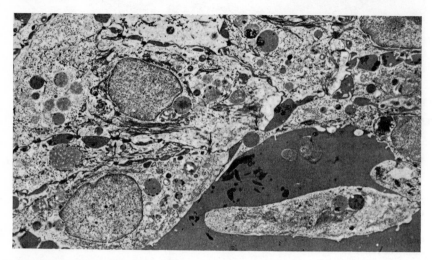

EPIDERMAL CELLS slightly to the rear of the leading edge of the epidermis contain round, dense vacuoles, filled with the serum protein and fibrin they have cleared from the wound by ingestion. Serum protein and somewhat darker patches of fibrin separate some of the cells from one another. These epidermal cells are strikingly different from those in undamaged tissue, which contain no vacuoles and have a different cytoplasmic structure.

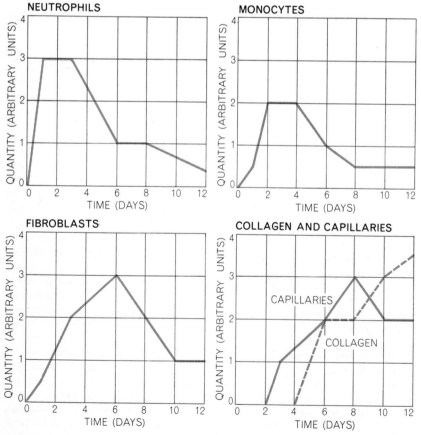

CELL POPULATIONS that occupy the wound at different times can be divided into three groups, each with its own special task to perform. The stages tend to overlap one another and may well be interdependent, since suppression of one can delay the start of the next.

20

The Immune System

by Niels Kaj Jerne
July 1973

*This diffuse organ has the assignment of monitoring
the identity of the body. Its basic constituents
are lymphocytes and antibody molecules, which
recognize both foreign molecules and one another*

The immune system is comparable in the complexity of its functions to the nervous system. Both systems are diffuse organs that are dispersed through most of the tissues of the body. In man the immune system weighs about two pounds. It consists of about a trillion (10^{12}) cells called lymphocytes and about 100 million trillion (10^{20}) molecules called antibodies that are produced and secreted by the lymphocytes. The special capability of the immune system is pattern recognition and its assignment is to patrol the body and guard its identity.

The cells and molecules of the immune system reach most tissues through the bloodstream, entering the tissues by penetrating the walls of the capillaries. After moving about they make their way to a return vascular system of their own, the lymphatic system [*see illustration on page 202*]. The tree of lymphatic vessels collects lymphocytes and antibodies, along with other cells and molecules and the interstitial fluid that bathes all the body's tissues, and pours its contents back into the bloodstream by joining the subclavian veins behind the collarbone. Lymphocytes are found in high concentrations in the lymph nodes, way stations along the lymphatic vessels, and at the sites where they are manufactured and processed: the bone marrow, the thymus and the spleen.

The immune system is subject to continuous decay and renewal. During the few moments it took you to read this far your body produced 10 million new lymphocytes and a million billion new antibody molecules. This might not be so astonishing if all these antibody molecules were identical. They are not. Millions of different molecules are required to cope with the task of pattern recognition, just as millions of different keys are required to fit millions of different locks.

The specific patterns that are recognized by antibody molecules are epitopes: patches on the surface of large molecules such as proteins, polysaccharides and nucleic acids. Molecules that display epitopes are called antigens. It is hardly possible to name a large molecule that is not an antigen. Let us consider protein molecules, which include enzymes, hormones, transport molecules such as hemoglobin and the great variety of molecules that are incorporated in cellular membranes or form the outer coat of viruses or bacteria.

Antigens and Antibodies

Each of the innumerable protein molecules is made up of polypeptide chains: linear strings of a few hundred amino acids chosen from a set of 20 amino acids. The number of amino acids in a large protein molecule is about equal to the number of letters in the column of text you are now reading, which is a linear string of letters chosen from an alphabet of 26 letters. Different protein molecules have different amino acid sequences just as different texts have different letter sequences. The string of letters in this column of text has been neatly "folded" into successive lines. The polypeptide chains of a protein molecule are also folded, although not so neatly. Their structure looks more like what you would obtain by haphazardly compressing a few yards of rope between your hands. There is nothing haphazard, however, about the folding of a particular polypeptide chain; the folding, and thus the ultimate conformation of the protein molecule, is precisely dictated by the amino acid sequence.

The parts of the folded chains that lie at the surface of a protein molecule make up its surface relief. An epitope (or "antigenic determinant") is a very small patch of this surface: about 10 amino acids may contribute to the pattern of the epitope. As Emanuel Margoliash of the Abbott Laboratories and Alfred Nisonoff of the University of Illinois College of Medicine showed for different molecules of cytochrome c, the replacement of just one amino acid by another in a polypeptide chain of a protein frequently leads to the display of a different epitope. The immune system recognizes that difference and is able to check on mutant cells that make mistakes in protein synthesis. Not only can an individual immune system recognize epitopes on any protein or other antigen produced by any of the millions of species of animals, plants and microorganisms but also it can distinguish "foreign" epitopes from epitopes that belong to the molecules of its own body. This recognition is a crucial event, since antibody molecules attach to the epitopes they recognize and thereby earmark the antigens (or the cells that carry them) for destruction or removal by other mechanisms available to the body.

Epitopes are recognized by the combining sites of antibody molecules. An antibody is itself a protein molecule consisting of more than 20,000 atoms. It is made up of four polypeptide chains: two identical light chains and two identical heavy chains. A light chain consists of 214 amino acids and a heavy chain of about twice as many. Antibody molecules are alike except for the amino acids at about 50 "variable" positions among the first 110 positions, which constitute what is called the variable region of both the light and the heavy chains. At the tip of each variable region there is a concave combining site whose three-dimensional relief enables it to recognize a complementary epitope and make the antibody molecule stick to the molecule displaying that epitope. Whether a combining site will recognize one epitope or a different one depends on which amino acids are located at the variable positions. If at each of 50 positions of both chains there were an independent choice between just two amino acids, there would be 2^{100} (or 10^{30}) potentially different molecules! The situation is not that simple, however. The chains fall into subgroups, within each of which there are far fewer than 50 variable positions. On the other hand, at some of those variable positions, clustered in so-called hot spots, the choice is actually among more than two alternative amino acids. There is general agreement that the differences in amino acid sequence among antibody molecules derive from mutations that have occurred in the genes encoding antibody structure.

The Recognition Problem

Smallpox being the nasty disease it is, one might expect nature to have designed antibody molecules with combining sites that specifically recognize the epitopes on smallpox virus. Nature differs from technology in its approach to problem solving, however: it thinks nothing of wastefulness. (For example, rather than improving the chance that a spermatozoon will meet an egg cell, nature finds it easier to produce millions of spermatozoa.) Instead of designing antibody molecules to fit the smallpox virus and other noxious agents, it is easier to make millions of different antibody molecules, some of which may fit. By way of analogy, suppose someone makes gloves in 1,000 different sizes and shapes: he would have a sufficiently well-fitting glove for almost any hand. Now imagine that hands were a great deal more variable; for example, the length of the fingers on a hand might vary independently from one inch to six inches. By making, say, 10 million gloves of different shapes the manufacturer would never-

theless be able to fit practically any hand—at the expense of efficiency, to be sure, since most of the gloves might never find a customer to fit them. Now be more wasteful still: have a factory with machines capable of turning out gloves of a billion different shapes, but turn off 99 percent of the machines, so that the factory actually turns out a random collection of 10 million of the potential billion shapes. You would still be doing all right. So would your colleague running a similar factory. Although the two sets of gloves you and he would make would show only a 1 percent overlap, each set would serve its purpose well enough.

That is how some of us think the immune system solves its recognition problem. By a more or less random replacement of amino acids in the hot-spot positions of the variable regions of antibody polypeptide chains, a set of millions of antibody molecules is generated with

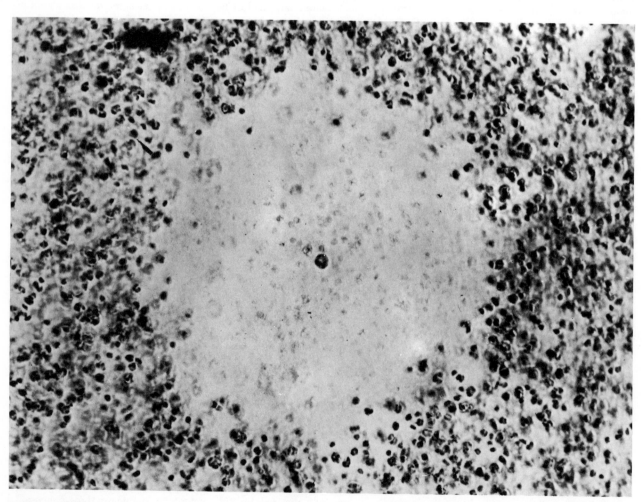

EFFECT OF ANTIBODY on an antigen is illustrated dramatically in a photomicrograph made by the author and Albert Nordin. The cell in the center is a plasma cell, an antibody-secreting lymphocyte of the immune system. It was embedded in a layer of culture medium along with millions of sheep red blood cells. The plasma cell is one that makes antibody against the sheep cells, specifically against epitopes, or small surface patches, on molecules in the surface membrane of the cells. Antibodies secreted by the plasma cell have destroyed the red blood cells in the area into which the antibodies have diffused; the radius of the area of destruction is about 1/5 millimeter. The technique illustrated here has become a standard one for measuring the immune response to an antigen.

different combining sites that will fit practically any epitope well enough. As has been demonstrated by Jacques Oudin of the Pasteur Institute and by Andrew Kelus and Philipp G. H. Gell of the University of Birmingham for rabbits and by Brigitte A. Askonas, Allan Williamson, Brian Wright and Wolfgang Kreth of the National Institute for Medical Research in London for mice, individual animals make use of entirely different sets of antibodies capable of recognizing a given epitope.

There is one serious snag in all of this, to which I alluded above: one's immune system does not seem to recognize the epitopes on molecules and cells that are part of one's own body. This property, which Sir Macfarlane Burnet called the discrimination between self and not-self, is often referred to as self-tolerance. You might think that self-tolerance derived from nature's being wise enough to construct the genes coding for your antibodies in such a way as not to give rise to combining sites that would fit epitopes occurring in your own body. It can easily be shown, however, that this is not so. For example, your father's antibodies could recognize epitopes occurring in your mother; some antibody genes inherited from your father should therefore code for antibodies recognizing epitopes inherited from your mother.

Self-tolerance, then, is not innate. It is something the immune system "learned" in embryonic life by either eliminating or "paralyzing" all lymphocytes that would produce self-recognizing antibodies. An original observation of this phenomenon by Ray D. Owen of the California Institute of Technology was generalized in a theoretical framework by Burnet and received experimental confirmation by P. B. Medawar in the 1950's, bringing Nobel prizes to Burnet and Medawar in 1960.

The Lymphocyte

Emil von Behring and Shibasaburo Kitazato discovered the existence of antibodies in Germany in 1890, but it was not until the 1960's that the structure of antibodies was determined, through the investigations initiated by R. R. Porter of the University of Oxford and Gerald M. Edelman of Rockefeller University [see "The Structure of Antibodies," by R. R. Porter, SCIENTIFIC AMERICAN Offprint 1083, and "The Structure and Function of Antibodies," by Gerald M. Edelman, SCIENTIFIC AMERICAN Offprint 1185]. The two men shared a Nobel prize last year for that work. Long before the structure of antibodies was

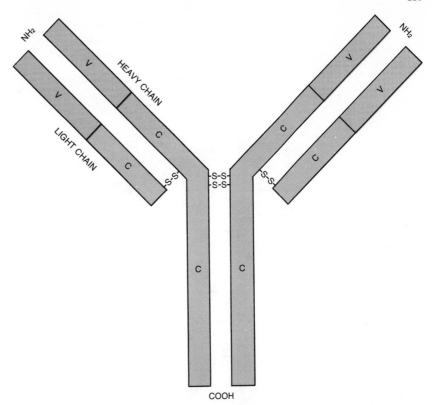

LINEAR STRUCTURE of an antibody molecule is shown schematically. The two heavy chains and two light ones are connected by disulfide bridges. Each chain has an amino end (NH₂) and a carboxyl end (COOH). Chains are divided into variable (V) regions (color), in which the amino acid sequence varies in different antibodies, and constant (C) regions.

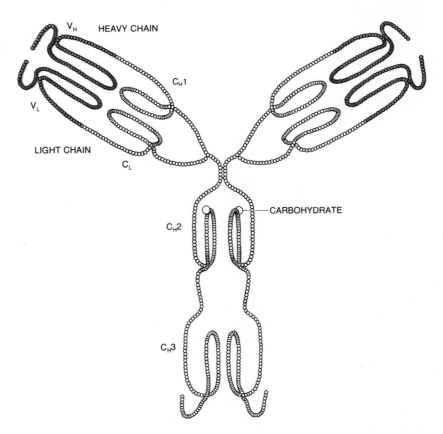

FOLDING OF THE FOUR CHAINS is suggested in this drawing based on a bead model of the antibody molecule made by Gerald M. Edelman and his colleagues. Each bead represents an amino acid, of which there are more than 1,200. The variable regions are in color.

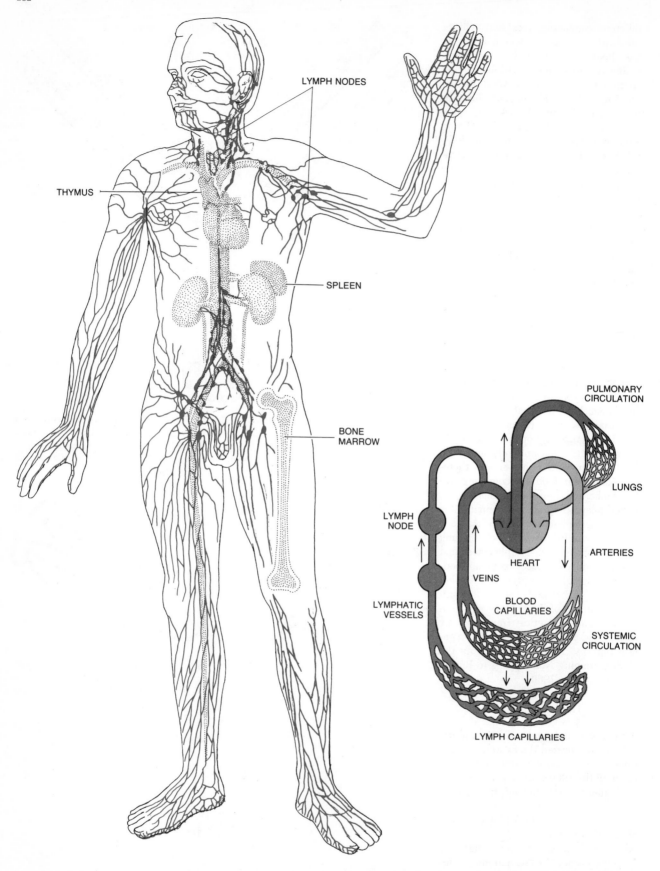

LYMPH NODES

THYMUS

SPLEEN

BONE
MARROW

PULMONARY
CIRCULATION

LUNGS

LYMPH
NODE

ARTERIES

LYMPHATIC
VESSELS

VEINS

HEART

BLOOD
CAPILLARIES

SYSTEMIC
CIRCULATION

LYMPH CAPILLARIES

IMMUNE SYSTEM consists of the lymphocytes and the antibody molecules they secrete. The cells and antibodies pervade most of the tissues, to which they are delivered by the bloodstream, but are concentrated in the tissues shown in color: the tree of lymphatic vessels and the lymph nodes stationed along them, the bone marrow (which is in the long bones, only one of which is illustrated), the thymus and the spleen. The lymphatic vessels collect the cells and antibodies from the tissue and return them to the bloodstream at the subclavian veins. Lymphocytes are manufactured in the bone marrow and multiply by cell division in the thymus, the spleen and the lymph nodes. The relation of the blood vessels and the lymphatic vessels is shown highly schematically in the illustration at right.

known, however, antibodies had been the subject of detailed studies. And yet it was not known that antibodies are produced by activated lymphocytes. Even 20 years ago lymphocytes were not thought to have anything to do with the immune system, something that seems odd now that they are known to constitute the immune system! It was only in the early 1960's that the involvement of lymphocytes was proved by James L. Gowans and Douglas McGregor of the University of Oxford.

Most lymphocytes (about 98 percent of them) do not actually secrete antibody. They are the "small" lymphocytes, spherical cells measuring about a hundredth of a millimeter in diameter, and they are said to be in a resting state. In order to secrete antibody a small lymphocyte must first become enlarged. In that state it can not only secrete antibody molecules but also divide and become two cells, which in turn can become four cells and so on. The offspring cells constitute the clone, or cell line, derived from one small lymphocyte.

As was originally postulated by Burnet in 1957, the antibody molecules produced by a lymphocyte and by the cells of its clone all have identical combining sites [see "The Mechanism of Immunity," by Sir Macfarlane Burnet; SCIENTIFIC AMERICAN, January, 1961]. G. J. V. Nossal, Burnet's successor as director of the Walter and Eliza Hall Institute of Medical Research in Melbourne, and his coworkers have accumulated much of the experimental evidence that now firmly supports this "single commitment" of the

lymphocyte [see "How Cells Make Antibodies," by G. J. V. Nossal; SCIENTIFIC AMERICAN Offprint 199]. The cells of one lymphocyte clone are committed to the expression of two particular genes coding for particular variants of the variable regions of the light chain and the heavy chain. Already in its resting, nonsecreting state a small lymphocyte produces a relatively small number of its particular antibody molecules, which it displays on the surface of its outer membrane. These antibody molecules are the "receptors" of the cell. A small lymphocyte displays about 100,000 receptors with identical combining sites, which are waiting, so to speak, for an encounter with an epitope that fits them.

When such an epitope makes contact, the lymphocyte can either become "stimulated" (respond positively) or become "paralyzed" (respond negatively), which is to say it is no longer capable of being stimulated. Investigations in progress by David S. Rowe of the World Health Organization, working in Lausanne, and Benvenuto Pernis at our Basel Institute for Immunology suggest that the distinction between excitatory and inhibitory signals may reside in differences in the constant regions of the lymphocyte's receptor antibody molecules. Whether a lymphocyte will choose to respond positively or negatively can be shown to depend on several conditions: the concentration of the recognized epitopes, the degree to which those epitopes fit the combining sites of the receptors, the way the epitopes are presented (for example whether they are presented

on molecules or on cell surfaces) and the presence or absence of other lymphocytes that can "help" or "suppress" a response. Much current experimentation aims at clarifying these complex matters.

A stimulated lymphocyte faces two tasks: it must produce antibody molecules for secretion and it must divide in order to expand into a clone of progeny cells representing its commitment. Progeny cells that go all out into the production and secretion of antibody molecules are called plasma cells. Each of them must transcribe its antibody genes into 20,000 messenger-RNA molecules that serve 200,000 ribosomes, enabling the cell to produce and secrete 2,000 identical antibody molecules per second. Other cells of the clone do not go that far; they revert to the resting state and represent the "memory" of the occurrence, ready to respond if the epitope should reappear. The immunological memory of what Stephen Fazekas de St. Groth of the University of Sydney, who is now working in our laboratory in Basel, has called "original antigenic sin" is remarkably persistent. People who are now 90 years old, for example, and had influenza in the 1890's still possess circulating antibodies to the epitopes of the influenza virus strains that were prevalent at that time.

If a lymphocyte that recognizes an epitope does not become stimulated, it may become paralyzed. Paralysis can occur when a lymphocyte is confronted by very high concentrations of epitope; this is called high-zone tolerance. David W. Dresser and N. Avrion Mitchison, who were working at the National Institute for Medical Research, have shown that paralysis can also result from the continuous presence of extremely small epitope concentrations, below the threshold required for stimulation; this is called low-zone tolerance. We need more knowledge of the mechanisms leading to paralysis, not only in order to understand how the immune system learns to tolerate self-epitopes but also to be able to induce the system to tolerate organ transplants.

Germ-Line and Soma Theories

The enormous diversity of antibodies raises the question of the origin of the genes that code for the variable regions of antibody molecules. Essentially two answers have been proposed to this question. They are the germ-line theory and the somatic theory. The argument of the germ-line theory is straightforward: All the cells of the body, including lym-

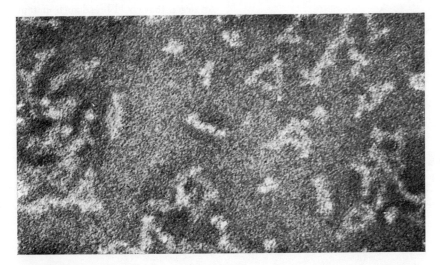

ANTIBODY MOLECULES are visible in an electron microscope when they are linked to antigens and one another in antigen-antibody complexes. In this micrograph, made by N. M. Green and the late Robin Valentine of the National Institute for Medical Research in London, rabbit antibodies are enlarged 500,000 diameters. The antigen is a short polypeptide chain with a dinitrophenyl group at each end; the antibodies are from a rabbit that was immunized against dinitrophenyl epitopes. The antigens (too small to be visible) link antibodies to form polygonal complexes whose geometry derives from antibody structure.

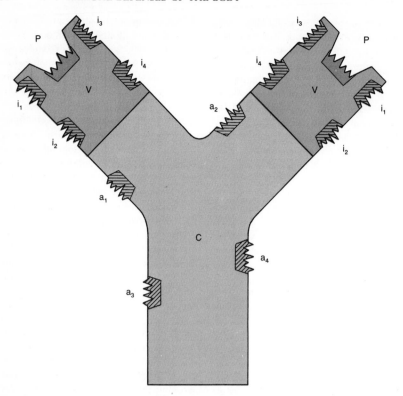

FUNCTIONAL TOPOGRAPHY of the antibody molecule is mapped. The end of each arm of the Y has a combining site (p) that recognizes epitopes on antigen molecules. The antibody also has its own epitopes, which can be recognized by other antibodies' combining sites. These include allotopes (a) in constant regions and idiotopes (i) in variable regions.

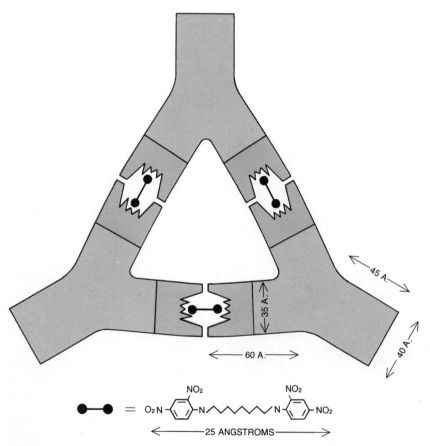

TRIANGULAR STRUCTURES in the micrograph on the opposite page are trimers, or complexes of three antibody molecules, linked by three double-ended dinitrophenyl antigens. The dimensions were worked out by Green and Valentine from electron micrographs.

phocytes, have the same set of genes, namely those in the fertilized egg from which the individual arose. Therefore genes for any antibody that an individual can make must already have been present in the fertilized egg cell. They are all transmitted to the individual's children through the germ-cell line: egg and spermatozoa and their precursors.

The somatic theory does not accept this approach. It is argued that the immune system needs millions of different antibodies for epitope recognition. Individual mice of an inbred strain, all having the same germ-line genes, have been shown to make use of entirely different sets of antibody molecules. The germ-line theory implies that the set of all these sets is represented in the genes of every single mouse of that strain. In that case, however, many of the genes would seem to have no survival value for the mouse, so that such a large number of genes cannot arise or be maintained in Darwinian evolution. Most antibody genes must therefore have arisen in the course of the somatic development of the individual by modification of a smaller number of germ-line genes. That is the point of departure for several variants of the somatic theory.

I have proposed that an inherited set of germ-line genes code for antibodies against certain self-epitopes. The clones of cells expressing these genes become suppressed except for mutant cells that, by an amino acid replacement, display new combining sites on their antibody receptor molecules. These mutant cells represent the enormous repertoire of antibodies that recognize foreign epitopes. An organ that could breed such mutant cells is the thymus gland. More than 10^{10} new lymphocytes arise in the thymus every day; the vast majority of these cells are killed in the thymus or immediately after they leave it.

It is not possible here to discuss the merits of these theories. That would require consideration of a large body of experimental results, such as the explorations of the genetics of immune responsiveness by Baruj Benacerraf of the Harvard Medical School, Hugh O. McDevitt of the Stanford University Medical Center and Michael Sela of the Weizmann Institute of Science in Israel.

T Cell and B Cell

All the lymphocytes that circulate in the tissues have arisen from precursor cells in the bone marrow. About half of these lymphocytes, the T cells, have passed through the thymus on their way to the tissues; the other half, the B cells, have not. This dichotomy was first discovered by Henry N. Claman of the University of Colorado Medical School and was characterized by Jacques F. A. P. Miller and Graham Mitchell, both of whom are now working with us in Basel. It has been the subject of thousands of investigations during the past five years. *T* cells and *B* cells cannot be distinguished by their form. Only *B* cells and their progeny cells secrete antibody molecules. One might think that this leaves little scope for *T*-cell function. On the contrary, *T* cells appear to be all-important. They too can recognize epitopes and must therefore, almost by definition, possess antibody molecules as surface

receptors, although these receptor molecules have been much harder to demonstrate experimentally than those on *B* cells.

T cells can kill other cells, such as cancer cells, and transplanted tissues that display foreign epitopes. *T* cells can also suppress *B* cells or alternatively can help *B* cells to become stimulated by epitopes. This "helper" function of *T* cells has been repeatedly demonstrated both in animal experiments and in experiments with cells in culture. In the cell-culture experiments, based on a technique developed by Richard W. Dutton and Robert I. Mishell at the University of California at San Diego, lymphocytes

taken from the spleen of an untreated animal are grown in a plastic dish together with molecules or cells that display foreign epitopes. After a few days' incubation lymphocytes that produce and secrete antibody molecules against the foreign epitopes can be shown to be present in the culture by the assay method for single antibody-producing cells [*see illustration on page 200*]. These antibody molecules are made by *B* cells, but the experiment will not work if only *B* cells are present. As soon as *T* cells are added to the culture dish, however, the *B* cells begin to respond and to produce antibody.

The dichotomy of the immune system into *T* and *B* lymphocytes adds a further dimension to the conceptual framework needed for the system's comprehension. That is not only an intellectual need but also a practical one, since the immune system is now known to be crucially involved in a vast number of diseases ranging from microbial infections and allergies to cancer, rheumatism, autoimmunity and many other degenerative disorders of aging.

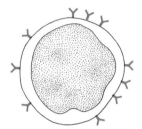
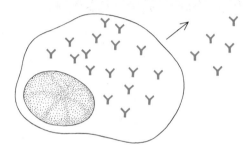

LYMPHOCYTES, the cells of the immune system, produce antibodies. Each cell is committed in advance to the production of one specific antibody. In its resting state, as a small lymphocyte (*left*), the cell displays such antibody molecules (*color*) on its surface as "receptors." The advent of an antigen with an epitope that fits the combining site of this particular antibody molecule may stimulate the lymphocyte to grow, change in structure and divide, eventually giving rise to a large number of plasma cells (*right*): lymphocytes specialized for the rapid synthesis and secretion of this cell line's specific antibody molecules.

The Lymphocyte Network

I have mentioned two striking dualisms within the immune system. One is the dichotomy of the lymphocytes into *T* cells and *B* cells, with functions that are partly synergistic and partly antagonistic. The second is the duality of the potential response of a lymphocyte when its receptors recognize an epitope: it can either respond positively (become stimulated) or respond negatively (become paralyzed). It is important to realize that the immune system displays a third dualism, namely that antibody molecules can recognize and can also be recognized. They not only have combining sites enabling them to recognize epitopes but also display epitopes enabling them to be recognized by the combining sites of other antibody molecules. That is true for the antibody molecules attached to the outer membranes of lymphocytes and serving as receptors as well as for the freely circulating antibody molecules, which can be regarded as messages released by lymphocytes.

Epitopes occur on both the constant and the variable regions of an antibody molecule. Since the patterns of the variable-region epitopes are determined by the variable amino acid sequences of the polypeptide chains, there are millions of different epitopes. The set of such epitopes on a given antibody molecule was named the idiotype of that molecule by Oudin. When antibodies produced by

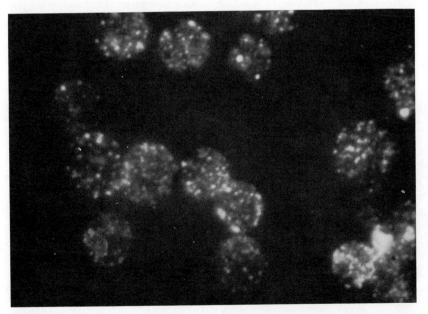

RECEPTOR ANTIBODY MOLECULES are demonstrated by a fluorescent stain in a photomicrograph made by Benvenuto Pernis. The cells are small lymphocytes from a patient with lymphocytic leukemia, in which a line of lymphocytes proliferates out of control. The receptor antibodies on the cell surfaces have epitopes in their constant regions (allotopes) characteristic of human antibody molecules. An antibody directed against those allotopes is prepared by injecting human serum into a rabbit. Rhodamine, a fluorescent dye, is coupled to the antibodies, which are added to a suspension of the lymphocytes. The bright spots on the cells represent fluorescent antibody bound to receptor molecules they "recognize."

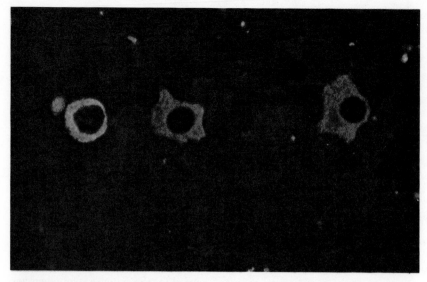

INTERNAL ANTIBODY MOLECULES, being produced by plasma cells for secretion, are stained in this photomicrograph and the one at the bottom of the page, also made by Pernis. The plasma cells are from a rabbit that is heterozygous for the structural gene that determines the constant region of the antibody molecule and therefore its set of allotopes, or its allotype; that is, the rabbit inherited paternal and maternal chromosomes containing two different determining genes. The plasma cells have been stained with two preparations of fluorescent antibodies: one, to which a green-fluorescing stain has been coupled, is directed against the paternal allotype (call it allotype A) and the other is directed against the maternal allotype (B). The internal antibodies in some of the plasma cells bind to the green-staining preparation, indicating that they are antibody molecules carrying allotype A.

animal A are injected into animal B, animal B will produce antibodies against the idiotypic epitopes ("idiotopes") of the injected antibody molecules. That is also true when A and B belong to the same animal species and even when they are of the same inbred strain, that is, when they are genetically identical. Evidence is emerging that, within one animal, the idiotopes occurring on one antibody molecule are recognized by combining sites on a set of other antibody molecules, and that the idiotopes on the receptor molecules of one lymphocyte are recognized by the combining sites of the receptor molecules of a set of other lymphocytes. We thus have a network of lymphocytes and antibody molecules that recognize other lymphocytes and antibody molecules, which in turn recognize still others.

I am convinced that the description of the immune system as a functional network of lymphocytes and antibody molecules is essential to its understanding, and that the network as a whole functions in a way that is peculiar to and characteristic of the internal interactions of the elements of the immune system itself: it displays what I call an eigen-behavior. (Eigen in German means peculiar to, or characteristic of. Eigen-behavior is analogous to such concepts as the eigenvalue or eigenfrequency of certain physical systems.) There is an increasing body of evidence for this view.

Antibody molecules are normally present in the blood in a concentration of about 5×10^{16} molecules per milliliter. The total concentration of combining sites and idiotopes is therefore of the order of 10^{17} per milliliter. If the immune system made use of 10 million different combining sites and 10 million different idiotopes, each single variant of these elements would be present, on the average, in a concentration of about 10^{10} per milliliter. Mitchison at the National Institute for Medical Research and Nossal, Gordon L. Ada and their colleagues at the Walter and Eliza Hall Institute and the Australian National University, experimenting with low-zone tolerance, have shown that epitope concentrations ranging for different antigens from a million to 10^{12} epitopes per milliliter suffice either to suppress or to paralyze lymphocytes that can recognize the epitopes. Nisonoff and his co-workers at the University of Illinois College of Medicine and Humberto Cosenza and Heinz Köhler at the University of Chicago have shown that injecting into an animal antibodies against an idiotype suppresses lymphocytes that have receptors with idiotopes recognized by those antibodies. Leonard A. Herzenberg of the Stanford University Medical Center and Ethel Jacobson in our laboratory in Basel find that T lymphocytes recognizing epitopes on the receptors of B lymphocytes can suppress those B lymphocytes.

What this adds up to is that lymphocytes are subject to continuous suppression by other lymphocytes and by antibody molecules with idiotopes or combining sites that fit. Some lymphocytes escape from suppression and divide. New lymphocytes emerge. Others remain suppressed or decay. The eigen-behavior is the dynamic steady state of the system as its elements interact. As the system expands in the course of development and later life, new idiotopes and new combining sites emerge. The "self"-epitopes of other tissues impinge on the

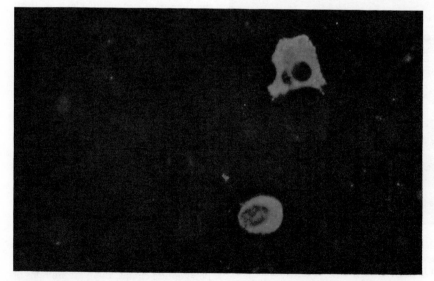

OTHER INTERNAL ANTIBODIES, in other plasma cells from the same field of the preparation as the micrograph at the top of the page, stain red. They are molecules with allotype B. Although all the rabbit's lymphocytes and plasma cells contain both the paternal and the maternal chromosomes, each cell expresses only one or the other constant-region gene.

network and cause certain elements to become more numerous and others less numerous. In this way each individual develops a different immune system.

Invading foreign antigens modulate the network; early imprints leave the deepest traces. A given foreign epitope will be recognized, with various degrees of precision, by the combining sites of a set of antibody molecules, and lymphocytes that are committed to producing antibody molecules of that set are then stimulated and become more numerous. That is not, however, the only imprint made by the foreign epitope. The set of combining sites that recognized the epitope also recognizes a set of idiotopes *within* the system, a set of idiotopes that constitutes the "internal image" of the foreign epitope. The lymphocytes representing the internal image will therefore be affected secondarily, and so forth in successive recognition waves throughout the network [*see illustration below*].

The structural properties of the immune system and its eigen-behavior reside in these complex ramifications.

Immune System and Nervous System

The immune system and the nervous system are unique among the organs of the body in their ability to respond adequately to an enormous variety of signals. Both systems display dichotomies: their cells can both receive and transmit signals, and the signals can be either excitatory or inhibitory. The two systems penetrate most other tissues of the body, but they seem to avoid each other: the "blood-brain barrier" prevents lymphocytes from coming into contact with nerve cells.

The nerve cells, or neurons, are in fixed positions in the brain, the spinal cord and the ganglia, and their long processes, the axons, connect them to form a network. The ability of the axon

of one neuron to form synapses with the correct set of other neurons must require something akin to epitope recognition. Lymphocytes are 100 times more numerous than nerve cells and, unlike nerve cells, they move about freely. They too interact, however, either by direct encounters or through the antibody molecules they release. These elements can recognize as well as be recognized, and in so doing they too form a network. As in the case of the nervous system, the modulation of the network by foreign signals represents its adaptation to the outside world. Both systems thereby learn from experience and build up a memory, a memory that is sustained by reinforcement but cannot be transmitted to the next generation. These striking analogies in the expression of the two systems may result from similarities in the sets of genes that encode their structure and that control their development and function.

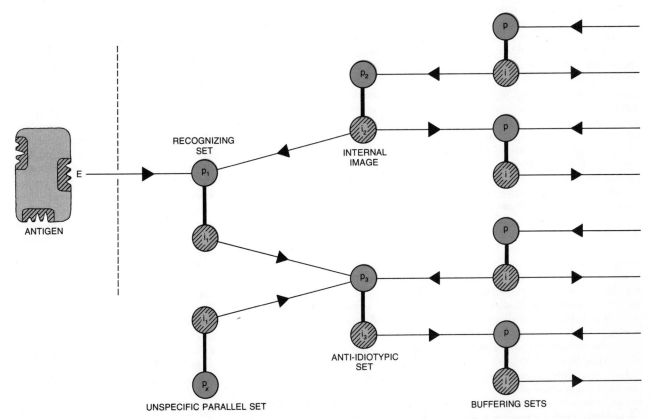

LYMPHOCYTE NETWORK is diagrammed in an effort to indicate how its steady-state ("eigen") behavior is established and how the network responds to an antigen. An epitope (E) on the antigen is recognized by a set (p_1) of combining sites on antibody molecules, both circulating antibody and cell-surface receptors. Cells with receptors of the recognizing set p_1 are potentially capable of responding to the antigenic stimulus (*arrowhead*) of epitope E, but there are constraints. The same molecules that carry combining sites p_1 carry a set of idiotopes (i_1). These are recognized within the system by a set of combining sites (p_3), called the anti-idiotypic set because they tend to suppress (*reverse arrowhead*) the cells of set i_1. (These idiotypes i_1 are also found on molecules with com-

bining sites that do not belong to the recognizing set p_1 but rather are unspecific with regard to epitope E.) On the other hand, the set p_1 also recognizes internal epitopes i_2, which therefore constitute an internal image of the foreign epitope E. In the steady state, molecules of the internal image tend to stimulate cells of set p_1 and thus to balance the suppressive tendency of the anti-idiotypic set. When the foreign antigen enters the system, its stimulatory effect on recognizing set p_1 allows cells of that set to escape from suppression. (The same thing happens to unspecific cells of the parallel set p_x.) The resulting immune response to the antigen is modulated by the buffering effects of many more sets of combining sites and idiotopes (*right*), which have a controlling influence on the response.

21

Markers of Biological Individuality

by Ralph A. Reisfeld and Barry D. Kahan
June 1972

The rejection of transplanted organs has focused attention on the body's ability to distinguish foreign cells from its own. The cells of each individual are uniquely marked with protein

It is often observed that every human being is unique. Our superficial differences, however, scarcely hint at the differences to be found at the level of the genes and the chemistry of the cell. It is now becoming clear that these differences play an important role in enabling living organisms to distinguish between "self" and "not self" and to reject cells that are not labeled with the appropriate recognition markers. The advances in surgery that make it possible to transplant organs from one person to another have stimulated an intensive study of the factors that cause the recipient to reject foreign tissue unless the donor is a close relative. Criteria for matching unrelated donors and hosts have proved extremely difficult to establish. The reason, we now see, is precisely the genetic and chemical uniqueness of each individual. Where organ transplants have been successful it is usually because ways have been found to suppress the rejection mechanism that normally operates. The techniques of suppression, however, still depend more on good luck than on fundamental understanding. Here we shall try to summarize what has recently been learned about the unique markers of individuality carried by each cell.

Individual differences are readily apparent in anatomy and physiology. "Normality" can only be defined by statistical methods involving many individuals, from which one can arrive at mean values with limits of variation. Any given individual, however, must possess many physical characteristics that are outside the normal range but that are not necessarily pathological. Functional differences are subtler; for example, individuals commonly exhibit large differences in the ratios of the electrolytes present in blood serum or in the structure and function of the various enzymes that mediate

specific biochemical reactions. It is clear that such differences are innate: they are demonstrable over periods of years and are only slightly influenced by the environment.

Individual peculiarities are sometimes tragically illustrated by the idiosyncratic reactions of patients to particular drugs. Fifty years ago Sir Archibald Garrod, in his pioneering book *Inborn Errors of Metabolism*, noted that "these [pathologic defects of enzyme chemistry] are merely examples of variations in chemical behavior which are probably everywhere present in minor degrees, and that just as no two individuals of a species are absolutely identical in bodily structure, neither are their chemical processes carried out on the same lines."

Since these individual differences can be preserved from generation to generation only by the maintenance of a complex polymorphic gene pool, that is, a pool of genes with many alternative expressions, they must serve some important purpose. Unless they contributed to the survival of the species they would have been eliminated long ago by the pressures of natural selection. Moreover, it is clear that individuality is not confined to higher organisms. As Thomas Humphreys and Aron A. Moscona of the University of Chicago have shown in their studies of specific aggregation of sponge cells, even primitive species can distinguish their own cells from others. Genetic variability enhances the ability of members of a species to survive changes in their external and internal environments. Thus by studying the factors that determine and express individuality one can hope to acquire insight into basic biological phenomena relevant not only to the role of isolated characteristics but also to the functioning of the organism as an integrated collection of cooperating processes.

One of the most valuable techniques for studying individuality pits the distinctive factors of one organism against those of another. It is easily achieved by transplanting a bit of tissue from a donor to a host. The host is then required to distinguish its own unique biological markers from those of the interloper. One might regard this as a test of parasitism: What prevents one individual from becoming an integral part of another?

Humans are not good subjects for transplantation studies because the uniqueness of each individual precludes getting a "typical" response. One would have to conduct many experiments with many individuals and interpret the results statistically, which would of course tend to obscure the individual factors one was setting out to study. The alternative is to use inbred strains of laboratory animals such as mice or guinea pigs, so that one has a better chance of isolating the chemical and biological determinants of individuality. In the 1920's Sewall Wright and Clarence Cook Little showed independently that by repeated brother-sister matings one can develop animal lines possessing a common pool of genes. This does not mean that each newborn individual contains the same set of genes but rather that each set is derived from the same limited pool. Wright and Little demonstrated that tissue grafts between members of the same line survived longer than grafts between members of two different lines.

Little's colleague George D. Snell, working at the Jackson Memorial Laboratory, went further and showed that tumors obtained from one line of mice are rejected by hosts that differ from the donor strain by a factor subsequently designated *H-2*. The gene controlling the *H-2* set of factors is transmitted by simple Mendelian laws [*see illustration below*]. Grafts of normal tissue exchanged between members of the same inbred line (isografts) are accepted. Grafts exchanged between members of two different lines (allografts) are uniformly rejected within 14 days. Grafts transferred from parental lines to the first-generation hybrids of those lines are accepted, but grafts transferred from the hybrid offspring back to the parents are not.

Moreover, if the grafts are tumors rather than normal tissue, the response generated by the *H-2* gene is strong enough to destroy the graft when the donor and the host possess different *H-2* genes. Evidently the *H-2* gene gives rise to strong markers on the surface of the cell that are readily recognized by cells of a different strain. In the absence of such a strong response tumors grow rapidly and outstrip the host's defenses. Snell also observed that when the grafts consist of normal tissue, they can be rejected by genetic factors weaker than those supplied by the *H-2* gene.

Furthermore, working with congenic lines of mice (inbred lines differing from one another at a single genetic locus), Snell found that a number of weaker factors, distributed as loci in the genetic

material different from the *H-2* locus, could initiate the rejection of normal-tissue grafts, although rejection could take as long as 200 days. Differences at a second genetic site are not capable, however, of stopping the growth of tumor grafts when the donor and the host possess the same *H-2* gene. The strong markers placed on cells by the *H-2* gene in mice have their counterpart in man in markers traceable to the human leukocyte locus *A* (HL-*A*) gene. Thus transplantation experiments have revealed a number of discrete genes that control the production of markers, or distinctive factors, that determine the fate of foreign grafts.

Genetic markers of this type have long been exploited in the typing and matching of red blood cells for transfusions. The individuality markers that have to be accounted for when tissues are transplanted from one organism to another, however, are much more numerous and harder to classify because many genes are involved. When studies of tissue transplantation were extended to outbred populations of mice, it was shown that the number of alternative gene expressions associated with the strong factors that control rapid rejection of transplants in mice is probably no smaller

than the number observed in man. A comparison of the data obtained in mice with observations in human populations led Jean Dausset of the University of Paris to suggest that the genes controlling the strong transplantation factors are collected in a single genetic region. This hypothesis was subsequently supported by the work of Ruggiero Ceppellini of the University of Turin and Jan van Rood of the University of Leiden. Ceppellini demonstrated that the rapid rejection of grafts exchanged between two human subjects depends on a single genetic locus; van Rood then found that the locus controls the production of factors that can be detected with HL-*A* antibodies. The region seems to be divided into two subregions, each of which determines a series of allelic, or alternative, factors. There are multiple alternatives in each subregion: the *D* and *K* regions in mice each have at least 10 alleles; in man the L-*A* region has at least 11 alleles and the region known as "Four" has at least 17. Thus a simple biological test, the acceptance or rejection of transplanted tissue, has uncovered a richly polymorphic genetic system in mice and men, providing a tool for attacking the problem of biological variability.

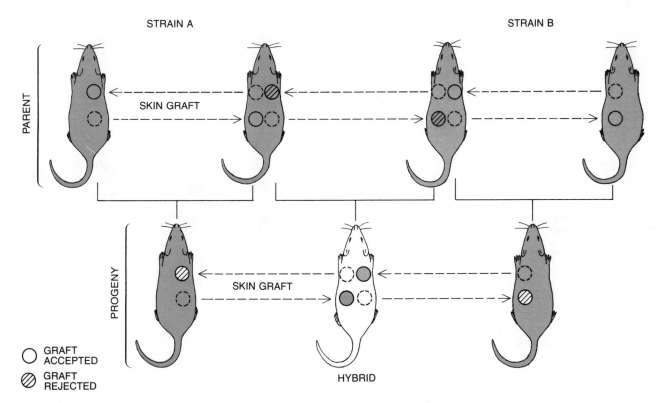

STRAIN A   STRAIN B

PARENT

SKIN GRAFT

PROGENY

SKIN GRAFT

○ GRAFT
  ACCEPTED

⊘ GRAFT
  REJECTED

HYBRID

STRONG RESPONSE TO TUMOR TRANSPLANTATION in mice depends on factors controlled by a specific gene known as *H-2*. George D. Snell of the Jackson Memorial Laboratory demonstrated in the 1940's that the *H-2* gene obeys simple Mendelian laws of inheritance. Tumor grafts exchanged between inbred lines of mice, such as between two mice of strain *A* or two of strain *B*, are accepted. Tumor transplants between mice of different strains, however, are vigorously rejected, being destroyed within two weeks. Hybrid offspring will accept grafts from either of the pure parental lines, but grafts in the reverse direction are destroyed.

The polymorphic marker system protects an individual from invaders that otherwise could not be distinguished from the self. Although data from human populations originally suggested that one out of every 10,000 unrelated individuals might possess the same strong transplantation genes, it has since been demonstrated that the frequency of identical genes is far rarer. In fact, Leo Loeb proposed many years ago that the possible combinations are infinite and that no exact matches can exist. The results of human transplantations offer little to refute Loeb's conjecture. Clinical success in organ transplants seems related less to genetic similarities between donors and recipients who are unrelated than to the impaired ability of the host to react to foreign markers.

What is the mechanism that enables the host to recognize and destroy foreign cells? There are many conceivable possibilities. For example, the host might provide a local environment in which foreign cells were deprived of essential nutrients. Plants are able to resist certain parasites by depriving them of substrates, for example polysaccharides, they need for growth. A parasite's success might depend on its ability to break up whatever such molecules may be provided by the host and to utilize their fragments. Another possibility, envisioned by Loeb, is that the growth of foreign tissue may result in the generation and release of specific substances that are toxic to the host. This might lead in turn to a nonspecific inflammatory response that would destroy the invader.

These and other hypotheses were finally ruled out some 20 years ago when P. B. Medawar and his colleagues at University College London demonstrated that the host's response to foreign tissue involves an immunological mechanism similar to the one that provides resistance to bacterial and viral infections. In his first group of experiments Medawar showed that the recipient of a graft develops a resistance that is specific for the donor: when the recipient is challenged with a second graft from the same donor, the transplanted tissue is destroyed even more rapidly than it was the first time. Medawar called this the second-set reaction [see "a" in illustration on opposite page]. Furthermore, the "memory" of the initial experience can be evoked by a second challenge graft no matter where it is placed on the recipient; in other words, the second-set response is system-wide, not local. Medawar then found that specific individuality markers are associated with every nucleated cell of an organism. Thus an animal can be "immunized" against foreign skin grafts by first injecting it with cells derived from other tissues of the same donor, for example cells from the spleen [see "b" in illustration on opposite page].

Finally, Medawar and two of his colleagues, Rupert E. Billingham and Leslie Brent, drawing on observations made by Ray D. Owen and a theory proposed by Sir Macfarlane Burnet, demonstrated that an animal of strain A can be made tolerant to tissue grafts from an animal of strain B by injecting the strain-A animal soon after birth with a suspension of spleen cells from the strain-B animal [see "c" in illustration on opposite page]. With this experiment Medawar and his colleagues proved that resistance or lack of resistance to foreign tissue is not a localized phenomenon but an immunological one. They later showed that the host's capacity to reject foreign tissues can be reconstituted by lymphoid grafts, which are known to restore immune reactions. This set of experiments left no doubt that the factors previously identified by genetic studies were those re-

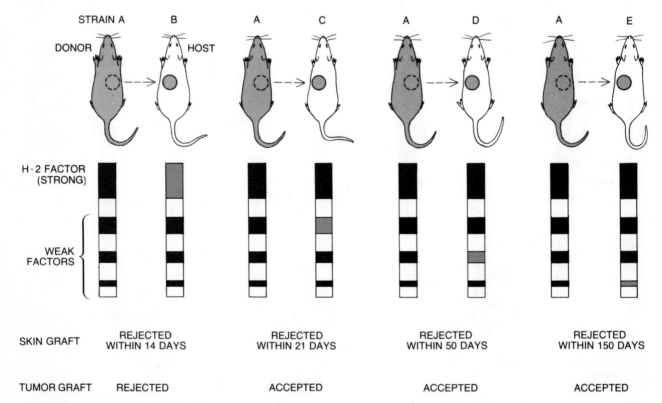

SKIN GRAFT — REJECTED WITHIN 14 DAYS — REJECTED WITHIN 21 DAYS — REJECTED WITHIN 50 DAYS — REJECTED WITHIN 150 DAYS

TUMOR GRAFT — REJECTED — ACCEPTED — ACCEPTED — ACCEPTED

ADDITIONAL GENETIC FACTORS determine the survival rate of normal skin grafts but not the survival rate of tumor grafts. If an animal of strain A and an animal of strain B differ in H-2 (that is, strong) individuality factors, each will vigorously reject tumor grafts as well as ordinary skin grafts from the other. Animals that share the same H-2 factors will accept tumor grafts from one another. Because animals C, D and E differ from A in weaker transplantation factors they reject normal skin grafts at various rates.

sponsible for graft rejection, since alteration of the host's responses toward these factors made it unable to react against grafts.

To account for these findings and related ones Medawar proposed that all nucleated cells possess surface markers that act as "antigens" when the cells are transplanted to another organism. The antigens trigger an immune response in the host that makes it specifically resistant to the donor's tissues. Evidently these markers are functional components of the outer membrane of every nucleated cell, although the role they play remains to be elucidated. Presumably the host becomes alerted to their presence when they perform their natural functions, perhaps as cell receptors or as factors involved in growth and development.

James L. Gowans of the University of Oxford has shown that foreign grafts are met by "wandering" lymphocytes, white scavenger cells produced in the lymph glands; the lymphocytes "check" cell surfaces like watchmen in order to ensure that all constituents of the host organism bear its own markers and not those of an invader [see illustration on next page]. When the wandering lymphocytes discover foreign markers in a graft, they carry the alarm back to the lymph nodes and stimulate an immune reaction. The reaction consists in the proliferation of lymphocytes with receptor sites that can engage the markers of foreign cells as a key engages a lock and thus interfere with their functioning, ultimately causing the death of the grafted cells. There is also evidence that whole cells or fragments from the graft tissue are detached and find their way into the drainage channels of the lymphatic system, where they eventually reach the lymph nodes and directly trigger the proliferation of suitably designed lymphocytes. Presumably the transplant markers have a role in the metabolism of the cell that requires them to occupy an exposed position on the cell's surface, so that a reaction against them is lethal to a cell transplanted in a foreign host.

Working with crude subcellular fractions, a number of investigators obtained evidence some years ago that nearly all the strong antigens of the cell are located on the surface of the cell membrane. Goran L. Möller of the Royal Caroline Institute in Stockholm found that serums produced in a host by challenge with intact foreign cells contain antibodies that react with cell surfaces [see illustration, page 215]. That these

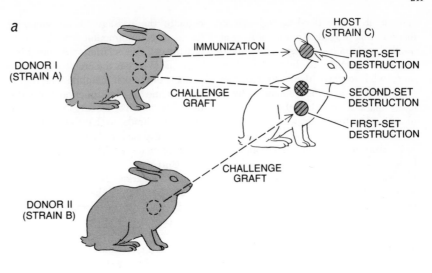

a

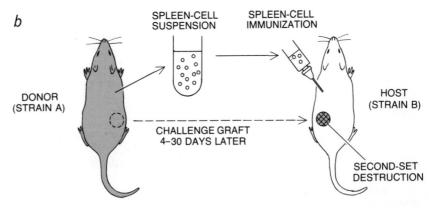

b

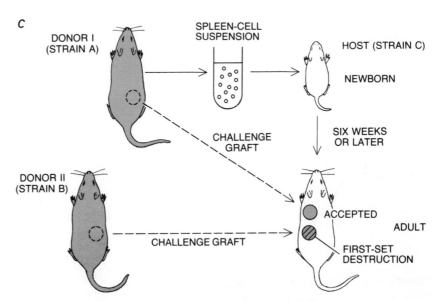

c

IMMUNOLOGIC NATURE OF TISSUE REJECTION was demonstrated by P. B. Medawar and his colleagues at University College London. His early studies showed (a) that a second graft from the same donor is destroyed more rapidly than the first. Medawar called this the second-set reaction; it shows that the host develops an "immunity" to a specific foreign donor. He then showed that immunization is not specific for the tissue involved (b). Animals exhibit a second-set reaction to a skin graft if they have previously been injected with cells derived from the donor's other tissues, such as the spleen. Finally Medawar and his group demonstrated that animals can develop immunologic tolerance (c). A mouse of strain C will accept a skin graft from a mouse of strain A if the strain-C mouse has been injected soon after birth with spleen cells from the mouse of strain A. The strain-C mouse, however, still retains its ability to reject challenge grafts from an animal of another strain, say strain B.

surface sites are indeed transplantation markers was proved in our laboratory at the Scripps Clinic and Research Foundation in La Jolla, Calif., by Soldano Ferrone, who recently demonstrated that solubilized and purified human individuality markers stimulate the production of antibodies that react with surfaces of intact cells.

Workers in a number of laboratories have been exploiting various techniques to isolate the individuality markers in a water-soluble form from the cell-membrane matrix with which they are associated. The task is complicated because the cell membrane consists of a lattice of lipoprotein molecules in which are embedded a variety of carbohydrates and proteins in addition to the individuality markers [*see top illustration on opposite page*]. Some of the proteins serve as structural components; others

regulate the passage of nutrients, respond to environmental stimuli and effect the cell's associations with neighboring cells. Whereas the architectural components of the cell membrane tend to be fixed, some of the functional proteins probably float like icebergs in a "sea" of membrane lipids, so that they are at least partially exposed to the external environment. Michael A. Edidin of Johns Hopkins University has inferred that transplantation markers are among those in a dynamic state from the way they spread rapidly between membranes when two unlike cells fuse.

One current model of the cell surface visualizes membrane components held together either by strong covalent chemical bonds or by weaker noncovalent interactions, such as those provided by hydrogen bonds and salt linkages. Since the individuality markers appear to be

mobile, it seems likely that they are associated with the architectural proteins and with other functional proteins through noncovalent interactions. One can only speculate on the role of these interactions. For example, by altering the expression of the individuality markers the interactions may protect the host against environmental agents that might alter the cell's identity. Fortunately it turns out that the antigenic activity associated with the individuality markers is independent of secondary interactions with other membrane components and depends solely on the marker's primary structure. Therefore it was possible for us to remove the markers from the membrane in water-soluble form and study their biological activity with a minimum of interference from other membrane components.

There are two general methods for re-

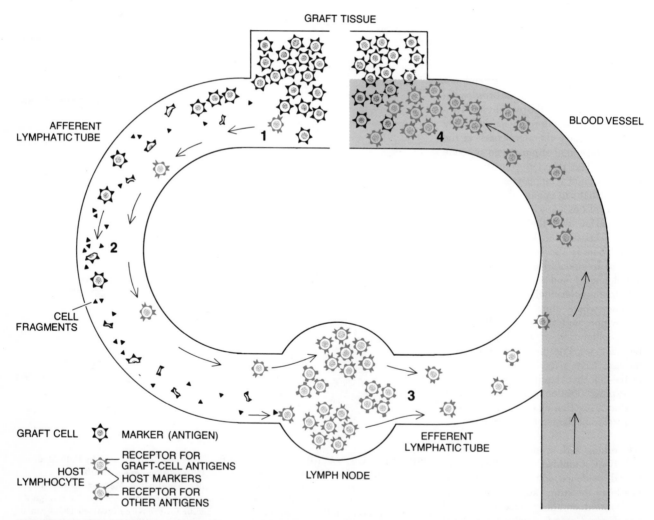

THEORY OF IMMUNOLOGIC RESPONSE to tissue transplants, proposed by Medawar, visualizes that cells carry individuality markers that another organism perceives as "antigens." Here the markers, or antigens, carried by the cells of a foreign graft are represented by black spikes. These are detected by the host's "wandering" lymphocytes, some of which are equipped with receptors that interact with the foreign markers in a lock-and-key fashion. These wandering lymphocytes (1) carry an alarm signal to the host's lymph nodes, stimulating the proliferation of more lymphocytes (3) equipped to destroy the invading cells (4). Cells detached from the graft, together with fragments carrying antigens, also enter the lymph drainage system (2) and carry the alarm message directly to the lymph nodes. Evidently lymphocytes are equipped with a variety of "keys" to engage any number of antigen "locks."

leasing markers from their membrane matrix [*see bottom illustration at right*]. One method uses proteolytic enzymes, for example papain, that randomly break the covalent bonds holding proteins together, thereby releasing an array of antigenic materials from membranes. Some of these fragments represent broken markers; others contain extraneous materials, such as carbohydrates, that apparently are not involved in the recognition of foreign cells in transplantation.

A gentler method of disruption, which leaves the markers more or less intact, involves exposing cells to low-intensity sonic energy. Noncovalent bonds are disrupted by a combination of cavitation (the rapid expansion and violent collapse of air bubbles trapped in the medium), mechanical agitation, foaming, shearing and local heating. The treatment produces a complex mixture of markers, other solubilized membrane components and soluble proteins that are released from the interior of the cell by the disintegration of its surface membrane.

Noncovalent bonds can also be dissociated by a simple salt, potassium chloride, which acts as a weak chaotropic agent, decreasing the orderly arrangement of water molecules in the medium surrounding the cell membrane. Irving M. Klotz and his colleagues at Northwestern University have postulated that a decrease in the ordered structure of water makes it possible for hydrophobic parts of membrane-embedded proteins to become detached from their lipid environment and to become dispersed in an aqueous one. Thus by subjecting cells to a strong solution of potassium chloride for about 16 hours one can obtain a crude extract containing a variety of cell components, including intact membrane proteins in solubilized form. Since the protein markers of cell individuality represent only a minor fraction of the entire crude mixture, their extraction and purification present a challenging task.

One separation technique that has proved successful is polyacrylamide-gel electrophoresis, originally developed by Leonard Ornstein and Baruch J. Davis of the Mount Sinai School of Medicine. Its usefulness has been considerably enhanced by a computer program based on theories developed by Andreas Chrambach and David Rodbard of the National Institutes of Health. With this program we have been able to exploit subtle differences in the molecular size and net charge of various components in the crude mixture of solubilized surface materials so that the transplantation mark-

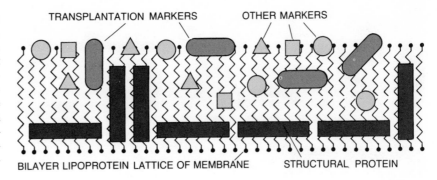

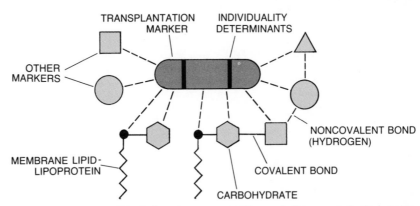

INDIVIDUALITY MARKERS that trigger a strong immunologic response in tissue transplantation are proteins (*dark color*) distributed in the surface membrane of virtually all nucleated cells. The black stripes represent determinants that vary from individual to individual. The membrane also contains other proteins that act as weaker surface markers. The matrix of the membrane is a bilayer of lipid molecules. The strong transplantation markers are bound to other membrane units (*bottom diagram*) largely through noncovalent bonds, which are more easily broken than the covalent bonds that hold molecules together.

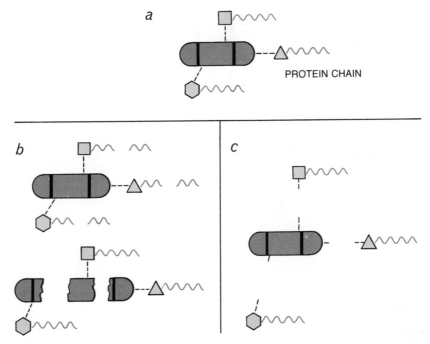

RELEASE OF MARKERS FROM CELL MEMBRANE can be accomplished by treating cells with proteolytic enzymes, which break covalent bonds, or by using milder methods that disrupt mainly noncovalent bonds. A marker is shown before treatment in *a*. Two possible results of treating cells with proteolytic enzymes are shown in *b*. In one case (*top*) the enzyme breaks the bonds of proteins loosely attached to the marker; in the other case (*bottom*) the enzyme breaks up the marker itself. The marker remains intact (*c*) if cells are subjected to low-energy sound or to solutions of certain simple salts, such as potassium chloride.

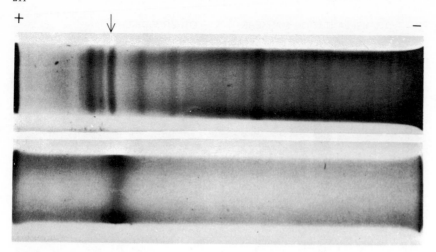

+   ↓   —

PURIFICATION OF MARKER ANTIGENS can be achieved by subjecting a crude membrane extract to polyacrylamide-gel electrophoresis. The separation exploits differences in size and charge affecting the rate at which proteins and other cell components move through a gel medium of defined pore size in response to an electric field. The image at the top shows the electrophoretic pattern of a crude extract containing a human transplantation marker (*arrow*) known as HL-*A* antigen. At the bottom the highly purified antigen isolated from a cultured human lymphoid cell line, designated RPMI 1788, is clearly separated.

ers separate clearly. In this way the markers carried by the cells of guinea pigs and the cells of humans have been extracted as homogeneous substances with high biological activity [*see illustration above*].

The purified markers turn out to belong to a family of proteins that differ from one another in their specific amino acid composition. It was exciting to find that individuality differences heretofore undetectable except by transplantation between two strains of guinea pigs or two unrelated human subjects could be predicted by specific differences in the amino acid composition between their isolated protein surface markers. Our findings supported earlier work by An-

drew A. Kandutsch of the Jackson Memorial Laboratory showing that the specificity of antigens arrayed on membrane fractions was destroyed by chemicals that attacked proteins. Thus antigenic activity resides entirely in membrane proteins and not, for example, in carbohydrates that might be associated with them. This was confirmed by Stanley Nathenson of the Albert Einstein College of Medicine in New York, who demonstrated that the carbohydrate fraction of solubilized *H*-2 antigens seems to be devoid of antigenic activity.

Transplantation markers are therefore simple proteins consisting of about 300 amino acid units strung together in a linear chain. Chemical studies show, however, that the markers of unrelated individuals have major differences in amino acid composition at five sites in the chain. As is well known from studies of the genetic code, a sequence of three bases in DNA, the genetic material, is needed to specify a particular amino acid in a protein molecule. In the case of at least four of the five variable amino acids a change in only a single base in one of the coding triplets is enough to account for the observed differences in the marker proteins.

In addition there are amino acid sequence differences involving the ordering of units in individual chains. When Michele Pellegrino in our laboratory subjected marker molecules to a proteolytic enzyme (trypsin) that attacked the covalent bonds adjacent to specific amino acids (lysine and arginine), he obtained 24 fragments, as determined by "fingerprinting" analysis [*see bottom illustration on this page*]. This is a technique in which a treated sample is placed in the corner of a square of filter paper; fragments are separated by chromatography in one direction and by electrophoresis in a direction at right angles to the first. The first separation reflects the rate at which fragments are carried across the paper by solvents; the second separation reflects the rate at which fragments move in an electric field. The resulting pattern provides a "fingerprint" of the protein.

The fingerprint of markers of two unrelated human subjects showed only six differences in the location of 24 fragments. This means that the amino acid composition of 18 of the fragments was the same in both subjects. The other six fragments exhibited amino acid differences. These findings suggest that transplantation markers are superimposed on a backbone of constant amino acid sequence but that interspersed along the

CHROMATOGRAPHY ⟶

X

—   ELECTROPHORESIS ⟶   +

FINGERPRINT OF HUMAN ANTIGEN is produced by a technique that combines electrophoresis with chromatography. The purified protein antigen isolated from cell line RPMI 1788 is treated with a proteolytic enzyme that breaks the molecular chain wherever it contains amino acid subunits of lysine or arginine. The chain is cleaved at 23 places, yielding a total of 24 fragments. The mixture is spotted at one corner (*X*) of a square of filter paper; the fragments are separated by chromatography in one direction and by electrophoresis in another. From such studies it appears that the amino acid differences in protein markers obtained from unrelated individuals are confined to six regions of the molecule.

backbone are small regions within at least some of which the amino acid sequence is varied to establish the organism's individuality.

Since transplantation markers are simple proteins, it should be possible in time to work out the complete base sequence in the structural gene that controls their synthesis. Direct gene-mapping is not possible for other strong individuality systems in mammals, such as blood-group substances, because their antigenicity depends on carbohydrates inserted in a supporting protein molecule by enzymes.

The limited variability of transplantation markers makes them much more tractable for detailed study of individuality than the immunoglobulins, the proteins that act as antibodies in blood serum, where they exhibit recognition and inactivation functions similar to those performed by cell-surface markers. The immunoglobulin molecule consists of a "constant" region whose amino acid composition varies among different individuals of the same species. This variation, known as allotypy, is not related to the molecule's recognition function in its role as an antibody. That difference is provided by the "variable" region of the molecule; each individual has the capacity to produce a vast number of different immunoglobulins (perhaps several hundred thousand) each containing a slightly different amino acid sequence in the variable region. It is not yet known how much of the far more limited variability detected in fingerprints of transplantation markers represents allotypy, or transplantation individuality differences, and how much represents functional variability.

The human marker studied by Pellegrino is known to be the product of the HL-A, or human leukocyte locus A, gene. The HL-A, gene product, which may be the primary determinant of whether a human graft will be tolerated or rejected, is now available for use in clinical transplantation.

The distribution of transplantation markers in the general population can be studied with the methods of "tissue-typing." Individuals can be immunized against foreign tissue in several ways: by skin grafting, by organ transplants or by the natural transplant represented by the fetus in pregnancy. Immunization produces antibody that reacts not only against the particular donor's cells but also against the cells of unrelated individuals. To perform the typing, blood serum containing antibody

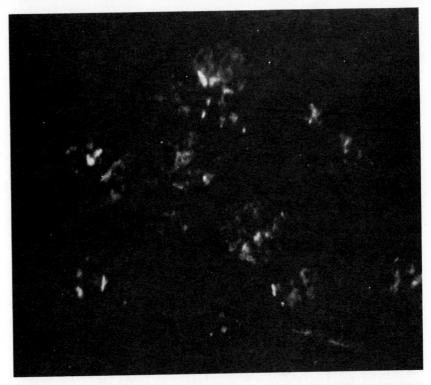

LOCATION OF INDIVIDUALITY MARKERS, the cellular proteins that help to establish each organism's identity, can be visualized by fluorescent microscopy. Target lymphocytes bearing the HL-A, or "strong," antigenic markers of one individual were first reacted with specific HL-A antibody from another individual. The resulting antigen-antibody complex on the lymphocyte surface was made visible by adding a fluorescein-labeled antibody to human gamma globulin produced in a goat. The micrograph, prepared in the authors' laboratory at the Scripps Clinic and Research Foundation by P. G. Natali, shows that the antibodies recognize and affix themselves to specific antigenic sites on the cell's surface.

is taken from the host and is mixed with white blood cells taken from another person [see illustration on next page]. Everyone whose cells are killed by the serum shares a factor foreign to the host and thus forms a tissue type.

Tissue-typing has been intensively studied during the past decade, but it is still in a period of flux. It has been found that human hosts can respond to multiple antigens in the challenge graft and not solely those related to transplantation. Furthermore, the reaction of a given individual against a set of determinants is highly complex. The reaction depends on the individual's capacity to respond to a given amount of marker, on his previous exposure to related materials that may have shared the same determinants, to say nothing of his own genetic constitution. As a result the superimposition of the individuality of immune responses on the vast genetic variety in transplantation markers and other markers yields a bewildering array of results, restricting the present value of typing in predicting the best unrelated donor for a given patient in need, say, of a kidney transplant.

An important advance in typing has recently come out of Ferrone's work in our laboratory. Ferrone finds that when solubilized human antigens (individuality markers released from cell membranes) are administered to rabbits, the animals' immune system can recognize individual differences it cannot discern when it is confronted with whole human cells. The explanation is that intact cells carry species-specific markers that are much stronger than the markers that vary from individual to individual. The purified soluble material evidently does not possess the species markers. As a result the rabbits respond to the individuality markers rather as another human being would. Studies involving many rabbits have shown that some animals respond to only a limited number of leukocyte locus A markers, and in some cases to only a single marker. These studies promise to improve tissue-typing and so lead to more successful organ-matching.

Other potential clinical applications of solubilized transplantation markers will be investigated in our transplantation unit at Northwestern University. The markers may provide an index to measure the patient's response to grafts so that the surgeon can regulate the amount of drugs necessary to suppress the patient's immune response and thereby enable the graft to survive. Another approach, which is now being tested in animals, is to pretreat the host with purified solubilized individuality markers obtained from the prospective donor. The host normally confronts such markers on the donor's intact cell surfaces; there is evidence that his immune system may be confounded and respond less vigorously or erratically when the markers are first presented in solubilized form. Under the circumstances his body may accept the markers as his own and fail to react at all, thus exhibiting immunological tolerance. He may develop a response that actually protects the subsequent graft rather than destroying it, a response termed immunological enhancement. Or his response may be an impotent "deviant" reaction that neither destroys the graft nor protects it. Whichever the case, experiments with animals demonstrate that the survival of a graft is significantly prolonged by pretreating the host with a solubilized individuality marker. Immunotherapy with these substances may well provide a major step in controlling the rejection of tissue transplants and break the barrier that has so far limited the clinical transplantation of organs.

As we have observed, individuality markers are present on virtually all nucleated cells and hence must play an essential role in the cell's economy. Since they spread freely across cell surfaces, it seems unlikely that they are involved in the cell structure itself. Conceivably they help to regulate the permeability of the cell membrane. Another possibility is that they may assist cells of the same type to form aggregates. Thus they may facilitate the adhesion of cells in the architecture of tissues or enable cells to exchange information. These functions might be a property of the structurally constant regions of molecules that also possess localized regions of individuality. It seems reasonable to suppose natural selection would have endowed marker molecules with one or more functions in addition to providing an identification label unique for each organism.

One puzzling observation is that cells from individuals never previously exposed to foreign markers in grafts often act as if they had encountered the markers before. H. Sherwood Lawrence of the New York University School of Medicine has suggested that such individuals may have encountered the markers, or close copies of them, in molecules carried by bacteria or viruses. This idea is supported by the fact that grafts, like intracellular bacterial and viral parasites, are destroyed by a cellular immunological mechanism. It seems entirely possible that each person is characterized not only by his innate individuality markers but also by an entire menagerie of infectious agents to which he has been exposed and whose markers he carries around throughout his life. This suggests in turn that a person's own markers may either help to protect him from certain disease processes or increase his susceptibility to them. In other words, in order to attack a cell successfully a bacterium or a virus might have to play a molecular game of wits with the individuality markers and with the immune potentials of the host that stand in its way. The hypothesis is supported by the observation that certain anti-HL-A antibodies that block leukocyte locus A marker sites also interfere with the infectivity of viral agents, thus suggesting that the agent shares the determinants.

There is evidence that various diseases are associated with leukocyte locus A factors, indicating that individuality markers are indeed related to the inception, development and pathogenic reaction to disease. On the other hand, the host's life might be prolonged if he were fortunate enough to harbor a para-

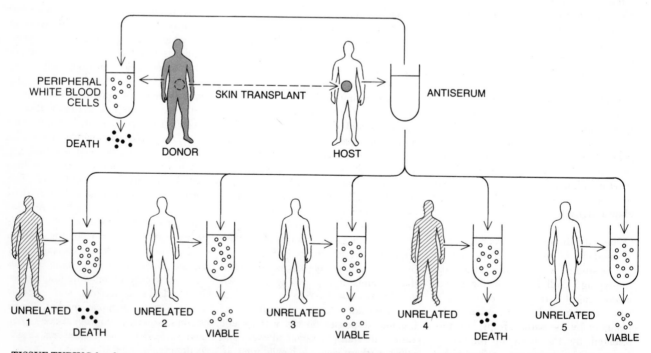

TISSUE-TYPING has been used to study the distribution of transplantation markers in the general population. An individual ("host," top) who has been immunized against an unrelated person ("donor"), for example by a skin graft, produces antiserum, a serum that contains antibody. The antiserum will destroy not only the donor's white blood cells but also the white cells of other unrelated individuals (1, 4). All the individuals whose cells are so destroyed are considered to belong to a common tissue type. Tissue-typing has been helpful in selecting related donors for organ transplants but is only modestly useful with unrelated individuals.

site that supplied markers he lacked. For example, it has been reported that leukemia and Burkitt's lymphoma have regressed after a patient had contracted measles. One might say that the perfect parasite is one that succeeds in lengthening the life of its host—but then should it still be called a parasite?

The individuality-marker system may protect the species against viral infections in another way. When some viruses that contain RNA rather than DNA as their genetic material infect a cell, they form buds and carry a portion of the cell's surface membrane along with them. If the cell's individuality markers thereby end up in the coat of the virus, they could elicit a "graft" rejection when the virus attacks another unrelated individual of the same species as the original host. The possibility that transplantation individuality markers can determine the host's response to viruses is potentially important in clinical diagnosis, prophylaxis and immunotherapy.

Recently Burnet has elaborated on a suggestion made earlier by Lewis Thomas of the Yale University School of Medicine to construct a comprehensive hypothesis of an immune surveillance system. Early in evolution primitive organisms developed a class of wandering cells capable of recognizing and destroying foreign cells or parasites. This mechanism, Burnet points out, would be able to act not only against agents of external origin but also against mutated cells of the host organism that exhibited new individuality markers. This would tend to make the individual resistant to new growths, that is, tumors and cancers in general. Functional surface markers on the host's own cells would serve as local recognition sites for the wandering surveillance cells; in addition markers released from cell membranes in the normal course of cell repair and replacement would telegraph subcellular messages to draining lymph nodes, keeping them informed of their constituency. Natural selection would favor such a system because it enhances the viability of the organism. Burnet suggests that the aging process involves a depletion in the number and vigilance of the "wanderers," allowing such abnormalities as cancer and immune responses to the host's own tissue to develop. Therefore markers of biological individuality provide an excellent tool for probing the most intimate functions of the cell. Knowledge so obtained cannot fail to be useful in meeting the challenges of human transplantation and disease.

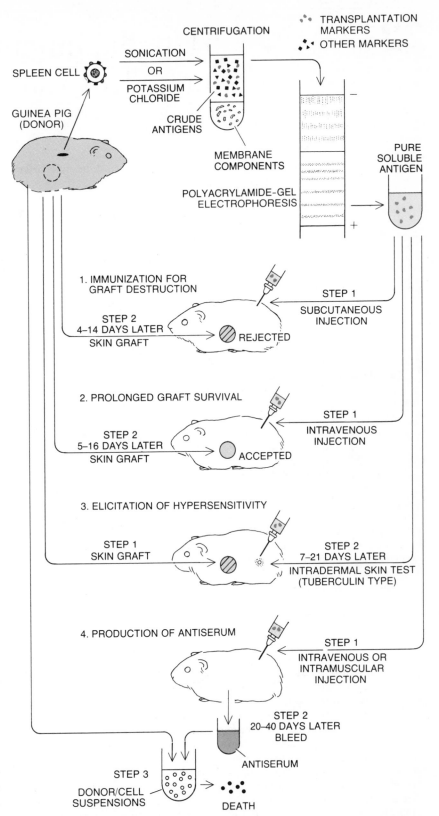

RESPONSE TO SOLUBLE INDIVIDUALITY MARKERS can take a variety of forms. Here pure markers, or antigens, obtained from guinea pig spleen cells are administered to four unrelated individuals. Subcutaneous injection of the antigens (1) will normally immunize an animal against a subsequent skin graft from the animal that supplied the antigens. On the other hand, an intravenous injection of antigens (2) will often prolong the survival of a skin graft made from five to 16 days later. One can tell if an animal is hypersensitive to the donor's tissue (3) by applying antigens from the donor in the form of an intradermal injection from one to three weeks after the skin graft. Finally, one can produce antiserum by injecting the antigens subcutaneously in an unrelated host (4). After a few weeks the host's serum will contain enough antibodies to destroy the donor's white cells.

22

How the Immune Response to a Virus Can Cause Disease

by Abner Louis Notkins and Hilary Koprowski
January 1973

*The body's defense mechanism may not always
be beneficial. In many cases the very process
that should combat a virus is itself a cause of
the damage associated with a viral disease*

In biology as in physics it is a truism that the deeper one goes in exploring elementary questions, the more one encounters paradoxical and puzzling phenomena. Such is the case in the investigation of the well-known animal defense mechanism called the immune response, whereby the body fights off infections and other invasions by foreign matter. The classic concept of this mechanism is quite simple: in response to invasion by the foreign substance, or antigen, the body produces specific antibodies that bind to it, thus neutralizing the invader so that it does not harm the invaded organism. Investigators are now learning, however, that this is far from the entire story, that the mechanism of immunity is much more complex than had been supposed.

This is true in particular in virus infections. In the simple case of direct attack by a virus (for example in poliomyelitis) the virus invades a cell and uses the cell's material to replicate, and soon the new crop of viruses bursts the cell and emerges to go on to infect other cells. The timely appearance of antibodies may prevent the spread of infection and the appearance of symptoms. In infections caused by other viruses, however, there is growing evidence that the cells are damaged not directly by the replicating virus but by a specific immune response that produces the symptoms of the disease. The complexities of the immune response to viruses are under exploration in a number of laboratories, including our own at the National Institutes of Health (Notkins) and at the Wistar Institute of Anatomy and Biology (Koprowski). Gradually an account of the immunity mechanism's diverse operations is being pieced together, and

what follows is a review of the emerging picture.

That the immunity system might sometimes be responsible for injurious effects was first suggested more than 60 years ago by Clemens von Pirquet, an Austrian pediatrician who was at one time also a professor at the Johns Hopkins School of Medicine. Von Pirquet noted that in "serum sickness," a disease that can follow injection of foreign blood serum, the patient's blood contained foreign proteins and antibody against them. He speculated that the combination of antibody with foreign proteins (antigen) perhaps produced a toxic substance that gave rise to the symptoms of the sickness: hives, rash, pain in the joints, shortness of breath and, in severe cases, death. He also conjectured that an interaction of antibodies with the viruses of such diseases as smallpox and measles might cause the skin eruptions characteristic of these diseases.

Von Pirquet's speculations that immune response to viruses might cause disease were not followed up at the time, but in the 1950's Wallace P. Rowe, a virus investigator at the National Institutes of Health, came on proof of the hypothesis in an ingenious series of experiments. Rowe was studying the pathology produced by a virus known as lymphocytic choriomeningitis (LCM), which infects rodents and occasionally

man and causes an inflammation of the membranes surrounding the brain (meningitis). He observed that although in infected mice the virus multiplied rapidly in many organs, the animals at first showed no sign of illness. On the sixth day, however, after the mice had begun to show an immune response to the virus, they developed meningitis and died. Was the disease caused by their immunological response to the virus rather than by the virus itself? Reasoning that if he inhibited the immune response, he might be able to prevent the disease, Rowe treated mice with X rays in doses known to suppress the immune response. He then infected both the treated mice and untreated control mice with LCM virus. The irradiated animals did not develop meningitis, although the virus replicated in their tissues just as rapidly as it did in the control mice, which died.

Later a group of investigators at the Johns Hopkins School of Hygiene and Public Health (Donald H. Gilden, Gerald A. Cole, Andrew A. Monjan and Neal Nathanson) took Rowe's experiments a step further. It was known by this time that the immune system responds to foreign substances in at least two ways, one mediated by antibody and the other mediated by a specific group of the cells known as lymphocytes. These "immune lymphocytes" recognize antigens on the surface of foreign cells

IMMUNE COMPLEXES are formed when an antiviral antibody combines with a virus and binds complement. The complexes are detected by a technique in which an antibody to complement is labeled with a fluorescent substance and incubated with tissue; if complexes are present, the antibody binds to them and fluoresces under ultraviolet radiation. The photomicrograph on the opposite page, made by David D. Porter of the University of California Center for Health Sciences, demonstrates the presence of complexes in the kidneys of mink that were infected with Aleutian virus and developed glomerulonephritis.

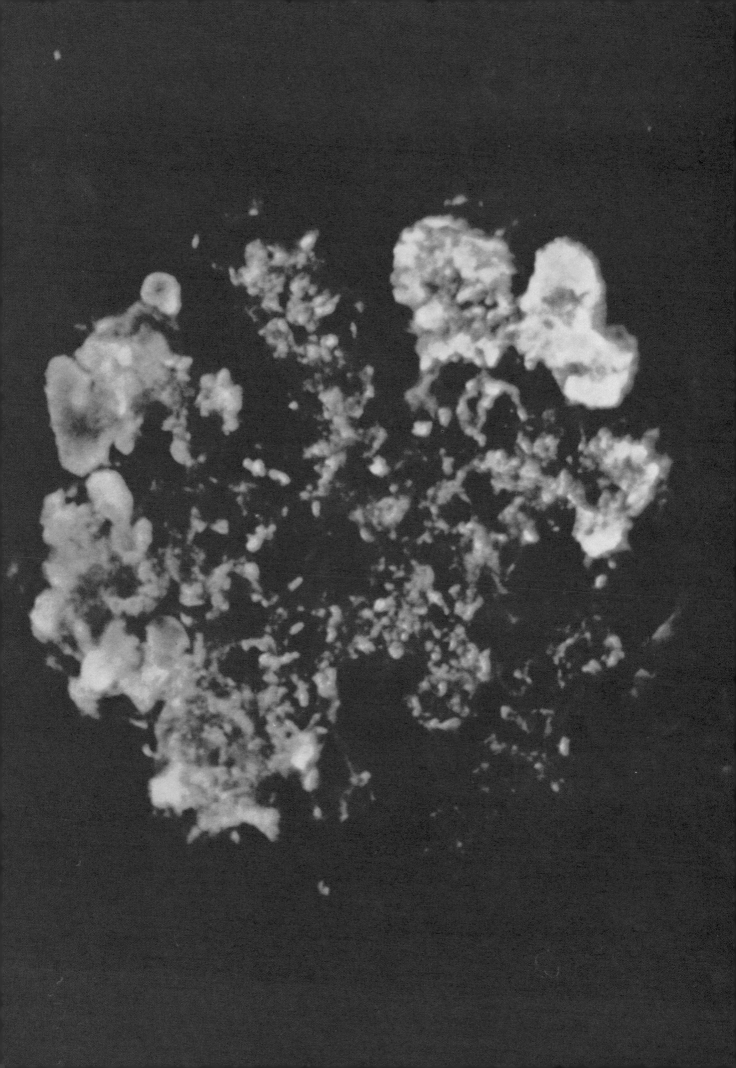

and thereby destroy tissue such as tumors or skin grafts [see the article "Markers of Biological Individuality," by Ralph A. Reisfeld and Barry D. Kahan, beginning on page 208]. Which of these factors was responsible for causing LCM disease in Rowe's experiments: antibodies or lymphocytes? The Johns Hopkins group used drugs to suppress the immunological response in mice, infected them with LCM virus and then divided the animals into three groups. One group received injections of anti-LCM antibody, the second was given anti-LCM lymphocytes and the third normal lymphocytes. The animals receiving the antibody or normal lymphocytes remained well but those given the immune lymphocytes developed the symptoms of LCM disease and died. Evidently in the case of LCM it was the combination of immune lymphocytes and the virus that produced the disease.

Extending their observations, the Johns Hopkins group found that in young rats LCM infection was not fatal but did damage the cerebellum, causing ataxia (inability to coordinate body movements). If the immune response was suppressed at the time of infection, however, the animals remained free of symptoms and cerebellar damage did not occur, even though the virus continued to replicate in the brain. As in the case of mice, development of the rats' disease was thus shown to be immunologically mediated. An interesting suggestion from these experiments is that perhaps other neurological disorders may arise from the immune response to viruses.

It was now time to look into the reasons why lymphocytes destroyed infected cells when the virus itself did not. In order to study this problem Duard L. Walker and his co-workers at the University of Wisconsin Medical School turned from experiments in animals to experiments in tissue culture. It was known that on infecting a cell some viruses induce the formation of viral antigens on the cell's surface. Walker reasoned that if these antigens are recognized by lymphocytes from animals immunized with the same virus, the lymphocytes might attack and destroy cells carrying the label of infection. To test this hypothesis he infected tissue-culture cells with mumps virus, which induced new antigens on the surface of the cells but did not destroy them. When he introduced into the infected cultures lymphocytes taken from animals that had been immunized with that virus, the lymphocytes did indeed destroy the infected cells. On the other hand, lymphocytes from animals that had not been immunized with the mumps virus did not attack the infected cells.

Other investigators soon obtained the same kind of result in tissue-culture experiments with LCM virus and the measles virus. A number of groups are now looking into the possibility that the interaction between immune lymphocytes and viral antigens formed on the surface of infected cells may account for some

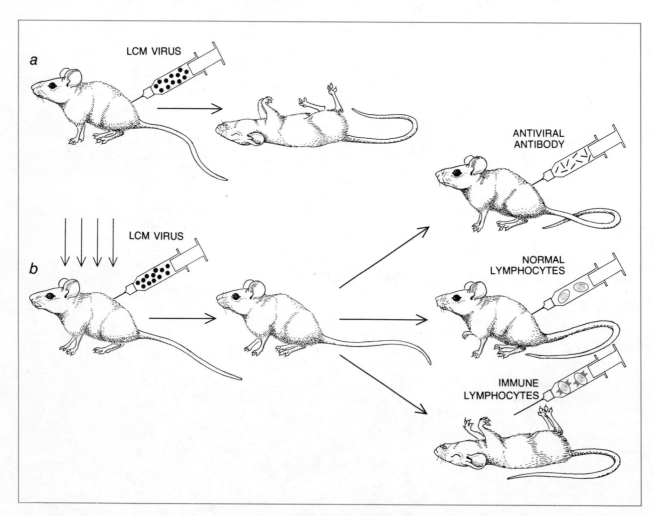

LCM VIRUS kills adult mice (a). If the immune response is suppressed by radiation or drugs (b), the mouse lives but develops a chronic infection. When immunological competence is restored by injecting lymphocytes from other animals immunized with LCM, test animal dies. Injection of anti-LCM antibody or of normal lymphocytes rather than anti-LCM lymphocytes does not cause death.

of the symptoms associated with viral diseases of man, including hepatitis.

Gary Rosenberg and Paul Farber in Notkins' laboratory at the National Institutes of Health undertook an even more detailed analysis of the behavior of lymphocytes in response to viral antigens. They used herpes simplex virus (HSV), which produces the familiar cold sores in man. They found that when lymphocytes from animals immunized with this virus were incubated in a test tube with HSV antigen, the antigen "turned on" the lymphocytes to replicate their DNA and divide. This reaction began within hours after exposure to the antigen and was quite specific: the anti-HSV lymphocytes were turned on only by the HSV antigen; they did not react at all to antigens of other viruses. In a follow-up study at the Wistar Institute with the rabies virus, Tadeusz J. Wiktor and Koprowski found that lymphocytes from rabbits immunized with that virus could be turned on not only by the complete virus but also by its subunits.

Further research showed that when lymphocytes are stimulated by exposure to viral antigens, they release potent chemical messengers, or mediators, that exhibit a variety of biological properties and are thought to be responsible for some of the inflammatory change and tissue injury associated with many viral infections. One of these mediators is known to attract inflammatory (white) cells and another can keep the inflammatory cells at the site of the infection. A third mediator, lymphotoxin, can destroy uninfected as well as infected cells, and a fourth is the now well-known substance interferon, which can inhibit the replication of viruses. Very likely a number of other mediators will be found to be released by the interaction of viruses with lymphocytes; mediators with at least a dozen different biological properties have been discovered in cultures of lymphocytes that are turned on by nonviral antigens.

If lymphocytes act as agents of tissue destruction and disease, might not antibodies also perform such a role? Mario Fernandas, Wiktor and Ernest Kuwert in Koprowski's laboratory began to explore the antibody phase of the immunity phenomenon. It had been known for some time that the attachment of antibody to antigens on the surface of cells could activate a group of proteins in the serum, known as complement, to break down cells. It was also known that under certain circumstances rabies virus could induce new antigens on the surface of tissue-culture cells without

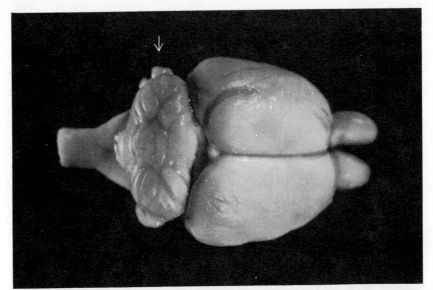

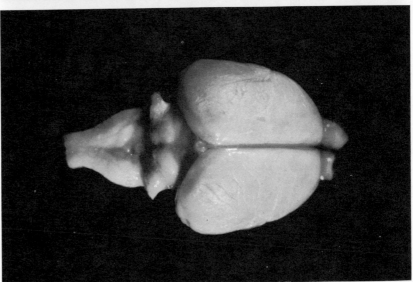

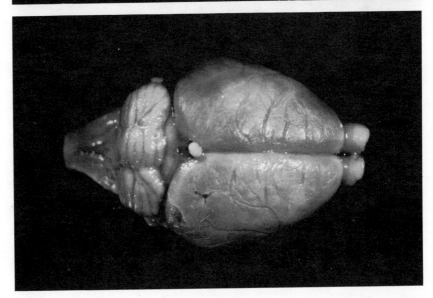

RAT BRAINS show the effects of an immune response to LCM infection demonstrated by Andrew A. Monjan, Gerald A. Cole and Neal Nathanson of the Johns Hopkins School of Hygiene and Public Health. The cerebellum is at the base of the brain (*top*). LCM infection produced severe cerebellar damage (*middle*), impairing the ability to coordinate movements. If the cellular immune response was suppressed, there was no brain damage (*bottom*).

destroying them. The Wistar Institute investigators added antirabies antibody or complement or both to the infected cultures. Neither the antibody nor the complement alone was injurious to the cells, but when they were added together the combination destroyed the rabies-infected cells.

Recent experiments suggest that antibody and complement also may contribute to the breakdown of cells infected with some of the well-known tissue-destroying viruses. Charles Wohlenberg, Arnold Brier and Joel Rosenthal of the National Institutes of Health laboratory showed that viral antigens on the surface

of cells infected with influenza, measles, vaccinia and HSV make the cells vulnerable to destruction by specific antibody and complement long before these cells break down as the direct result of viral replication. It seems highly likely, therefore, that in the body the symptoms and other effects of these diseases are produced by a collaboration between the immunological process and the virus.

How does the interaction of complement with antibody bring about tissue damage and inflammation in the infected animal? Ralph Snyderman of the National Institutes of Health laboratory conducted an experiment that suggested

a likely answer. He found that when complement and antibody to HSV were added to a culture containing cells infected with that virus, a mediator was released from one of the components of complement. This mediator was identified as one that had previously been shown to increase the permeability of blood vessels and to attract white cells. Although white cells are known to be important in the defense against certain infections, they contain potent tissue-destroying enzymes and so they may also act as agents of cell destruction in some viral infections. In fact, several groups of investigators have found that the injection of specific antibody into virus-infected animals has the effect of increasing the number of white cells and the amount of tissue damage in the infected organs. Now it appears that the sometimes fatal shock syndrome associated with dengue fever, a viral disease in Southeast Asia, may be mediated by antibody and complement. Scott B. Halstead of the University of Hawaii Medical School and Philip K. Russell of the Walter Reed Army Institute of Research, who originally proposed the idea, believe mediators released from complement may be one of the factors responsible for the increase in permeability of blood vessels and the consequent shock that marks this disease. Several workers are now attempting to gather proof for this hypothesis.

U p to this point we have discussed primarily mechanisms by which the immune response to viral antigens on the surface of infected cells can cause tissue injury. Von Pirquet's studies in the early 1900's on serum sickness suggested a different mechanism, one involving the combination of antigen and antibody. Again, it was not until the 1950's that firm evidence began to come into view. Frederick G. Germuth, Jr., and his co-workers at the Johns Hopkins Hospital and Frank J. Dixon and his colleagues at the University of Pittsburgh School of Medicine found that they could produce serum sickness in rabbits by injecting combinations, prepared in the test tube, of foreign protein and the antibody to it. They also ascertained that some of these injected complexes, circulating in the rabbits' blood, became trapped in the capillaries of the kidneys and led to inflammation, loss of kidney functions and symptoms characteristic of the human disease known as glomerulonephritis.

In the light of the evidence that the kidney disease was caused by an antigen-antibody complex, it was called "immune complex" disease. Although

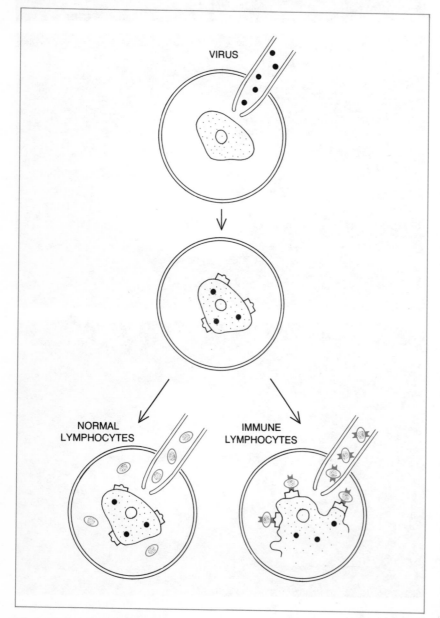

VIRUS-INFECTED CELLS are destroyed by an immune response as shown here. The infecting virus induces the formation of new antigens on the cell surface. Addition of lymphocytes to a culture of infected cells has no effect. Lymphocytes from an animal immunized with the same virus recognize the viral antigen, however, and cell destruction results.

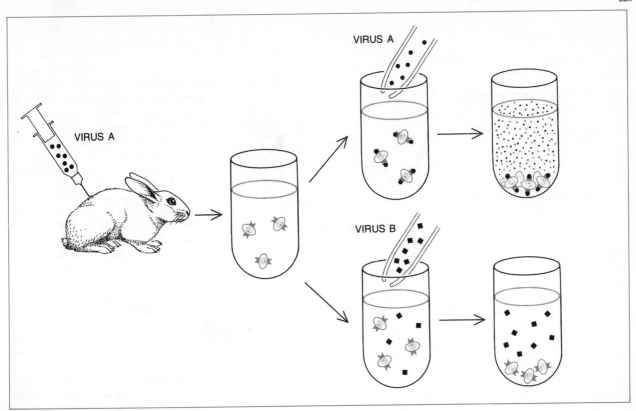

SPECIFICITY of the effect of immune lymphocytes was demonstrated by immunizing rabbits with a virus and then exposing their lymphocytes to the same virus and to another virus. The lymphocytes exposed to the same virus were stimulated to synthesize new DNA and to divide; the others were not. The stimulated lymphocytes produce mediators, some of which destroy normal tissue.

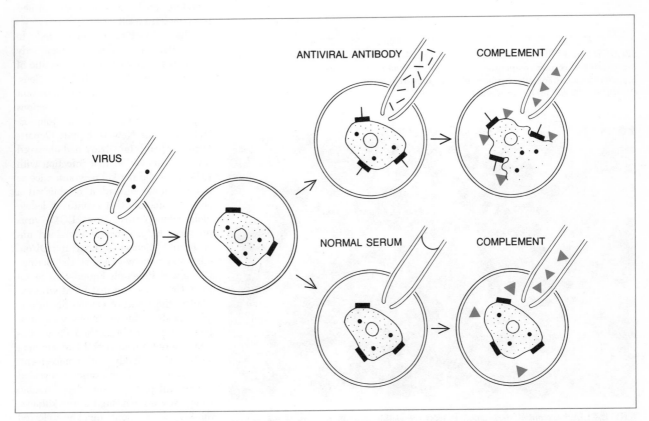

ANTIBODY-MEDIATED IMMUNE RESPONSE is demonstrated in a tissue culture. In the first step antiviral antibody recognizes and binds to virus-induced antigens on the surface of infected cells. Then complement interacts with the newly formed antigen-antibody complex and the cells are destroyed. Neither the antiviral antibody nor the complement alone destroys the infected cells.

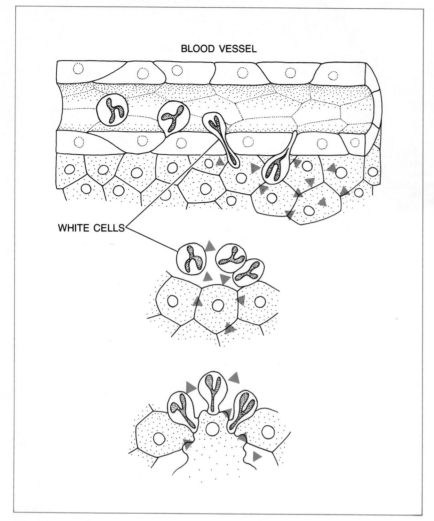

SECONDARY EFFECT of the interaction of complement with antigen-antibody complexes is activation of certain components of complement (*color*) that increase the permeability of blood-vessel walls and attract white cells. Inflammation results, and enzymes released from the white cells can injure uninfected tissue, contributing to the symptoms of the infection.

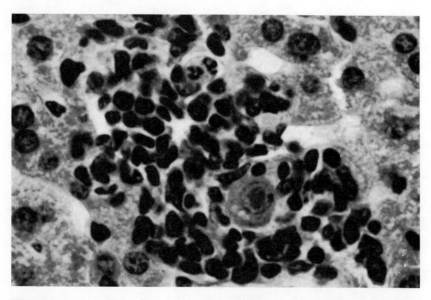

WHITE CELLS (*rounded, dark-staining cells*) are visible in a photograph of mouse-liver tissue made by Donald Henson of the National Institutes of Health. They were attracted to the site of a virus-induced lesion, presumably in the manner diagrammed at the top of the page: the large cell at the center of the clustered cells was infected with cytomegalovirus.

the immune-complex syndrome could readily be produced in animals in the laboratory, the fact that it could also occur in response to viral infections under natural conditions was not demonstrated until a decade later. Virus-antibody complexes were not easy to identify, and the problem became much more manageable when a new technique for doing so was developed.

For a number of years the National Institutes of Health laboratory had been studying an unusual virus called lactic dehydrogenase virus (LDV) because it elevates the level of that enzyme in the blood. Inoculation of mice with the virus produced large amounts of infectious virus in the blood and a chronic infection without any indication of an immune response. It was supposed that the animals were unable to make antibody against the virus, a situation known as immunological tolerance. The laboratory devised a highly sensitive technique for detecting virus-antibody combinations, however, and with this technique [*see top illustration on opposite page*] discovered that the infected mice were indeed making antibody to LDV. The reason it had not been recognized before was that although antibody had combined with the virus, the resulting complex remained infectious and therefore could not be distinguished from the virus itself.

Other investigators proceeded to show that infectious virus-antibody complexes were in fact characteristic of several of the chronic viral infections. What is more, it soon became apparent that some of these chronic infections ended in glomerulonephritis. John E. Hotchin of the New York State Department of Health in Albany had observed several years earlier that infection with LCM virus did not kill newborn mice as it did adults; instead it established a chronic infection that eventually led to glomerulonephritis. How LCM virus produces kidney disease remained unclear until Michael Oldstone and Dixon, who was now working at the Scripps Clinic and Research Foundation in La Jolla, Calif., showed that LCM virus existed in the blood of chronically infected animals as an infectious virus-antibody complex and that the kidneys contained large amounts of LCM antigen, anti-LCM antibody and complement. Moreover, microscopic studies revealed the typical pattern seen when immune complexes are deposited in the kidneys. Similar observations quickly followed with other chronic viral infections, in some of which the inflammatory changes were not confined to kidneys but ap-

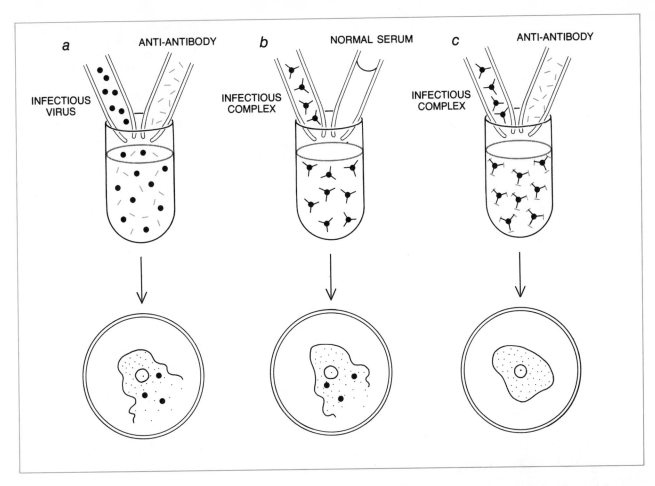

INFECTIOUS VIRUS-ANTIBODY COMPLEXES can be detected by a technique utilizing an antibody that recognizes an antiviral antibody as an antigen: an "anti-antibody." When the anti-antibody is incubated with untreated virus, the virus retains its infectivity (*a*). When virus with antiviral antibody on its surface (infectious virus-antibody complex) is treated with normal serum, the complex remains infectious (*b*). When the infectious complex is treated with the anti-antibody, however, the latter attaches itself to the antibody of the complex, neutralizing the complex (*c*). Thus the test distinguishes between virus and infectious complex.

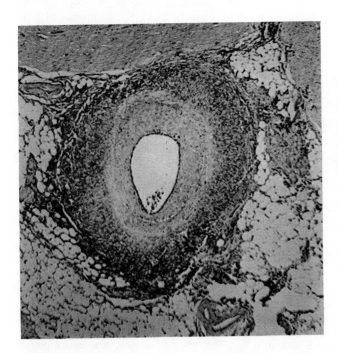

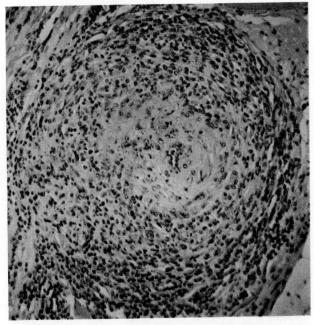

ARTERITIS occurs in some viral immune-complex diseases. Photomicrographs of the coronary artery of mink infected with Aleutian virus, made by Porter, show mild arteritis (*left*) and, at higher magnification, severe arteritis with infiltration of white cells (*small dark-staining cells*) and obstruction of the vessel (*right*). Damaged arteries contained viral antigen, antibody and complement.

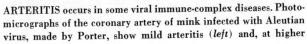

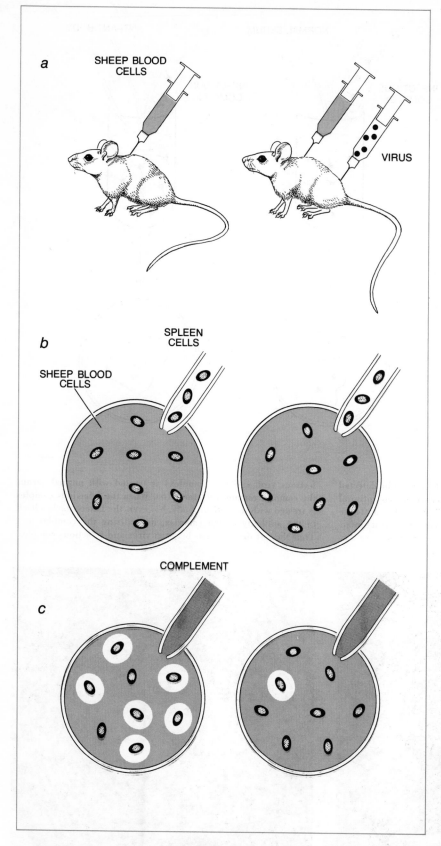

a SHEEP BLOOD CELLS VIRUS

b SPLEEN CELLS SHEEP BLOOD CELLS

COMPLEMENT

c

INFECTION of mice with leukemia virus depressed their ability to make antibody. Mice, some of which had been infected with virus, were immunized with sheep red blood cells (*a*). Spleen cells from the mice were spread over layers of sheep red blood cells in laboratory dishes (*b*) and complement was added (*c*). Only spleen cells in which antibody to the sheep cells had been induced were able, in the presence of complement, to destroy surrounding sheep cells. Counting the patches of dead sheep cells (*white*) showed spleens of infected mice contained many fewer antibody-producing cells than spleens of healthy mice.

peared in blood vessels and organs in other parts of the body.

The precise steps involved in the production of immune-complex disease by virus-antibody complexes are still only partly understood. Presumably when the complexes are trapped in the kidneys or on blood-vessel walls, they activate the components of complement; mediators are generated, inflammation results and the tissue-injuring enzymes are released from white cells. There are still many unanswered questions, however. Why does LCM virus produce a severe glomerulonephritis and LDV only a mild one? Why do some animals develop glomerulonephritis when they are exposed to a given virus whereas other animals do not? Are there genetic factors governing susceptibility to the virus or the immunological response to it?

Be that as it may, thousands of people develop glomerulonephritis each year, and so the finding that virus-antibody complexes produce the disease in animals has intensified research interest in such complexes and their possible involvement in human disease. Recently evidence has begun to accumulate that in man the hepatitis virus circulates in the blood as an immune complex and that some of the manifestations of this disease, including the associated high incidence of arthritis, may be due to these immune complexes. If it turns out that the immune response to viruses is actually responsible for glomerulonephritis or other immune-complex diseases in man, then controlling the adverse effects of the immune response may be essential for therapy. In animals many manifestations of these chronic infections, including glomerulonephritis, can be prevented or reduced by drugs that suppress the immune response.

The possibility that viruses and the immune response may also be involved in autoimmune diseases such as rheumatoid arthritis and lupus erythematosus is being investigated in a number of laboratories. An autoimmune disease is one in which the body treats its own tissue as an alien antigen and produces antibodies that attack the tissue. What causes the host suddenly to turn against its own tissues? It has long been suspected that viral infections may be a triggering factor. Several hypotheses about how a virus might bring the immunological system into play against the host's own cells have been suggested. The viral infection may unmask or release a potential antigen that normally is hidden within the cells, out of contact with the immune system. Or a viral antigen may combine with an indigenous protein on the cell's surface and thus form a new "foreign" substance. Another possibility is that a viral infection may activate genes in the cell whose information is ordinarily repressed, thereby causing them to begin producing "new" substances that act as antigens. Wanda Baranska and Wojciech Sawicki at the Wistar Institute found support for this idea in experiments with mouse ova and embryos. They showed that an antigen that was present in the animals' earliest embryonic stage could not be detected in adults, but that when adult cells were transformed into tumor cells by the virus known as SV-40, the embryonic antigen reappeared on the cell surface. Apparently in these transformed mouse cells the genes were again able to redirect formation of the "embryonic antigen."

Still another possibility is that a viral infection may cause the cells of the immune system to behave abnormally and so produce antibodies against some of the host's own tissue. Although proof for this hypothesis is still lacking, there is considerable evidence, particularly from animal studies, that certain viral infections depress the function of the cells of the immune system. Again, it was von Pirquet who first observed that the reactivity to the tuberculin test (an immune response) was depressed in patients infected with measles virus. The effect of viruses on immune function received little attention, however, until 1963, when Robert A. Good and his colleagues at the University of Minnesota College of Medical Sciences showed that mice infected with a leukemia-producing virus were markedly depressed in their ability to make antibody against foreign substances. Other studies, notably by Walter Ceglowski and Herman Friedman at the Albert Einstein Medical Center in Philadelphia, showed that not only was the amount of antibody in the blood reduced but also the number of cells capable of making antibody was curtailed by as much as 99 percent [see illustration on opposite page]. Moreover, the immune response was depressed within a few days after infection, long before the animals developed visible signs of leukemia.

It soon became apparent that non-leukemia-producing viruses also could impair immune function. Richard J. Howard and Stephan E. Mergenhagen of the National Institutes of Health laboratory showed this in tests of the reactivity of LDV-infected mice to foreign skin grafts. These animals rejected grafts at a lower rate than uninfected animals. Moreover, there is evidence that the re-

MECHANISMS	PATHOLOGY

1 VIRUS-INDUCED ANTIGENS ON CELL SURFACE:
   a INTERACTION WITH IMMUNE LYMPHOCYTES
   b INTERACTION WITH ANTIVIRAL
     ANTIBODY AND COMPLEMENT
	DESTRUCTION
	OF INFECTED CELLS

2 ACTIVATION OF MEDIATORS FROM IMMUNE LYMPHOCYTES OR COMPONENTS OF COMPLEMENT	INFLAMMATION, ALLERGIC REACTIONS, DESTRUCTION OF CELLS
3 FORMATION OF CIRCULATING VIRUS-ANTIBODY COMPLEXES	IMMUNE-COMPLEX DISEASE
4 IMMUNE RESPONSE TO HOST-CELL ANTIGENS ALTERED OR DEREPRESSED BY VIRUS	AUTOIMMUNE REACTIONS
5 INFECTION OF CELLS OF IMMUNE SYSTEM	INHIBITION OR ENHANCEMENT OF IMMUNE FUNCTION

MECHANISMS of the various immune-response disorders that are caused by viral infection and are discussed in the article are summarized, together with the associated pathology.

jection of transplanted tumors also is slowed by viral infection of the immune system. In fact, it seems possible that the virus-induced depression of antibody-mediated and lymphocyte-mediated immunity may be a factor in the initiation and development of tumors and also may account for the chronic nature of certain viral infections.

A curious twist in the already complicated story of immunity and viruses is the finding that lymphocytes, which are the body's major defense against tumors, can actually act as agents for the induction of tumors. It has been known for some time that an unusually high incidence of lymphomas (tumors of the lymphoid glands) occurs in animals or patients undergoing chronic stimulation of these glands as a result of autoimmune disorders or rejection of a graft such as a kidney transplant. It is also known that certain viruses, including the leukemia virus of mice and the mononucleosis virus of man, exist in lymphoid cells in a "latent" state. On the basis of these observations, groups headed by Martin S. Hirsch and Paul H. Black of the Harvard Medical School and Robert S. Schwartz of the Tufts University School of Medicine conducted experiments to see if stimulation of the lymph-

oid elements of the immune system might arouse the latent virus. In order to stimulate the immune system they exposed mice to foreign grafts, and they found that this activated the leukemia virus that had previously been latent in the lymphoid cells. The findings suggest that such activation of leukemia virus from immunologically stimulated lymphocytes may be responsible for the high incidence of lymphomas associated with autoimmune disorders and graft rejection.

All in all, it is now obvious that the interrelations encompassing viruses, immunity and disease are indeed complex. It appears that the immune response to viral infection can have both beneficial and deleterious effects on the host. On the one hand, it may be the chief or only weapon against the infection; on the other hand, it may be responsible for some of the noxious symptoms and even the fatal effects of the disease. Probably the immune response makes some contribution, large or small, to the pathologic picture in most viral infections. Although we must recognize that the immune system is not an unmixed blessing, it is encouraging to know that by learning more about it we may eventually find new approaches to the treatment of viral diseases.

AGING

INTRODUCTION

Old age is not an illness; it is a continuation of life with decreasing capacities for adaptation." Frederic Verzar's view of aging in terms of a progressive failure of the body's various homeostatic adaptive responses has gained wide acceptance only recently, for there had been a strong tendency to confuse what we now recognize as a distinct aging process with those diseases frequently associated with aging. But, as Alexander Leaf points out in "Getting Old" and as we have emphasized in this anthology, these diseases—notably atherosclerosis and cancer—are not necessary accompaniments of old age, even though their incidence does increase with age; rather, they seem to interact with the aging process in a positive feedback cycle, each accelerating the other. For example, one theory of aging focuses on the fact that the collagen molecules of connective tissue develop increased cross-linkages with age, and that the resulting rigidity decreases tissue functioning. It is not difficult to imagine how this aging change in collagen, were it to occur in the large arteries (which contain connective tissue), would also enhance the rate of development of atherosclerosis, which, in turn, might enhance the rate of cross-linkage formation by altering the chemical environment of the collagen.

Given that the deterioration and decreased adaptability characteristic of aging can be distinguished from the diseases usually associated with it in industrialized societies, a series of questions can be generated: first, if these diseases are not really "caused" by aging what are they due to? This question, of course, has been a dominant theme throughout the anthology, and Leaf's study of three peoples who remain relatively free of these diseases as they age provides a beautiful summary of several environmental factors which might be implicated. These include nutrition (total calories, protein, animal fat, and other specific nutrients), level of physical activity, and the psychosocial environment. However, as the reader should be able to predict, Leaf is unable to draw any solid conclusions about the quantitative contributions of these specific factors.

A next logical question is: what is the nature of the aging process itself (again, in contrast to the "diseases of aging")? As described by Shock in "The Physiology of Aging," the physiological manifestations of aging are a gradual deterioration in function and in the capacity to respond to environmental stresses. Thus, such quantitative parameters of function as glomerular filtration rate and basal metabolic rate decrease, as does the ability to maintain the internal environment constant in the face of changes in temperature, diet, oxygen supply, etc. These manifestations of aging are related both to a decrease in the actual number of cells in the body (for example, we lose an estimated 100,000 brain cells each day) and to the disordered functioning of many of the cells which remain.

In thinking of how the total number of cells in the body diminishes, it is crucial to recognize that this number reflects the balance between new cell generation (by cell division) and the death of old cells. It had been known for many years that, in the adult human, certain specialized cells—notably nerve and muscle—lose their ability to divide, but it was not until the so-called "Hayflick experiment" (described by Leonard Hayflick in "Human Cells and Aging") that a limitation on cell division of other cell types was firmly established. Hayflick and his co-workers demonstrated that cells, when grown outside of the body, divided only a certain number of times and then stopped; moreover, the number of divisions correlated with the age of the donor. The fact that the number of divisions also correlated with the normal life-span of the different species from which the cells were obtained is strong evidence for the idea that cessation of mitosis is a normal, genetically programmed event. However, more recent experiments have demonstrated that manipulation of the chemical environment of the cells (in this case, by the addition of large quantities of vitamin E) can result in the cells dividing 120 times (rather than the 50 usually observed) before mitosis ceases. Thus, as in any gene-environment interaction, it is likely that environmental factors determine just how many divisions actually occur, within the limits set by the genetic program.

In addition to "inherent degeneration" of macromolecules (to use Hayflick's phrase) and external-environmental factors ("wear-and-tear," radiation, chemicals, etc.), it is very likely that cell death may be the result of the influence of one tissue or organ upon another. Perhaps the most exciting possibility in this last regard concerns the role of the immune system in aging. The concept of autoimmunity was discussed in an earlier section of this anthology; as emphasized there, the immune system all too frequently attacks the body's own cells and it seems likely that, over time, forbidden clones might arise, or the macromolecules of the body's cells might undergo subtle changes such that they would no longer be recognized as "self," with the failure of recognition leading to their destruction by the immune system.

Hayflick suggests that the ultimate failure of mitosis is due to an "accumulation of copying errors" in a cell's DNA molecules. This emphasis on the importance of DNA in aging applies not only to cell division but to all aspects of cell function. As mentioned above, aging is expressed not only by a decrease in total number of cells but also by the deterioration of the functional capacity of those cells which remain. There is fairly general agreement that the immediate cause of this deterioration is an interference in the function of the cells' macromolecules—not just DNA, but RNA, cell proteins, and the flow of information between these macromolecules as well. There are probably many factors responsible for these macromolecular disturbances; as one recent reviewer put it, this is a field in which there are as many theories as there are investigators.

In ending "Human Cells and Aging," Hayflick raises a question of profound biological importance: how effective are the mechanisms for correcting or repairing errors in DNA as they arise? Is it possible, he asks, that humans have a longer life-span than other mammals because they have more effective repair mechanisms? Ten years before his article was written, this question would not have been raised, since we were unaware even of the existence of cellular mechanisms for repairing DNA. It is now clear that such repair mechanisms do exist in mammals, and recent evidence suggests that, as Hayflick surmised, there may well be a correlation between an organism's life-span and the effectiveness of its DNA repair mechanisms. The implications of these mechanisms go well beyond their possible role in the aging process, for they are very likely of critical importance in determining the incidence of birth defects, cancer, and disease arising as a result of mutation, since these three have as their common denominator an alteration of DNA.

23

Getting Old

by Alexander Leaf
September 1973

*Everyone ages, but some seem to age less quickly
than others. In search of clues to the phenomenon the
author visits three communities where vigorous
oldsters are remarkably numerous*

"The patient, Mr. *X*, is 81 years old. A resident of the Dunhill Nursing Home, he has had two strokes, the first three years ago and the second a year ago. Since the last stroke he has been bedridden, incontinent and senile. He no longer recognizes members of his own family. For the past two months he has been eating poorly and failing generally. He was brought in last night by ambulance to our Emergency Service, where we found pulmonary congestion from a failing arteriosclerotic heart and pneumonia. Treatment was started with diuretics and digitalis for his congestive heart failure and high doses of penicillin for his pneumonia. This morning his fever is gone and he is breathing quietly."

I listen to this familiar story related by my intern at the Massachusetts General Hospital and mentally fill in the remainder of the picture. Sometimes the patient is 65, sometimes 70 or 90. Sometimes there is an underlying cancer. Usually, however, cardiovascular or respiratory problems dominate the clinical situation but are superimposed on a substrate of debility, wasting and senility. In the past pneumonia usually terminated such stories with some degree of dignity. Today modern medicine, with its antibiotics, intravenous infusions, cardiac

pacemakers, respirators, diuretics and the like can often resolve the immediate problem (pneumonia and congestive heart failure in this instance) and return the patient to his nursing home again.

In a large metropolitan teaching hospital some 40 percent of the medical (as opposed to surgical) beds are occupied by patients over 65, many of them in a condition similar to Mr. *X*'s. More than 20 million citizens are 65 or older in the U.S. today. They are the major reservoir of illness and medical needs. When I chose a career in medicine 35 years ago, I did so with the conviction that regardless of my eventual specialization my work would promote health and relieve human suffering. Today I contemplate Mr. *X* and wonder if my initial conviction was right. Of course, medicine does more than treat the Mr. *X*'s, but are we doing the best we can for contemporary society? The proportion of the population over 65 is between 10 and 11 percent now and will come close to 15 percent in the next few decades. What is medicine doing about it? Are we applying our resources wisely? Should we devote proportionately more of our efforts to trying to learn how we can prevent the infirmities of old age?

In the U.S. in 1969 the life expectancy at birth for white males was 67.8 years

and for white females 75.1 years. That is some 23 years longer than life expectancy was in 1900. However, an adult who had reached 65 at the turn of the century could expect to live another 13 years and an adult who reaches 65 today can expect to live another 15 years, or only two years longer than in 1900. The seemingly large increase in life expectancy at birth actually reflects the great reduction in infant mortality. Little progress has been made in controlling the major causes of death in adults: heart disease, cancer and stroke. Accidents, of which nearly half are caused by motor vehicles, are a fourth major cause of adult deaths.

At the Ninth International Congress of Gerontology in July, 1972, M. Vacek of Czechoslovakia discussed the increase in mean age and the rising proportion of old people in the population. He pointed out the fact that the rise in the number of people over 65 toward 15 percent in the populations of stable, industrial countries could be attributed to a falling birthrate. When at the conclusion of his remarks someone in the audience timidly asked if the achievements of the medical sciences had played a role in the rise in mean age, he responded that indeed they had, but that the simultaneous deterioration of the environment had canceled out any positive effect from medicine.

Mere length of life, however, may be a poor concern on which to focus. Most would agree that the quality of life, rather than its duration, should be the prime issue. The active life, so warmly espoused by Theodore Roosevelt, has been the American model. If one can extend the period of productive activity as did Verdi and Churchill, so much the better.

Most students of the field would agree

NINE REMBRANDT SELF-PORTRAITS, painted between the ages of 27 or 28 and 63, document the progress of aging in the case of a generally prosperous Dutch burgher of the mid-17th century. The painter was born in July, 1606, and died in October, 1669. His self-portrait of 1633 or 1634 (*top left on opposite page*) is now in the Uffizi Gallery in Florence. Next (*top center*) is his self-portrait of 1652, when he was 46 years old. This is one of three portraits reproduced here that are in the Kunsthistorisches Museum in Vienna. Next (*top right*) is a portrait at age 49, also in Vienna. The second row begins (*left*) with a portrait in the National Gallery of Scotland in Edinburgh, painted at the age of 51. Next is the third of the Vienna paintings, which may show Rembrandt one year older (*center*). At the right is his portrait at age 54, now in the Louvre. The bottom row begins (*left*) with a portrait at age 55, now in the Rijksmuseum in Amsterdam. The next portrait (*center*) may have been painted in 1664 when Rembrandt was 58; it is also in the Uffizi. The last portrait (*right*) Rembrandt painted in the year of his death; it is at the Mauritshuis in The Hague.

with the statement of Frederic Verzár, the Swiss dean of gerontologists. "Old age is not an illness," says Verzár. "It is a continuation of life with decreasing capacities for adaptation." The main crippler and killer, arteriosclerosis, is not a necessary accompaniment of aging but a disease state that increases in incidence with age; the same is true of cancer. If we could prevent arteriosclerosis, hypertension and cancer, the life-span could be pushed back closer to the biological limit, if such a limit in fact exists.

In order to observe aged individuals who are free of debilitating illness I recently visited three remote parts of the world where the existence of such individuals was rumored. These were the village of Vilcabamba in Ecuador, the small principality of Hunza in West Pakistan and the highlands of Georgia in the Soviet Caucasus.

A census of Vilcabamba taken in 1971 by Ecuador's National Institute of Statistics recorded a total population of 819 in this remote Andean village. Nine of the 819 were over 100 years old. The proportion of the population in this small village over 60 is 16.4 percent, as contrasted with a figure of 6.4 percent for rural Ecuador in general. The valley that shelters Vilcabamba is at an altitude of some 4,500 feet. Its vegetation appears quite lush. The people live by farming, but the methods are so primitive that only a bare subsistence is extracted from the land.

A team of physicians and scientists from Quito under the direction of Miguel Salvador has been studying this unique population. Guillermo Vela of the University of Quito, the nutritionist in the group, finds that the average daily caloric intake of an elderly Vilcabamba adult is 1,200 calories. Protein provided 35 to 38 grams and fat only 12 to 19 grams; 200 to 250 grams of carbohydrate completed the diet. The contribution of animal protein and animal fat to the diet is very low.

The villagers of Vilcabamba, like most inhabitants of underdeveloped countries, live without benefit of modern sanitation, cleanliness and medical care. Cleanliness is evidently not a prerequisite for longevity. A small river skirts the village and provides water for drinking, washing and bathing. When we asked various villagers how long it had been since they had last bathed, the responses showed that many had not done so for two years. (The record was 10 years.) The villagers live in mud huts with dirt floors; chickens and pigs share their quarters. As one might expect in such surroundings, infant mortality is high, but so is the proportion of aged individuals. By extrapolation it stands at 1,100 per 100,000, compared with three per 100,000 in the U.S. Statistically, of course, such an extrapolation from so small a number is unwarranted; nonetheless, it shows how unusual the age distribution is in this little village.

The old people of Vilcabamba all appeared to be of European rather than Indian descent. We were able to validate the reported ages of almost all the elderly individuals from baptismal records kept in the local Catholic church. Miguel Carpio, aged 121, was the oldest person in the village when we were there. José Toledo's picture has appeared in newspapers around the world with captions that have proclaimed his extreme old age; actually he is only 109. All the old people were born locally, so that what is sometimes called the Miami Beach phenomenon, that is, an ingathering of the elderly, cannot account for the age distribution.

Both tobacco and sugarcane are grown in the valley, and a local rum drink, zuhmir, is produced. We did not, however, witness any drunkenness, and although most villagers smoke, there is disagreement among visiting physicians about how many of them inhale. The villagers work hard to scratch a livelihood from the soil, and the mountainous terrain demands continuous and vigorous physical activity. These circumstances are hardly unique to Vilcabamba, and one leaves this Andean valley with the strong suspicion that genetic factors must be playing an important role in the longevity of this small enclave of elderly people of European stock.

The impression that genetic factors are important in longevity was reinforced by what we saw in Hunza. This small independent state is ruled over by a hereditary line of leaders known as Mirs. In 1891 the Mir of that day surrendered control over defense, communications and foreign affairs to the British when they conquered his country. In 1948, after the partition of Pakistan and India,

IN SOVIET GEORGIA, one of three widely separated parts of the world visited by the author, a 105-year-old man has the place of honor (*at head of table, left*) at a party held near the village of Gurjanni. The centenarian was the oldest guest but the man and the two women on his left are all over 80. The average diet of those over 80 in the Caucasus is relatively rich in proteins (70 to 80 grams a day) and fats (40 to 60 grams), but the daily caloric intake, 1,700 to 1,900 calories, is barely half the U.S. average.

ON THE PAKISTAN FRONTIER old men of the principality of Hunza, another area visited by the author, winnow threshed wheat in a mountain village. Ibriham Shah (*right*) professes to be 80 years old but Mohd Ghraib (*left*) says he is only 75. Both vigorous work and a low-calorie diet are factors in the fitness of Hunza elders. An adult male's daily diet includes 50 grams of proteins, 36 grams of fats and 354 grams of carbohydrate. This combination provides a somewhat higher daily total than in the Caucasus: 1,923 calories.

the present Mir yielded the same right to Pakistan. Hunza is hidden among the towering peaks of the Karakorum Range on Pakistan's border with China and Afghanistan and is one of the most inaccessible places on the earth. After a day's travel from Gilgit, first by jeep and then on foot from the point where a rockslide had cut the mountain road, we found ourselves in a valley surrounded on all sides by peaks more than 20,000 feet high, blocked at the far end by Mount Rokaposhi, which rises to 25,500 feet in snow-clad splendor.

The valley is arid, but a system of irrigation canals built over the past 800 years carries water from the high surrounding glaciers, converting the valley into a terraced garden. The inhabitants work their fields by primitive agricultural methods, and the harvests are not quite sufficient to prevent a period of real privation each winter. According to S. Maqsood Ali, a Pakistani nutritionist who has surveyed the diet of 55 adult males in Hunza, the daily diet averages 1,923 calories: 50 grams are protein, 36 grams fat and 354 grams carbohydrate. Meat and dairy products accounted for only 1 percent of the total. Land is too precious to be used to support dairy herds; the few animals that are kept by the people are killed for meat only dur-

ing the winter or on festive occasions.

As in Vilcabamba, everyone in Hunza works hard to wrest a living from the rocky hills. One sees an unusual number of vigorous people who, although elderly in appearance, agilely climb up and down the steep slopes of the valley. Their language is an unwritten one, so that no documentary records of birth dates are available. In religion the Hunzakuts are Ismaili Moslems, followers of the Agha Khan. Their education is quite limited. Thus, although a number of nonscientific accounts have attributed remarkable longevity and robust health to these people, no documentation substantiates the reports.

Unfortunately there is no known means of distinguishing chronological age from physiological age in humans. When we asked elderly Hunzakuts how old they were and what they remembered of the British invasion of their country in 1891, their answers did little to validate their supposed age. If, however, they are as old as both they and their Mir, a well-educated and worldly man, maintain, then there are a remarkable number of aged but lean and fit-looking Hunzakuts who can climb the steep slopes of their valley with far greater ease than we could. Putatively the oldest citizen was one Tulah Beg,

who said he was 110; the next oldest, also a male, said he was 105.

In addition to their low-calorie diet, which is also low in animal fats, and their intense physical activity, the Hunzakuts have a record of genetic isolation that must be nearly unique. Now, there do not seem to be "good" genes that favor longevity but only "bad" genes that increase the probability of acquiring a fatal illness. One may therefore speculate that a small number of individuals, singularly lacking such "bad" genes, settled this mountain valley centuries ago and that their isolation has prevented a subsequent admixture with "bad" genes. This, of course, is mere speculation.

The possible role of genetic factors in longevity seemed of less importance in the Caucasus. In Abkhasia on the shores of the Black Sea and in the adjoining Caucasus one encounters many people over 100 who are not only Georgian but also Russian, Jewish, Armenian and Turkish. The Caucasus is a land bridge that has been traveled for centuries by conquerors from both the east and the west, and its population can scarcely have maintained any significant degree of genetic isolation. At the same time, when one speaks to the numerous

IN THE ANDES OF ECUADOR an 85-year-old woman of the village of Vilcabamba works in a cornfield, continuing to contribute to the economic welfare of the community at an age when many women no longer work. Of the areas visited by the author, Vilcabamba afforded the elderly the most sparse diet. The daily intake was less than 40 grams of proteins, 20 grams of fats and 250 grams of carbohydrate for a total intake of 1,200 calories.

BLACK SEA VILLAGER from the coast near Sukhumi is over 100 years old. To live to be 100 or older in the Caucasus gains one a kind of elder statesman's place in local society.

centenarians in the area, one invariably discovers that each of them has parents or siblings who have similarly attained great age. The genetic aspect of longevity therefore cannot be entirely dismissed.

In contrast to the isolated valleys of Vilcabamba and Hunza, the Caucasus is an extensive area that includes three Soviet republics: Georgia, Azerbaijan and Armenia. The climate varies from the humid and subtropical (as at Sukhumi on the Black Sea, with an annual rainfall of 1,100 to 1,400 millimeters) to drier continental conditions, marked by extremes of summer heat and winter cold. The population extends from sea-level settlements along the Black and Caspian seas to mountain villages at altitudes of 1,000 to 1,500 meters. More old people are found in the mountainous regions than at sea level, and the incidence of atherosclerosis among those who live in the mountains is only half that in the sea-level villages. G. Z. Pitzkhelauri, head of the Gerontological Center in the Republic of Georgia, told me that the 1970 census placed the number of centenarians for the entire Caucasus at between 4,500 and 5,000. Of these, 1,844, or an average of 39 per 100,000, live in Georgia. In Azerbaijan there are 2,500 more, or 63 per 100,000. Perhaps of more pertinence to my study was the record of activity among the elderly of the Caucasus. Of 15,000 individuals over 80 whose records were kept by Pitzkhelauri, more than 70 percent continue to lead very active lives. Sixty percent of them were still working on state or collective farms.

I returned from my three surveys convinced that a vigorous, active life involving physical activity (sexual activity included) was possible for at least 100 years and in some instances for even longer.

Longevity is clearly a multifactorial matter. First are the genetic factors; all who have studied longevity are convinced of their importance. It is generally accepted that the offspring of long-lived parents live longer than others. Yet a long life extends beyond the period of fertility, at least in women, so that length of life can have no direct evolutionary advantage. It has been suggested that living organisms are like clocks: their life-span ends when the initial endowment of energy is expended, just as the clock stops when its spring becomes unwound. L. V. Heilbrunn estimated in the 1900's that the heart of a mouse, which beats 520 to 780 times per minute, would contract 1.11 billion times during the 3.5 years of the mouse's normal life-span. The heart of an elephant, which beats 25 to 28 times per minute, would have beaten 1.021 billion times during the elephant's normal life-span of 70 years. The similarity between these two figures seems to suggest some initial equal potential that is gradually dissipated over the animal's life-span. Such calculations, however, are probably more entertaining than explanatory of longevity.

As I have mentioned, genes evidently influence longevity only in a negative way by predisposing the organism to specific fatal diseases. Alex Comfort of University College London has pointed out the possible role of heterosis, or "hybrid vigor." The crossbreeding of two inbred strains—each with limited growth, size, resistance to disease and longevity—can improve these characteristics in the progeny. This vigor has been manifested in a wide variety of species, both plant and animal, but its significance in human longevity is not known.

Next are the factors associated with nutrition. Since we are composed of what we eat and drink, it is not surprising that students of longevity have emphasized the importance of dietary factors. Indeed, the only demonstrable means of extending the life-span of an experimental animal has been the manipulation of diet. The classic studies of Clive M. McCay of Cornell University in the 1930's showed that the life-span of albino rats could be increased as much as 40 percent by the restriction of caloric intake early in life. Rats fed a diet otherwise balanced but deficient in calories showed delayed growth and maturation until the caloric intake was increased. At the same time their life-span was extended. The significance of these experiments with respect to human longevity is not known, but they raise questions about the current tendency to overfeed children.

The role of specific dietary factors in promoting longevity remains unsettled. When the old people we visited were asked to what they attributed their long life, credit was usually given to the local alcoholic beverage, but this response generally came from the more vocal and chauvinistic males. Much publicity is given today to the possible role of animal fats in the development of arteriosclerosis. Saturated fatty acids and cholesterol have also been suggested as the causal agents of coronary atherosclerosis. Since the atheromas themselves—the deposits that narrow and occlude the blood vessels of the heart, the brain and other organs—contain cholesterol, the suspicion

ELDER OF HUNZA is examined by the author. A resident of the village of Mominabad, the man professed to be 110 years old. Birth records are not kept in Hunza, however, and evidence supporting his claim and other Hunza elders' claims of extreme age did not exist.

of complicity comes easily. It is also well documented that the level of cholesterol in the blood serum has a positive prognostic relation to the likelihood of heart attacks.

Since the diet of affluent societies has a high content of animal fats, which are rich in cholesterol and saturated fatty acids, the suspicion further arises that these dietary constituents are contributing to the increase in the number of heart attacks affecting young adult males. The marked individual variability in tolerance to the quantity of fat and cholesterol in the diet, however, makes it difficult to ascribe prime importance to this single factor. We have seen that in both Vilcabamba and Hunza the diet was not only low in calories but also low in animal fats. In the Caucasus an active dairy economy allows cheese and milk products to be served with every meal. Perhaps it is significant that in this area the cheese is low in fat content and the total fat intake is only some 40 to 60 grams per day. This level is in sharp contrast to figures from a recent U.S. Department of Agriculture report stating that the average daily fat intake for Americans of all ages is 157 grams. The best-informed medical opinion today

generally agrees that the Americans' average daily intake of 3,300 calories, including substantial quantities of fat, is excessive and conducive neither to optimal health nor to longevity.

Let us now consider physical activity. There is increasing awareness that early heart disease is a price we are paying for our largely sedentary existence. In the three areas I visited physical fitness was an inevitable consequence of the active life led by the inhabitants. My initial speculation that their isolated lives protects them from the ravages of infectious diseases and from acquiring "bad" genes is probably incorrect. A simpler explanation for their fitness is that their mountainous terrain demands a high level of physical activity simply to get through the day.

A number of studies have examined the effects of physical activity on the incidence and severity of heart attacks among various sample populations. For example, among British postal workers it was found that those who delivered the mail had a lower incidence of heart attacks and, when attacks did occur, a lower mortality rate than their colleagues who worked at sedentary jobs. This same study also compared the relative incidence of heart attacks among London bus drivers and conductors. The conductors, who spent their working hours climbing up and down the double-deck buses collecting fares, had less heart trouble than the sedentary drivers. It has been reported recently that the weekend athlete who engages in vigorous physical activity is only one-third as prone to heart disease as his age-matched sedentary neighbor. It is well known that exercise increases the oxygenation of the blood and improves the circulation. Exercise will also improve the collateral circulation of blood to the heart muscle. When the exercises are carefully graded and performed under appropriate supervision, they are undoubtedly the best means of rehabilitating the heart muscle after a heart attack.

Exercise improves circulation to nearly all parts of the body. Circulation is increased to the brain as well as to the heart and skeletal muscles. Recently it has been asserted that improvement in the oxygen supply to the brain will actually improve thinking. In a resting sedentary individual, however, the blood supplied to the brain contains more oxygen than the brain is able to extract. It is difficult to explain how an added overabundance of a constituent that is normally not a rate-limiting factor in

cerebral function would enhance that function. It may nonetheless be that the sense of well-being that the exercising individual enjoys is likely to increase self-confidence and as a result improve both social and intellectual effectiveness.

Finally, physical activity helps to burn off excess calories and dispose of ingested fats. Exercise may thus counteract the deleterious effects of a diet that includes too many calories and too much fat. Indeed, it may well be one factor that helps to account for the great individual differences in tolerance for dietary factors.

Psychological factors must also be considered. It is characteristic of each of the areas I visited that the old people continue to be contributing, productive members of their society. The economy in all three areas is agrarian, there is no fixed retirement age and the elderly make themselves useful doing many necessary tasks around the farm or the home. Moreover, increased age is accompanied by increased social status. The old people, with their accumulated experience of life, are expected to be wise, and they respond accordingly. In Hunza the Mir rules his small state with the advice of a council of elders, who sit on the ground in a circle at his feet and help him with his decisions and pro-

nouncements. When we met Temur Tarba in the Caucasus just three weeks after his 100th birthday, it was clear from his manner that he was delighted to have at last "arrived." He proudly displayed his Hero of Labor medal, the highest civilian award of the U.S.S.R. He had won it only seven years earlier for his work in hybridizing corn. The cheerful centenarian still picked tea leaves, rode his horse and worked on a collective farm.

People who no longer have a necessary role to play in the social and economic life of their society generally deteriorate rapidly. The pattern of increasingly early retirement in our own society takes a heavy toll of our older citizens. They also find that their offspring generally have neither any room nor any use for them in their urban apartment. These are economic determinants that cannot be reversed in our culture today. Their devastating effect on the happiness and life-span of the elderly could be countered at least in part by educational programs to awaken other interests or avocations to which these people could turn with zest when their contribution to the industrial economy is no longer needed. The trend toward shorter working hours and earlier retirement makes the need for such education urgent. It seems a corruption of the very purpose of an economy that instead of freeing us from drudgery and need and allowing us to enjoy a better life, it holds us slaves to its dictates even though affluence is at hand.

In their remote farms and villages the old people I visited live oblivious to the pressures and strains of modern life. Such American controversies as the fighting in Southeast Asia, conflict in the Middle East, environmental pollution, the energy crisis and the like that fill our news media were unknown to them. Of course, most of the stresses and tensions to which mankind is subject arise from more personal social interactions, such as a quarrel with a spouse or misbehavior by an offspring. As a result it is impossible for any social group to entirely escape mental stress and tensions. Nonetheless, in societies that are less competitive and less aggressive even these personal stresses can be less exhausting than they are in our own. The old people I met abroad showed equanimity and optimism. In the Caucasus I asked Gabriel Chapnian, aged 117, what he thought of young people, of his government and of the state of the world in general. When he repeatedly responded "Fine, fine," I asked with some annoy-

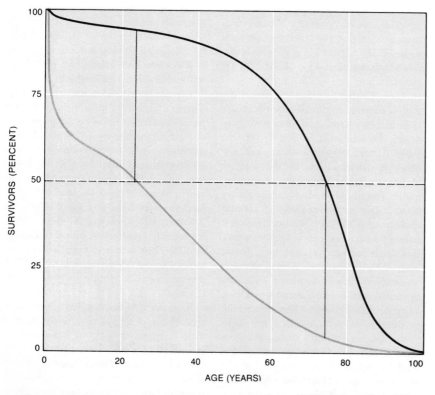

FEW LIVE TO GROW OLD in underdeveloped nations, whereas the opposite is true in the developed nations. For example, half of the women who died in India (*color*) from 1921 to 1930 were 24 or younger; only 5 percent of the women who died in New Zealand from 1934 to 1938 were that young. Half of the New Zealand women who died in the same period were 74 or older. In India only 5 percent of women's deaths were in this age group.

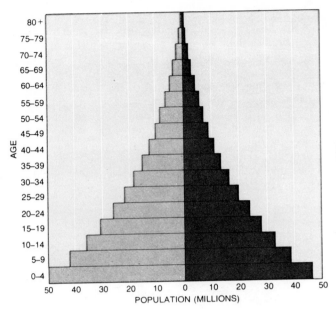

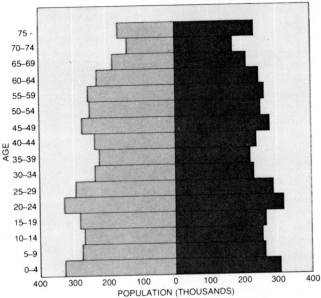

EXAGGERATED PYRAMID appears when the males (*left*) and females (*right*) of a population with a high growth rate are counted in ascending five-year steps. The graph shows the 1970 population of India; many more Indians are below the age of 20 than are above that age. In such an expanding population the elderly, although numerous, comprise a small percentage of the total.

UNEVEN COLUMN, rather than a pyramid, appears when the number of males (*left*) and females (*right*) in a slow-growing population are similarly displayed. The graph shows the population of Sweden in 1970; about as many Swedes are above the age of 35 as are below that age. In this population, noted for its low birthrate, the elderly comprise an increasingly large percentage of the total.

ance, "Isn't there anything today that disturbs you?" He replied cheerfully, "Oh yes, there are a number of things that are not the way I would want them to be, but since I can't change them I don't worry about them."

One cannot discuss the quality of life at advanced age without considering sexual activity. In most societies the combination of male boasting about sexual prowess and taboos against discussing the subject makes it difficult to collect reliable information about sexual activity in the elderly. Women's ovaries stop functioning at menopause, usually in the late 40's or early 50's, but this has little influence on the libido. Similarly, aging is associated with a gradual decrease in the number of cells in certain organs, including the male testes. The cells that produce sperm are the first to be affected, but later the cells that produce the male hormone testosterone may also diminish. in number and activity. Sexual potency in the male and libido in the female may nonetheless persist in advanced old age. Herman Brotman of the Department of Health, Education, and Welfare reports that each year there are some 3,500 marriages among the 20 million Americans over the age of 65, and that sexual activity is cited along with companionship as one reason for these late unions.

Research on aging is proceeding in two general directions. One aims at a better understanding of those disease states that, although they are not an integral part of the aging process, nevertheless increase in incidence with age and constitute the major cripplers of the aged. Here both arteriosclerosis and cancer are prominent. Knowledge of the causes and prevention of these diseases is still very limited, and much more work is necessary before there can be sufficient understanding to allow prevention of either. If both arteriosclerosis and cancer could be prevented, it should be possible to extend man's life-span close to its as yet undetermined biological limit.

There is reason to believe that a limit does exist. Even if one could erase the cumulative wear and tear that affects the aging organism, it seems probable that the cells themselves are not programmed for perpetual activity. The differences in the life-spans of various animals suggest some such natural limit. The changes that are observed in the tissues of animals that undergo metamorphosis—for example insects—are indicative of the programmed extinction of certain tissues. So is the failure of the human ovaries at menopause, long before other vital organs give out. Leonard Hayflick of the Stanford University School of Medicine has reported that the cells in cultures of human embryonic connective tissue will divide some 50 times and then die. If growth is interrupted by plunging the culture into liquid nitrogen, the cells will resume growth on thawing, continue through the remainder of their 50 di-

visions and then die. Although one may question whether the cells are not affected by adverse environmental influences, Hayflick's studies support the notion of programmed death. Only cancer cells seem capable of eternal life in culture. One familiar line of cancer cells, the "HeLa" strain, has often been the subject of cell-culture studies; it originated as a cervical cancer 21 years ago and has been maintained in culture ever since.

As investigations and speculations seek an answer to the question of the natural limits of life, other workers have noted that the giant molecules of living matter themselves age. Collagen, the main protein of connective tissue, constitutes approximately 30 percent of all the protein in the human body. With advancing age collagen molecules show a spontaneous increase in the cross-linking of their subunits, a process that increases their rigidity and reduces their solubility. Such a stiffening of this important structural component of our bodies might underlie such classic features of aging as rigidity of blood vessels, resistance to blood flow, reduced delivery of blood through hardened arteries and, as a final consequence, the loss of cells and of function. Other giant molecules, including the DNA molecule that stores the genetic information and the RNA molecule that reads out the stored message, may also be subject to spontaneous cross-linking that would eventually prevent the normal self-renewal of tissues.

A new area in aging research appears to be opening up in studies on the immune system of the body. In addition to providing antibodies against bacteria or foreign substances introduced into the body, the immune system recognizes and destroys abnormal or foreign cells. When these functions of the immune system diminish, as happens with advancing age, there may be errors in the system that result in antibodies that attack and destroy normal body cells or impair their function. The net result of this disturbed activity of the immune system, like the cross-linking of collagen, would be what is recognized as the aging process.

Takashi Makinodan of the Baltimore City Hospitals thinks that methods of rejuvenating the immune system are possible. The hope that some medicine will be discovered that can block the fundamental process of aging, however, seems to me very remote until the nature of the aging process is far better understood.

Much research is needed before such an understanding can be attained. It is nonetheless encouraging that aging research is finding increased interest and support in several countries. It is also encouraging to perceive a parallel interest in the aged among scientists, physicians, economists, politicians and others. To consider any extension of the human life-span without a serious effort to anticipate and plan for the impact of increased longevity on society would be entirely irresponsible.

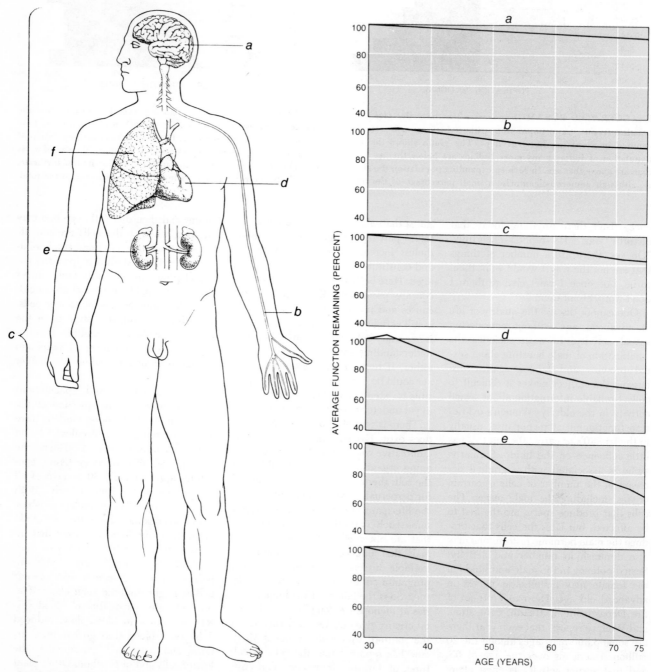

LOSS OF FUNCTION with increasing age does not occur at the same rate in all organs and systems. Graphs (*right*) show loss as a percentage, with the level of function at age 30 representing 100 percent. Thus brain weight (*a*) has diminished to 92 percent of its age-30 value by age 75 and nerve-conduction velocity (*b*) to about 90 percent. The basal metabolic rate (*c*) has diminished to 84 percent, cardiac output at rest (*d*) to 70 percent, filtration rate of the kidneys (*e*) to 69 percent and maximum breathing capacity (*f*) to 43 percent. Diseases, however, rather than gradual diminution of function, are at present the chief barrier to extended longevity.

The Physiology of Aging

by Nathan W. Shock
January 1962

The processes of aging operate during all of adult life. The decline in function that accompanies these processes is apparently due to the progressive loss of body cells

With the virtual conquest of want and infectious disease in technologically advanced countries, men and women in increasing numbers are living out the promised Biblical life span of three score years and ten. The diseases of age and the fundamental process of aging are moving to the center of interest in the practice of medicine and in medical research. Few people die of old age. Mortality increases rapidly with age—in precise logarithmic ratio to age in the population as a whole—because the elderly become more susceptible to diseases that kill, such as cancer and cardiovascular disease. The diseases of old age are the province of the relatively new medical specialty known as geriatrics. A still younger discipline called gerontology deals with the process of aging itself. This is a process that continues throughout adult life. It goes on in health as well as in sickness and constitutes the primary biological factor underlying the increase in susceptibility to the diseases that are the concern of geriatrics.

Gerontology is still in the descriptive stage. Investigators have only recently developed objective standards for measurement of the decline in the performance and capacity of the body and its organ systems, and they have just begun to make such measurements on statistically significant samples of the population. The first general finding in gerontology is that the body dies a little every day. Decline in capacity and function over the years correlates directly with a progressive loss of body tissue. The loss of tissue has been shown to be associated in turn with the disappearance of cells from the muscles, the nervous system and many vital organs. To get at the causes of death in the cell gerontology has entered the realm of cellular physiology and chemistry.

The ideal way to study the aging of the human body would be to apply the same battery of tests to a large group of people at repeated intervals throughout their lives. Such a program would require dedicated subjects and a scientific staff organized for continuity of operation over a period of perhaps 50 years. Obviously some compromise must be made. Instead of starting observations on a group of subjects at age 30 and following them for 50 years, it is possible to begin with subjects of various ages and follow them for 20 years. At our laboratories in the Gerontology Branch of the Baltimore City Hospitals we started such a study on 400 men in 1958. Until this and similar undertakings have had time to yield results, gerontologists must rely on data accumulated from one-time tests of rather large numbers of different individuals ranging in age from 20 or 30 up to 80 or 90. Although subjects of any specific age differ widely, the average values for many physiological characteristics show a gradual but definite reduction between the ages of 30 and 90. Individual differences become quite apparent, for example, in studies of the amount of blood flowing through the kidneys. Whereas this function generally declines markedly with age, it is the same in some 80-year-old men as it is in the average 50-year-old man.

One of the most obvious manifestations of aging is the decline in the ability to exercise and do work. In order to measure the extent of the change it is necessary to set up laboratory experiments in which the rate of output and the amount of work done can be precisely determined along with the responses of various organ systems. Subjects may be put to walking a treadmill or climbing a certain number of steps at a specified rate. In our laboratory the subject lies on his back and turns the crank of an ergometer, an apparatus for measuring the work done. When our purpose is to measure the subject's maximum output in a given time, the crank can be adjusted to turn more stiffly or more easily. With the subject lying supine it is easier to make the necessary measurements of blood pressure, heart rate and heart output (blood pumped) and to collect the respiratory gases through a face mask for the measurement of oxygen consumption and carbon dioxide production. These measurements are customarily made before, during and after exertion in order to establish the subject's norms, his capacity and the rate at which vital functions recover their normal or resting rates.

As a common denominator of capacity we seek to determine the maximum amount of work a subject can do and have his heart return to normal within two minutes after he stops working. Men 30 years old achieve an output of 500 kilogram-meters per minute (the equivalent of lifting 500 kilograms one meter in one minute), whereas 70-year-old men on the average reach only 350 kilogram-meters per minute. Thus at age 70 a man's physical capacity as defined by this test has declined by 30 per cent. Over the years from 35 to 80 the maximum work rate for short bursts of crank-turning falls almost 60 per cent, from about 1,850 kilogram-meters per minute for young men to 750 kilogram-meters for the 80-year-old.

Physical performance, of course, reflects the combined capacity of the different organ systems of the body working together. The ability to do work depends on the strength of the muscles, the co-ordination of movement by the nervous system, the effectiveness of the heart in propelling blood from the lungs to the working muscles, the rate at which air moves in and out of the lungs, the efficiency of the lung in its gas-exchange function, the response of the kidneys to the task of removing excess waste materials from the blood, the synchronization of metabolic processes by the endocrine glands and, finally, the constancy with which the buffer systems in the blood maintain the chemical environment of the body. In order to determine the causes of the decline in overall capacity, it is necessary to assess the effects of aging on each of the organ systems.

Tests of the strength of the hand serve in our laboratory to isolate one aspect of muscle function. The subject simply squeezes a grip-measuring device as hard as he can for a moment. In a group of 604 men the strength of the dominant hand dropped from about 44 kilograms of pressure at age 35 to 23 kilograms at age 90. Although the dominant hand is stronger at all ages, it loses more of its strength over the years than the subordinate hand. Endurance, measured by the average grip pressure exerted for one minute, drops from 28 kilograms at age 20 to 20 kilograms at age 75. That muscle performance is not the only factor involved in maximum work rates is indicated by the fact that the decrease in muscular strength over the years is less than the decline in work rates.

The nerve fibers that connect directly with the muscles show little decline in function with age. The speed of nerve impulses along single fibers in elderly people is only 10 to 15 per cent less than it is in young people. Simple neurological functions involving only a few connections in the spinal cord also remain virtually unimpaired. It is in the

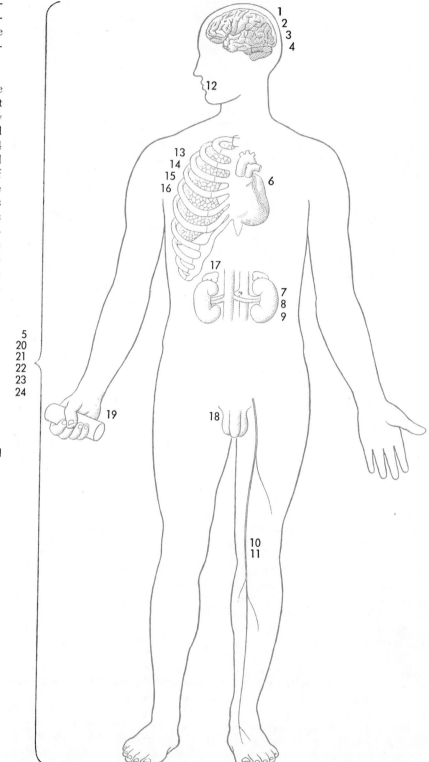

1 BRAIN WEIGHT [56]
2 MEMORY LOSS
3 SLOWER SPEED OF RESPONSE
4 BLOOD FLOW TO BRAIN [80]
5 SPEED OF RETURN TO EQUILIBRIUM OF BLOOD ACIDITY [17]
6 CARDIAC OUTPUT (AT REST) [70]
7 NUMBER OF GLOMERULI IN KIDNEY [56]
8 GLOMERULAR FILTRATION RATE [69]
9 KIDNEY PLASMA FLOW [50]
10 NUMBER OF NERVE TRUNK FIBERS [63]
11 NERVE CONDUCTION VELOCITY [90]
12 NUMBER OF TASTE BUDS [36]
13 MAXIMUM OXYGEN UPTAKE (DURING EXERCISE) [40]
14 MAXIMUM VENTILATION VOLUME (DURING EXERCISE) [53]
15 MAXIMUM BREATHING CAPACITY (VOLUNTARY) [43]
16 VITAL CAPACITY [56]
17 LESS ADRENAL ACTIVITY
18 LESS GONADAL ACTIVITY
19 HAND GRIP [55]
20 MAXIMUM WORK RATE [70]
21 MAXIMUM WORK RATE FOR SHORT BURST [40]
22 BASAL METABOLIC RATE [84]
23 BODY WATER CONTENT [82]
24 BODY WEIGHT FOR MALES [88]

PHYSIOLOGICAL DECLINE ACCOMPANYING AGE appears in many measurements throughout the body. Changes are great in some cases, small in others. The figures in brackets following most of the labels in the key at left are the approximate percentages of functions or tissues remaining to the average 75-year-old man, taking the value found for the average 30-year-old as 100 per cent.

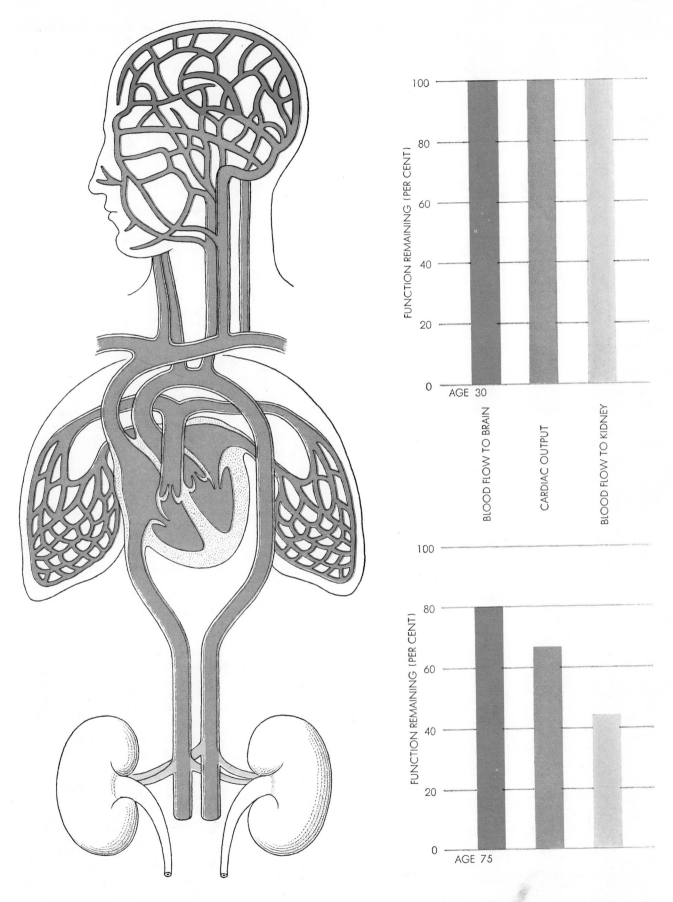

DECLINE IN HEART FUNCTION by age 75 means that both brain and kidney receive less blood than in the younger man. At age 75 the heart pumps only about 65 per cent as much blood as at age 30. The brain receives 80 per cent as much blood; the kidney, only 42 per cent as much. Large decrease in flow to kidney may be an adaptive mechanism permitting greater flow to brain.

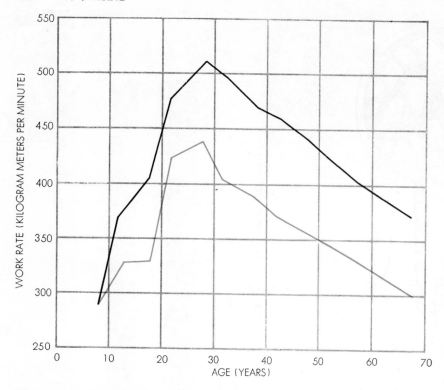

CHANGES IN WORK RATE with age are striking. The function reaches a peak at age 28 and declines steadily thereafter. Rate for males is shown in black, for females in color.

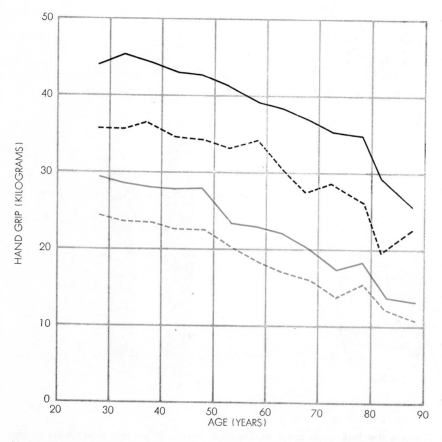

HAND GRIP GROWS WEAKER with age. Rate for men is shown in black, for women in color. Solid lines represent dominant hand; broken lines represent subordinate hand.

central nervous system, where complex connections are made, that aging takes its toll. Memory loss, particularly for recent events, often plagues the elderly. The older person requires substantially more time for choosing between a number of possible responses to a situation, although with enough time he arrives at the correct decision. Certain routine mental activities, on the other hand, hardly change with age. Vocabulary comprehension, for example, remains strong in most people. Experienced proofreaders maintain a high degree of accuracy even at advanced ages.

Because muscles engaged in sustained exercise require extra oxygen and other nutrients and produce more waste to be carried away, the heart must work harder to move more blood through the system. During exercise the heart pumps more blood at each stroke, at a faster rate and at higher pressure. Although the resting blood pressure in healthy individuals increases only slightly with age, a given amount of exercise will raise the heart rate and blood pressure in old people more than it will in young. And when subjects exert themselves to the maximum, the heart of the older person cannot achieve as great an increase in rate as that of the younger. During exercise, therefore, the cardiac output, or amount of blood pumped per minute, is less in the old than it is in the young. This, of course, imposes limits on the amount of work the elderly can do.

Cardiac output can be measured directly and, in subjects at rest, quite easily. (The measurements are difficult during exercise.) In one procedure a known amount of blue dye is injected into a vein of one arm and blood samples are then taken periodically from a small catheter in the large artery of the opposite leg. The dilution of the dye provides a measure of heart output. The amount of blood pumped falls from an average of 3.75 liters per minute (a liter is slightly more than a quart) per square meter of body surface in 20-year-olds to two liters per minute in 90-year-olds.

The lung plays as important a role in exercise as the heart. We have studied the two aspects of lung function—the maximum amount of oxygen that can be taken up from inspired air during exercise and the ability of the lung to move air in and out. The amount of oxygen that the blood takes up from the lung and transports to the tissues during exercise falls substantially with age. The blood of 20-year-old men takes up, on the average, almost four liters of oxygen per minute, whereas at age 75 the rate is only 1.5 liters per minute. This function has been tested in several individuals over many years. D. Bruce Dill, a physiologist now at Indiana University, found that his own maximum oxygen uptake declined from 3.28 liters at age 37 to 2.80 liters at age 66.

Another measurement reveals that in order to double the level of oxygen uptake during exercise the older individual must move about 50 per cent more air in and out of his lungs. No doubt the decline in oxygen absorption reflects in part the reduced heart output, for less blood flows through the lungs of the older person in a given time. But the great difference in oxygen uptake between young and old shows that the lung tissue too has changed.

The decline in respiratory function also reflects a loss in simple mechanical efficiency. In normal respiration less air turns over and the amount of dead air space in the lungs increases, although total lung volume remains almost unchanged. Even the "vital capacity" (the amount of air that can be forcibly expired from the lung) diminishes with age. The nature of this impairment becomes clear when one measures the subject's maximum breathing capacity—the amount of air he can move through his lungs in 15 seconds. The chart for this test shows a decline of about 40 per cent between the ages of 20 and 80. Since the older person expels about as much air at each breath as the younger person does, it is clear that his capacity is less because he cannot maintain as fast a rate of breathing. The impairment is an expression of the general decline in neuromuscular capacity.

Exercise produces acids and other metabolic waste products that are excreted primarily by the kidney. Because the heart pumps less blood with advancing age, less blood flows through the kidney in a given time. Changes within the kidney itself further reduce the flow of blood as well as the efficiency with which the kidney processes the wastes. The kidney puts the blood through a delicate, multistaged process. First it filters the blood, flushing the waste products out of the bloodstream in a filtrate from which it withholds the red cells and larger molecules; then it processes this filtrate, recovering the smaller useful molecules, such as those of glucose, that get through the filter; and finally the kidney actively excretes waste molecules, some of them too large to pass through the filter. The active

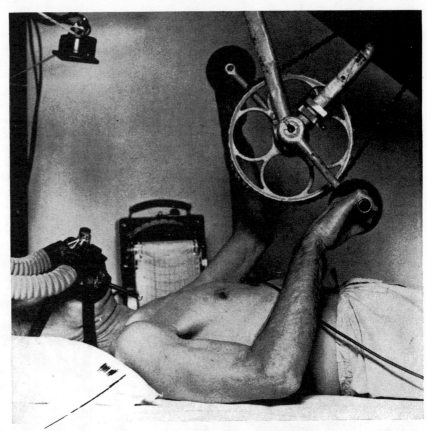

ABILITY TO DO WORK is measured by ergometer. Subject turns crank while breathing through face mask. Oxygen consumption, work rate and other functions can be measured.

TEST OF GRIP STRENGTH serves to isolate muscle function. The subject squeezes the grip and the dial registers the amount of pressure, in kilograms, that he is able to exert.

functions take place in the tubular lining of the nephrons, the functional units of the kidney. A full test of kidney performance involves measurement of the amount of filtrate formed per minute, the quantity of blood plasma (the liquid portion of the blood) passing through the kidney per minute and the maximum excretory capacity of the nephrons. The measurements are made by infusing the blood with substances that the kidney removes in different ways; one substance, the metabolically inert polysaccharide inulin, goes out through the filter, whereas para-aminohippuric acid must be actively excreted. Analysis of blood samples and urine during the infusion shows how efficiently the kidney is working. Such tests show that between the ages of 35 and 80 the flow of blood plasma through the kidney declines by 55 per cent. The filtration rate and the maximum excretory capacity, as well as glucose reabsorption, decline to the same extent.

Intravenous administration of the substance pyrogen increases the flow of blood through the kidney in young and old. Apparently reduction in blood flow through the kidney is an adaptive mechanism of the aging body. It seems to result from constriction of kidney blood vessels, which makes more blood available to other organs. Because the kidney of the older person has less blood to work on it cleanses the blood of waste more slowly; given enough time, however, it will do the job.

The endocrine glands regulate a wide variety of physiological processes, ranging from cellular metabolism to regulation of the diameter of small blood vessels and consequently the amount of

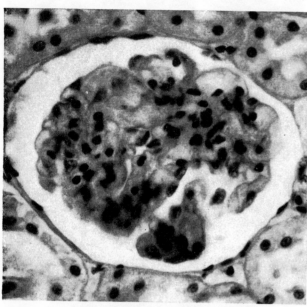
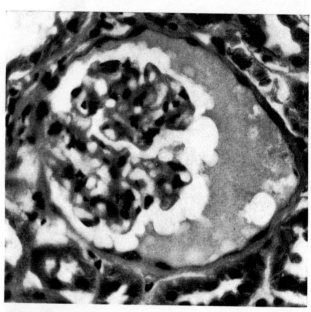

DEGENERATION OF KIDNEY accompanies aging in the rat. At left is a section through the kidney of a normal adult rat. Section at right is from senile rat. "Colloid" has stagnated in a glomerulus (filtration unit) of kidney, causing atrophy. Magnification is 570 diameters. These four photomicrographs were made by Warren Andrew of the Indiana University School of Medicine.

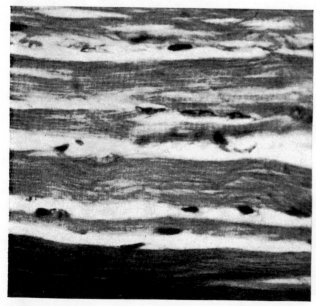
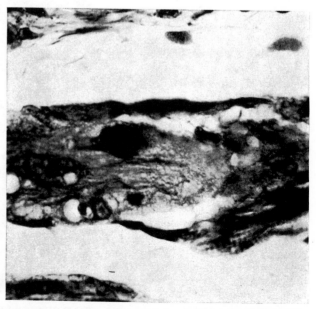

DETERIORATION OF LEG MUSCLE in rat shows plainly in these photomicrographs, each enlarged 710 diameters. Section at left is from normal adult rat, that at right from senile rat. Connective tissue has replaced many muscle fibers that have died.

blood reaching various tissues. At the center of the endocrine system is the pituitary gland, which secretes hormones that stimulate the adrenal glands, the thyroid, the ovaries, the testes and other glands to release their hormones. There is no way to test the performance of the master gland in human subjects, but the responsiveness of other endocrine glands can be tested by administering the appropriate pituitary hormone. Sometimes the level of activity of a given gland can be estimated from the amount of its hormone or of the breakdown products of the hormone that appear in the urine. Adrenal activity is customarily measured by this means and shows a decline with age. Administration of the pituitary hormone that stimulates the cortex of the adrenals produces a smaller elevation of adrenal activity in older people. Since the adrenal hormones are the "stress" hormones, this indicates a reduction in the capacity to respond to stress. On the other hand, the pituitary hormone that stimulates the thyroid gland produces the same result in the old as it does in the young. Even at advanced ages the thyroid retains its ability to manufacture and release thyroxine, which regulates the basal metabolism, or the rate at which the resting subject consumes oxygen.

Normal function in the cells of the body requires that the chemical composition of the fluids surrounding them be closely regulated. Because the intercellular fluids cannot be sampled directly, estimates of the internal environment must come from analyses of the blood. Such factors as total blood volume, acidity, osmotic pressure, protein content and sugar content remain constant in both young and old subjects at rest. But when these variables are deliberately altered, the older person needs a much longer time to recover internal chemical equilibrium. The acidity of the blood, for example, can be increased by oral administration of ammonium chloride; a younger subject recovers normal acidity within six to eight hours, whereas the 70-year-old requires 36 to 48 hours to recover.

Although the average blood-sugar level remains quite constant even into advanced age, the rate at which the system removes extra glucose drops significantly in older people. Insulin, normally secreted by glands in the pancreas, greatly accelerates the removal of sugar from the blood. When we administer insulin intravenously along with extra glucose, the glucose disappears from the blood of the young person at a much

higher rate than it does from the old.

It may well be that subtle changes in the chemical composition of the blood and other body fluids account for certain physiological changes in the elderly. So far as the gross chemical characteristics go, however, the old animal or human easily maintains a constant, normal internal environment when completely at rest. But increasing age is accompanied by a definite reduction in the capacity to readjust to changes that accompany the stresses even of daily living. In other words, a key characteristic of aging is a reduction in the reserve capacities of the body—the capacities to return to normal quickly after disturbance in the equilibrium.

Another important element in the aging process shows up in those functions and activities that involve a high degree of co-ordination among organ systems. Co-ordination breaks down, in the first place, because the different organ systems age at different rates. Conduction of the nerve impulse, for instance, hardly slows with age, whereas cardiac output and breathing capacity decline considerably. Thus in those functions that involve the simultaneous output of several organ systems—sustained physical exercise, for example—the performance of the body shows marked impairment. Most of the debilities of age apparently result from a

loss of tissue, particularly through the death and disappearance of cells from the tissues. The wrinkled and flabby skin so apparent in elderly people offers mute testimony to this loss. Body weight declines, especially after middle age. In a large sample of the male population of Canada average weight showed a decline from 167 pounds at ages 35 to 44 to an average of 155 pounds for men 65 and over. A sample of men in the U.S., all of them 70 inches tall, averaged 168 pounds in weight at ages 65 to 69 and 148 pounds at ages 90 to 94. Women 65 inches tall weighed, on the average, 148 pounds at ages 65 to 69 and 129 pounds at ages 90 to 94.

Individual organs also lose weight after middle age. For example, the average weight of the brain at autopsy falls from 1,375 grams (3.03 pounds) to 1,232 grams (2.72 pounds) between ages 30 and 90. The same striking loss in total weight and in the weight of specific organs shows up also in the senile rat: the total weight of certain muscle groups drops 30 per cent.

The microscope shows that in many tissues a decrease in the number of cells accompanies the weight loss. Connective tissue replaces the lost cells in some cases, so that the loss of cells is even greater than the reduction in weight would indicate. In the senile rat the

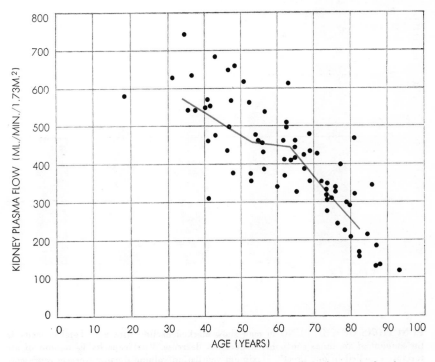

LARGE INDIVIDUAL DIFFERENCES in aging show up when, for example, the rate of flow of blood plasma through the kidney is plotted against age for some 70 men. The plasma flow is measured in milliliters per minute per 1.73 square meters of body-surface area.

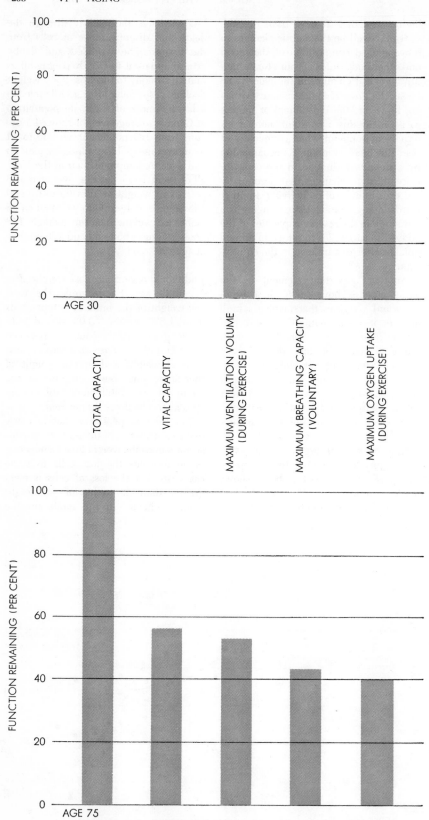

FUNCTIONS OF THE LUNG in man show marked decline with age. Total capacity is the amount of air lungs can hold; it does not decrease. Vital capacity is amount of air forcibly expelled in one breath. Maximum ventilation volume during exercise represents involuntary movement of air. Maximum breathing capacity is the amount of air that can be moved in and out of the lungs voluntarily in 15 seconds. Oxygen uptake is the quantity of oxygen absorbed by the blood from the lungs for transportation to the body cells.

muscle fibers show degenerative changes with replacement by connective tissue and an increase in the spaces between fibers. Such loss of muscle fibers no doubt accounts in large measure for the lower muscular strength of elderly humans as well.

The nervous system shows a similar decline. The number of nerve fibers in a nerve trunk decreases by 27 per cent at advanced ages. In the kidney the disappearance of cells is accompanied by a reduction in the number of nephrons. According to counts made by Robert A.

Moore of the State University of New York Downstate Medical Center, the number drops over the life span from 800,000 to 450,000. A final example: The number of taste buds per papilla of the tongue falls from an average of 245 in young adults to 88 in subjects aged 70 to 85 years.

In early life the body is endowed with tremendous reserve capacities. The loss of a few hundred or a few thousand cells hardly affects the performance of an organ. As age advances and losses accumulate, however, impairments develop; eventually the stress of daily living or disease imposes demands beyond the reserve capacity of the organism.

Understanding of the process of aging will ultimately require discovery of the factors that cause the death and disappearance of individual cells. It may be that the death of individual cells is merely a chance event that occurs on a statistical basis in most tissues. It is more likely that changes in the internal metabolism of a cell damage its capacity for self-repair and reproduction. Biochemical study of the life processes in tissue cells is in its infancy. Within the past few years investigation of tissues from the rat and lower organisms such as the hydra have begun to reveal the effects of age on enzyme activity in body cells.

One key to understanding aging and particularly to taking action that might extend the human life span can be found in the differences in the rate of aging observed in different individuals. These differences indicate that many factors play a role in aging. When we know why some people age less rapidly than others, we may be able to create conditions that will minimize the loss of functioning cells and tissues, thereby enabling many more people to live as long as those who live longest today.

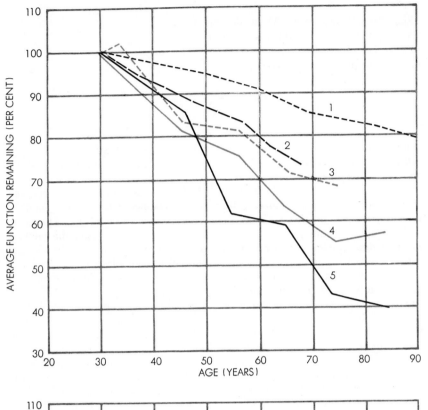

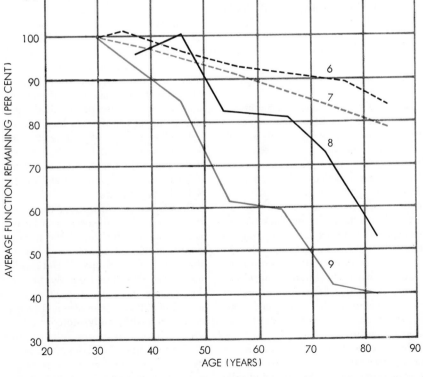

1 BASAL METABOLIC RATE
2 WORK RATE
3 CARDIAC OUTPUT (AT REST)
4 VITAL CAPACITY OF LUNGS
5 MAXIMUM BREATHING CAPACITY (VOLUNTARY)
6 NERVE CONDUCTION VELOCITY
7 BODY WATER CONTENT
8 FILTRATION RATE OF KIDNEY
9 KIDNEY PLASMA FLOW

PERCENTAGE CHANGES WITH AGE for nine different physiological functions are shown in these two diagrams. The average value for each function at age 30 is taken as 100 per cent. Small drop in basal metabolism (1) is probably due simply to loss of cells.

25

Human Cells and Aging

by Leonard Hayflick
March 1968

When normal cells are grown outside the body, their capacity to survive dwindles after a period of time. This deterioration may well represent aging and an ultimate limit to the span of life

The common impression that modern medicine has lengthened the human life-span is not supported by either vital statistics or biological evidence. To be sure, the 20th-century advances in control of infectious diseases and of certain other causes of death have improved the longevity of the human population as a whole. These accomplishments in medicine and public health, however, have merely extended the *average* life expectancy by allowing more people to reach the upper limit, which for the general run of mankind still seems to be approximately the Biblical fourscore years. Aging and a limited life-span apparently are characteristic of all animals that stop growing after reaching a fixed, mature size. In the case of man, after the age of 30 there is a steady, inexorable increase in the probability of death from one cause or another; the probability doubles about every eight years as one grows older. This general probability is such that even if the major causes of death in old age—heart disease, stroke and cancer—were eliminated, the average life expectancy would not be lengthened by much more than 10 years. It would then be about 80 years instead of the expectancy of about 70 years that now prevails in advanced countries.

Could man's life-span be extended—or is there an inescapable aging mechanism that restricts human longevity to the present apparent limit? Until recently few biologists ventured to attempt to explore the basic processes of aging; obviously the subject does not easily lend itself to detailed study. It is now receiving considerable attention, however, in a number of laboratories [see the article "The Physiology of Aging," by Nathan W. Shock, beginning on page 243]. In this article I shall discuss some new findings at the cellular level.

No doubt many mechanisms are involved in the aging of the body. At the cell level at least three aging processes are under investigation. One is a possible decline in the functional efficiency of nondividing, highly specialized cells, such as nerve and muscle cells. Another is the progressive stiffening with age of the structural protein collagen, which constitutes more than a third of all the body protein and serves as the general binding substance of the skin, muscular and vascular systems [see "The Aging of Collagen," by Frederic Verzár; SCIENTIFIC AMERICAN Offprint 155]. In our own laboratory at the Wistar Institute we have addressed ourselves to a third question: the limitation on cell division. Our studies have focused particularly on the structural cells called fibroblasts, which produce collagen and fibrin. These cells, like certain other "blast" cells, go on dividing in the adult body. We set out to determine whether human fibroblasts in a cell culture could divide indefinitely or had only a finite capacity for doing so.

Alexis Carrel's famous experiments more than a generation ago suggested that animal cells per se (that is, cells removed from the body's regulatory mechanisms) might be immortal. He apparently succeeded in keeping chick fibroblasts growing and multiplying in glass vessels for more than 30 years—a great deal longer than a chicken's life expectancy. Later experimenters reported similar successes with embryonic cells from laboratory mice. It has since been learned, however, through improved techniques and a better understanding of cell cultures, that the conclusions drawn from those early experiments were erroneous.

In the case of chick fibroblasts it has been repeatedly demonstrated that, if care is taken not to add any living cells to the initial population in the glass vessel, the cell colony will not survive long. The early cultures, including Carrel's, were fed a crude extract taken from chick embryos, and it is now believed these feedings must have contained some living chick cells. That is to say, in all probability the reason the cultures continued to grow indefinitely was that new, viable fibroblasts were introduced into the culture at each feeding.

Restudy of the experiments in culturing mouse cells has brought to light a highly interesting fact. It has been found that when normal cells from a laboratory mouse are cultured in a glass vessel, they frequently undergo a spontaneous transformation that enables them to divide and multiply indefinitely. This type of transformation takes place regularly in cultures of mouse cells but only rarely in cultures of the fibroblasts of man or other animals. These transformed cell populations have several abnormal properties, but they are truly immortal: many of the mouse-derived cultures have survived for decades. Similarly, the famous line of transformed human cells called HeLa, originally derived from cervical tissue in 1952 by George O. Gey of the Johns Hopkins University School of Medicine, is still growing and multiplying in glass cultures.

On microscopic examination the transformed cells show themselves to be indeed abnormal. Instead of the normal number of 46 chromosomes in a human diploid cell, the "mixoploid" HeLa cells may have anywhere from 50 to 350 chromosomes per cell. They differ from normal chromosomes in size and shape and also stain differently. Moreover, they often behave like cancer cells: inoculated into a suitable laboratory animal, they can grow as tumors. This property of transformed cells has become an important tool in investigations of the

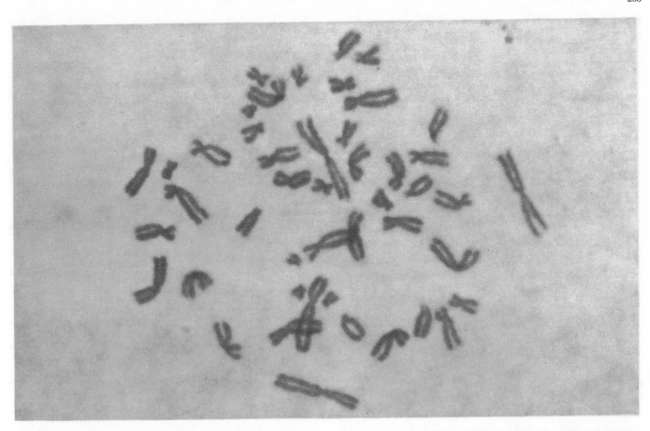

HUMAN CHROMOSOMES are seen magnified 3,000 diameters in a normal diploid cell grown in culture. The chromosomes assume this compact form during mitotic cell division. When grown in culture, human cells display a limited capacity to divide; their finite longevity may be related to the finite span of human life. The chromosomes in the picture are stained with a dye called Giemsa.

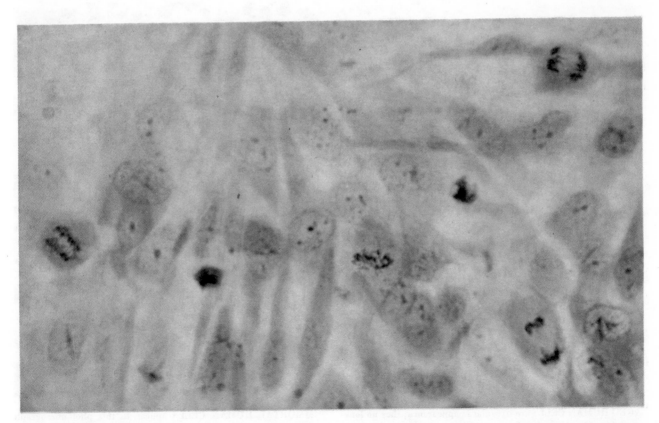

STAGES OF CELL DIVISION appear in this photomicrograph of human cells. The chromosomes are seen as dark clusters. Those in the nucleus at the center have not begun to divide. At left is a nucleus in which the chromosome cluster has separated into two parts; at right, top and bottom, are two nuclei in a later stage of the division cycle. The culture is stained with aceto-orcein and green counterstain. The magnification is 750 diameters. Both photomicrographs were made by Paul S. Moorhead of the Wistar Institute.

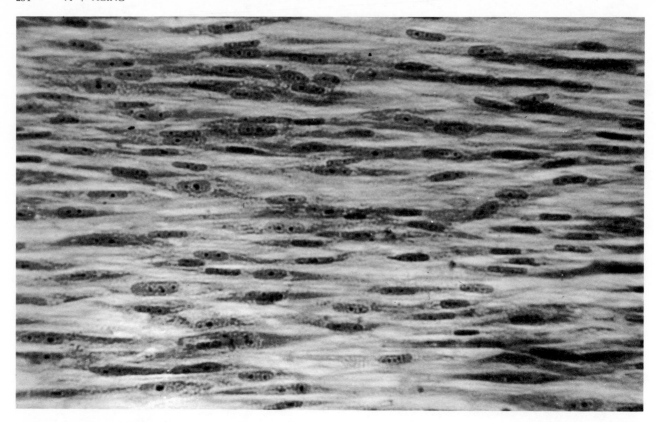

GROWING HUMAN CELLS cover the surface of the glass vessel in which they were planted. They form a layer one cell deep. Under proper conditions this normal cell population will continue to proliferate; it will not, except in unusual cases, multiply indefinitely.

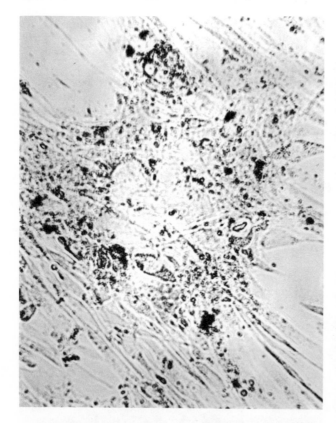

AGED HUMAN CELLS are irregular in appearance; they no longer divide. The aging of such normal cell populations is apparently due to an intrinsic process, not a deficiency in growing conditions. The cells were taken from lung tissue and grown in glass. This photomicrograph and the one at top were made by the author.

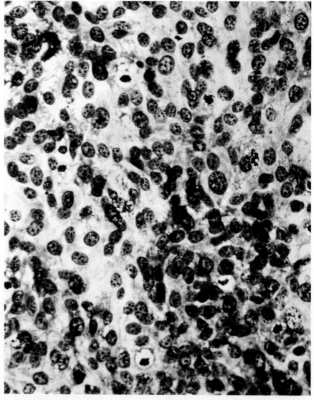

IMMORTAL HUMAN CELLS appear in this photomicrograph made by Fred C. Jensen of the Wistar Institute. The cells, once normal like those at top, have been treated with the monkey virus SV-40; thus transformed, they can apparently multiply indefinitely. The magnification in these photomicrographs is 300 diameters.

genesis of cancer. Although the spontaneous transformation of human cells is rare, investigators of cancer are making use of the recent discovery that normal human cells can be routinely transformed into cancer cells by exposing them to the monkey virus known as SV-40.

A crucial consideration for the relevance of cells in culture to aging, of course, is that they are normal cells. Our interest is therefore directed not to abnormal cells but to the observation that normal cells do not divide indefinitely. Whereas a population of transformed cells will proliferate and survive for decades in cell culture, no one has succeeded in perpetuating a culture of normal animal cells. The same is true of cells implanted in a living animal. Transformed cells will go on growing indefinitely in a series of tissue transplants from animal to animal, but normal cells will not. Peter L. Krohn of the University of Birmingham has shown, for example, that normal mouse skin can survive only a limited number of serial grafts from one mouse to another in the same inbred strain.

Over the past seven years in our laboratory Paul S. Moorhead and I have been studying cell cultures of normal human fibroblasts. Unlike highly specialized cells, fibroblasts (which serve as the structural bricks for most body tissues) will grow and multiply in a nutrient medium in glass bottles. We have used lung tissue as our principal source of the cells. We break down the tissue into separated cells by means of the digestive enzyme trypsin, then remove the trypsin by centrifugation and seed the cells in a bottle containing a suitable growing medium. After a few days of incubation at 99 degrees Fahrenheit the fibroblasts have spread out on the glass surface and begun to divide. In a week or so they cover the entire available surface with a layer one cell deep. Since they will proliferate only in a single layer, we strip off the layer that has covered the bottle surface, again separate the cells with trypsin and plant half of the cells in each of two new bottles with fresh medium. In three or four days the inoculated fibroblasts grow over all the available surface in each bottle, thus doubling the number of cells taken from the original bottle. As this procedure is repeated at four-day intervals, the fibroblasts continue to proliferate in new bottles, doubling in number each time.

We found that fibroblasts taken from four-month-old human embryos doubled in this way about 50 times (the limit ranged between 40 and 60 doublings).

After reaching this limit of capacity for division the cell population died. It could therefore be concluded that human fibroblasts derived from embryonic tissue and grown in cell culture have a finite lifetime amounting to approximately 50 population doublings (which in our culture covered a span of six to eight months).

Further study reinforced this conclusion. It turned out, for example, that if cell division was interrupted and then resumed, the total number of population doublings was not altered. In our experiments we did not, of course, double the number of bottles at each step; after only 10 doubling passages we would have had 1,024 bottles, and 50 doublings of the original seeding of fibroblasts could have produced about 20 million tons of cells! To keep the yield within reasonable bounds we set aside most of the cells from the subcultivations and put them in cold storage. We found they could be kept in suspended animation for apparently unlimited periods; even after six years in storage they proved to be capable of resuming division when they were thawed and placed in a culture medium. They "remembered" the doubling level they had reached before storage and completed the course from that point. For example, cells that had been stored at the 30th doubling went on to divide about 20 more times.

The geometric rate of increase of the cells in culture has made it possible to provide essentially unlimited supplies of the cells for experiments. Samples of one of our strains of normal human fibroblasts (WI-38), "banked" after the eighth doubling in liquid nitrogen at 190 degrees below zero centigrade, have been distributed to hundreds of research laboratories around the world. The stored cells presumably would be available for study far in the future. For instance, well-protected capsules containing frozen cells might be buried in Antarctica or deposited in orbit in the cold of outer space for retrieval many generations hence. Investigators might then be able to use them to study, among other things, whether or not time had brought about any evolutionary change in the aging of man or other animals at the cellular level.

We found further confirmation of the finite lifetime of human fibroblasts when we cultured such cells from adult donors. Samples of lung fibroblasts taken at the time of death from eight adults, ranging in age from 20 to 87, underwent 14 to 29 doublings in cell culture afterward. The number of doublings in these

tests did not show a clear correlation with the age of the donor, but presumably this was because our method of measuring the doubling of cell populations in bottles is not sufficiently precise to disclose such a correlation in detail. Our current experiments do suggest, however, that consistent differences can be found between broad age groups. It appears that fibroblasts from human embryos will divide in cell culture 50 ± 10 times, those from persons between birth and the age of about 20 will divide 30 ± 10 times and those from donors over 20 will divide 20 ± 10 times.

We have tested fibroblasts from several human embryonic tissues besides lung tissue and found that they too are limited to a total of about 50 divisions in culture. The doubling lifetimes of fibroblasts from animals other than man have also been studied. As one would expect, the cells of the shorter-lived vertebrates show less capacity for division. For example, normal fibroblasts from embryos of chickens, rats, mice, hamsters and guinea pigs usually double no more than 15 times in cell culture, and cells that have been taken from adults of the same species undergo considerably fewer than 15 divisions.

Early in our experiments it became evident that we had to examine the possibility that a lethal factor in the culture medium or a defect in the culture technique might be responsible for the limitation of division and ultimate death of the cells. Did the cells stop dividing because of some lack in the nutrient mixture or the presence of contaminating microorganisms such as viruses? We explored these possibilities by various experiments, one of which consisted in culturing a mixture of normal fibroblasts taken from male and female donors. Female cells can be distinguished from male cells either by the presence of special chromatin bodies (found only in female cells) or by the visible difference between the XX female sex chromosomes and the XY male chromosomes. As a consequence we were able to use these cell "markers" as a label for following the progeny of the respective original parents.

We seeded a bottle with a certain number of fibroblasts from a male population that had undergone 40 doublings and with an equal number of fibroblasts from a female population that had gone through only 10 doublings. If some inadequacy of the culture or the accumulation of a killing factor in the medium were the primary cause of cell death in our cultures, then in this experiment the number of doublings in culture should have been the same for both the male and the female cells after we had mixed them together; there is no reason to suppose that a nutritional inadequacy or a lethal factor such as a virus would act preferentially on the cells of a particular sex. Actually the male and female cells composing the mixture, presumably because of their difference in age, proved to have sharply different survival rates. Most of the male population, having already undergone 40 doublings before the mixture was made, died off after 10 more doublings of the mixed culture, whereas the "young" female population of cells in the same mixed culture, with only 10 previous doublings, went on dividing for many more than the male. After 25 doublings all the male cells had disappeared and the culture contained only female cells. These results appear to confirm in an unambiguous way that the life-span of fibroblast populations in our cell cultures is determined by intrinsic aging of the cells rather than by external agencies.

Does this aging result from depletion or dilution of the cells' own chemical resources? We considered the possibility that the eventual death of the cells might be attributable to the exhaustion of some essential metabolite the cells could not synthesize from the culture medium. If that were the case, however, the original

store of this substance must be very large indeed to enable the cells to multiply for 50 generations. Simple mathematics showed that in order to provide at least one molecule of the hypothetical substance for each cell by the 50th doubling, the original cell would have to have at least three times its known weight even if the substance in question were the lightest element (hydrogen) and

the original parent cell were composed entirely of that element!

By isolating individual cells and developing clones (colonies) from them we have been able to establish that each human fibroblast from an embryo is capable of giving rise to about 50 doubling generations in cell culture. As the cell proliferation proceeds there is a gradual decline in the capacity for reproduction.

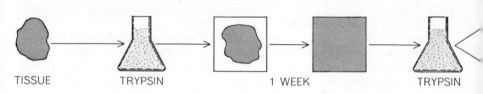

CULTIVATION OF HUMAN CELLS in the author's laboratory begins with the breakdown of lung tissue into separate cells. This is accomplished by means of the digestive enzyme

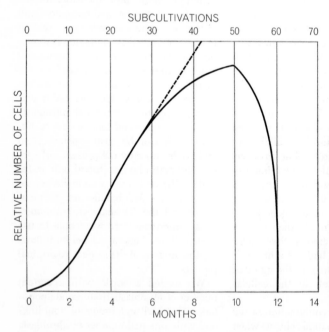

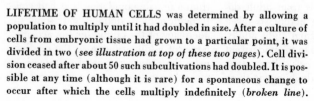

LIFETIME OF HUMAN CELLS was determined by allowing a population to multiply until it had doubled in size. After a culture of cells from embryonic tissue had grown to a particular point, it was divided in two (see illustration at top of these two pages). Cell division ceased after about 50 such subcultivations had doubled. It is possible at any time (although it is rare) for a spontaneous change to occur after which the cells multiply indefinitely (broken line).

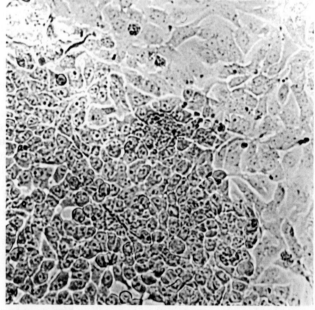

TRANSFORMED HUMAN CELLS are distinguished by their morphology and reaction to staining. The darker amnion cells, forming a large island at lower left, have undergone a spontaneous transformation. They will continue to divide after the neighboring cells have died. Transformed, or "mixoploid," cells have more chromosomes than diploid cells do; they are utilized in cancer research. The magnification of this photomicrograph is about 180 diameters.

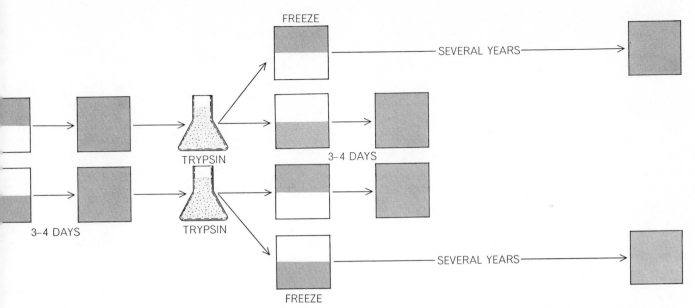

FREEZE

SEVERAL YEARS

TRYPSIN

3–4 DAYS

3–4 DAYS

TRYPSIN

SEVERAL YEARS

FREEZE

trypsin. After the cells, seeded in a bottle, have multiplied to cover its surface, they are again treated with trypsin, then divided into two halves and replanted. Most cells thus grown are placed in cold storage. Thawed and planted years later, they resume division.

Investigators in several laboratories have found that in successive generations of the multiplying cells a larger and larger fraction of the progeny becomes incapable of dividing until, by the 45th or 50th doubling generation, the entire population has lost this ability.

Curiously, as the population of human fibroblasts approaches the end of its lifetime, aberrations often crop up in the chromosomes. Chromosome aberrations and cell-division peculiarities related to age have also been observed in human leucocytes and in the liver tissue of living mice. The question of whether or not cell abnormalities are a common accompaniment of human aging remains a moot point, however; there is no clear clinical or laboratory evidence on the matter.

Our information on the aging of cells so far is limited to what we can observe in cell cultures in glassware. It has not yet been established that fibroblasts behave the same way in living animals as they do in an artificial culture. However, in man several organs lose weight after middle age, and this is directly attributable to cell loss. The human brain weighs considerably less in old age than it does in the middle years, the kidney also shows a large reduction in nephrons accompanying cell loss and the number of taste buds per papilla of the tongue drops from an average of 245 in young adults to 88 in the aged. We cannot be certain that fibroblasts stop dividing or divide at a lower rate as an animal ages or that the bodily signs of aging can be explained on that basis. It is well known, however, that certain cell systems in animals do stop dividing and die in the normal course of development. Familiar examples are the larval tissues of insects, the tail and gills of the tadpole and some embryonic kidney tissues (the pronephros and mesonephros) in the higher vertebrates.

Reviewing these phenomena, John W. Saunders, Jr., of the School of Veterinary Medicine of the University of Pennsylvania has suggested that the death of cells resulting in the demise of specific tissues is a normal, programmed event in the development of multicellular animals. By the same reasoning we can surmise that the aging and finite lifetime of normal cells constitute a programmed mechanism that sets an overall limit on an organism's length of life. This would suggest that, even if we were able to checkmate all the incidental causes of human aging, human beings would inevitably still succumb to the ultimate failure of the normal cells to divide or function.

In this connection it is interesting to consider what engineers call the "mean time to failure" in the lifetime of machines. Every machine embodies built-in obsolescence (intentional or unintentional) in the sense that its useful lifetime is limited and more or less predictable from consideration of the durability of its parts. By the repair or replacement of elements of a machine as they fail its lifetime can be extended, but barring total replacement of all the elements eventual "death" of the machine as a functioning system is inevitable.

What might determine the "mean time to failure" of an animal organism? I suggest that animal aging may result from deterioration of the genetic program that orchestrates the development of cells. As time goes on, the DNA of dividing cells may become clouded with an accumulation of copying errors (analogous to the "noise" that develops in the serial copying of a photograph). The coding and decoding system that governs the replication of DNA operates with a high degree of accuracy, but the accuracy is not absolute. Moreover, there is some experimental evidence that, as Leslie E. Orgel of the Salk Institute for Biological Studies has suggested, certain enzymes involved in the transcription of information from DNA for the synthesis of proteins may deteriorate with age. At all events, since the ability of cells to divide or to function is controlled by the inherited information-containing molecules, it seems likely that some inherent degeneration of these molecules may hold the key to the aging and eventual death of cells.

Pursuing the machine analogy, we might surmise that man is endowed with a longer life-span than other mammals because human cells have evolved a more effective system for correcting or repairing errors as they arise. Such an evolution would account for the generally progressive lengthening of the fixed life-span from the lower to the higher animals; presumably the march of evolution has developed improvements in the cells' error-repairing mechanisms. It is clear, however, that even in man this system is far from perfect. In the idiom of computer engineers we might say that man, like all other animals, has a "mean time to failure" because his normal cells eventually run out of accurate program and capacity for repair.

EPILOGUE

The Present Evolution of Man

by Theodosius Dobzhansky
September 1960

Man still evolves by natural selection for his environment, but it is now an environment largely of his own making. Moreover, he may be changing the environment faster than he can change biologically

Any discussion of the evolution of the human species deals with a natural process that has transcended itself. Only once before, when life originated out of inorganic matter, has there occurred a comparable event.

After that first momentous step, living forms evolved by adapting to their environments. Adaptation—the maintenance or advancement of conformity between an organism and its surroundings—takes place through natural selection. The raw materials with which natural selection works are supplied by mutation and sexual recombination of hereditary units: the genes.

Mutation, sexual recombination and natural selection led to the emergence of *Homo sapiens*. The creatures that preceded him had already developed the rudiments of tool-using, toolmaking and cultural transmission. But the next evolutionary step was so great as to constitute a difference in kind from those before it. There now appeared an organism whose mastery of technology and of symbolic communication enabled it to create a supraorganic culture. Other organisms adapt to their environments by changing their genes in accordance with the demands of the surroundings. Man and man alone can also adapt by changing his environments to fit his genes. His genes enable him to invent new tools, to alter his opinions, his aims and his conduct, to acquire new knowledge and new wisdom.

The authors of the preceding articles have shown how the possession of these faculties brought the human species to its present biological eminence. Man has spread to every section of the earth, bringing high culture to much of it. He is now the most numerous of the mammals. By these or any other reasonable standards, he is by far the most successful product of biological evolution.

For better or worse, biological evolution did not stop when culture appeared. In this final article we address ourselves to the question of where evolution is now taking man. The literature of this subject has not lacked for prophets who wish to divine man's eventual fate. In our age of anxiety, prediction of final extinction has become the fashionable view, replacing the hopes for emergence of a race of demigods that more optimistic authorities used to foresee. Our purpose is less ambitious. What biological evolutionary processes are now at work is a problem both serious and complex enough to occupy us here.

The impact of human works on the environment is so strong that it has become very hard to make out the forces to which the human species is now adjusting. It has even been argued that *Homo sapiens* has already emancipated himself from the operation of natural selection. At the other extreme are those who still assume that man is nothing but an animal. The second fallacy is the more pernicious, leading as it does to theories of biological racism and the justification of race and class prejudice which are bringing suffering to millions of people from South Africa to Arkansas. Assuming that man's genetic endowment can be ignored is the converse falsehood, perhaps less disastrous in its immediate effects, but more insidious in the long run.

Like all other animals, man remains the product of his biological inheritance. The first, and basic, feature of his present evolution is that his genes continue to mutate, as they have since he first appeared. Every one of the tens of thousands of genes inherited by an individual has a tiny probability of changing in some way during his generation. Among the small, and probably atypical, sample of human genes for which very rough estimates of the mutation frequencies are available, the rates of mutation vary from one in 10,000 to one in about 250,000. For example, it has been calculated that approximately one sex cell in every 50,000 produced by a normal person carries a new mutant gene causing retinoblastoma, a cancer of the eye affecting children.

These figures are "spontaneous" frequencies in people not exposed to any special agents that can induce mutation. As is now widely known, the existence of such agents, including ionizing radiation and certain chemicals, has been demonstrated with organisms other than man. New mutagens are constantly being discovered. It can hardly be doubted that at least some of them affect human genes. As a consequence the members of an industrial civilization have increased genetic variability through rising mutation rates.

There is no question that many mutations produce hereditary diseases, malformations and constitutional weaknesses of various kinds. Some few must also be useful, at least in certain environments; otherwise there would be no evolution. (Useful mutants have actually been observed in experiments on lower organisms.) But what about minor variations that produce a little more or a little less hair, a slightly longer or a slightly shorter nose, blood of type O or type A? These traits seem neither useful nor harmful. Here, however, we must proceed with the greatest caution. Beneficial or damaging effects of ostensibly neutral traits may eventually be discovered. For example, recent evidence indicates that people with blood of type O have a slightly higher rate of duodenal ulcer than does the general population. Does it follow that O blood is bad? Not necessarily; it is the most frequent type in many populations, and it may conceivably confer some advantages yet undiscovered.

Still other mutants that are detrimental when present in double dose (the so-called homozygous condition, where the same type of gene has been inherited from both parents) lead to hybrid vigor in single dose (the heterozygous condition). How frequently this happens is uncertain. The effect surely operates in the breeding of domestic animals and plants, and it has been detected among X-ray-induced mutations in fruit flies. Only one case is thus far known in man. Anthony C. Allison of the University of Oxford has found that the gene causing sickle-cell anemia in the homozygous condition makes its heterozygous carriers relatively resistant to certain forms of malaria. This gene is very frequent in the native population of the central African lowlands, where malaria has long been endemic, and relatively rare in the inhabitants of the more salubrious highlands. Certainly there are other such adaptively ambivalent genes in human populations, but we do not know how many.

Despite these uncertainties, which cannot be glossed over, it is generally agreed among geneticists that the effects of mutation are on the average detrimental. Any increase of mutation rate, no matter how small, can only augment the mass of human misery due to defective heredity. The matter has rightly attracted wide attention in connection with ionizing radiation from military and industrial operations and medical X-rays. Yet these form only a part of a larger and more portentous issue.

Of the almost countless mutant genes that have arisen since life on earth began, only a minute fraction were preserved. They were preserved because they were useful, or at least not very harmful, to their possessors. A great majority of gene changes were eliminated. The agency that preserved useful mutants and eliminated injurious ones was natural selection. Is natural selection still operating in mankind, and can it be trusted to keep man fit to live in environments created by his civilization?

One must beware of words taken from everyday language to construct scientific terminology. "Natural" in "natural selection" does not mean the state of affairs preceding or excluding man-made changes. Artificially or not, man's environment has altered. Would it now be natural to try to make your living as a Stone Age hunter?

Then there are phrases like "the struggle for life" and "survival of the fittest." Now "struggle" was to Darwin a metaphor. Animals struggle against cold by growing warm fur, and plants against dryness by reducing the evaporating leaf surface. It was the school of so-called social Darwinists (to which Darwin did not belong) who equated "struggle" with violence, warfare and competition

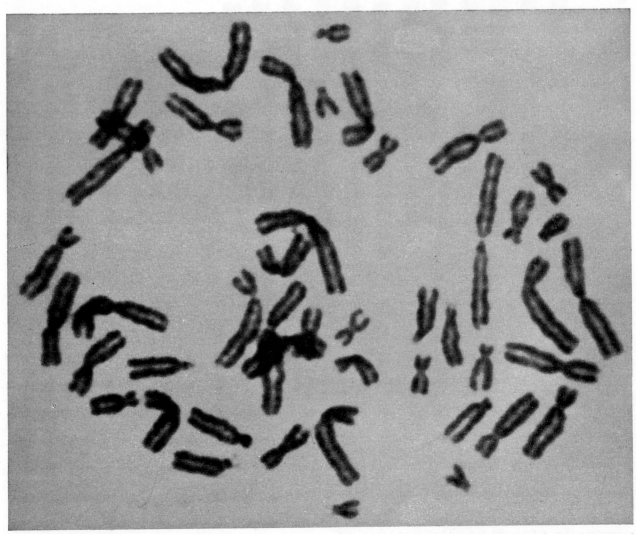

HUMAN CHROMOSOMES are enlarged some 5,000 times in this photomicrograph made by J. H. Tjio and Theodore T. Puck at the University of Colorado Medical Center. The photomicrograph shows all of the 23 pairs of chromosomes in a dividing body cell.

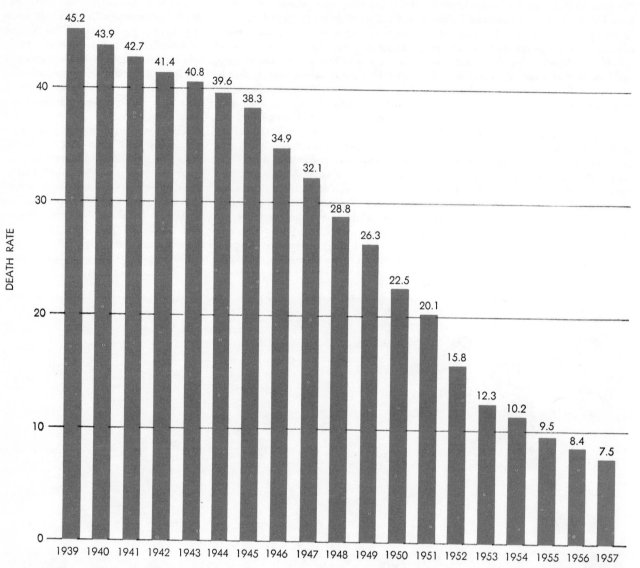

GENERAL DISTRIBUTION OF PENICILLIN
INTRODUCTION OF STREPTOMYCIN IN HOSPITALS

GENERAL DISTRIBUTION OF STREPTOMYCIN

INTRODUCTION OF PENICILLIN IN HOSPITALS

INTRODUCTION OF ISONIAZID

TUBERCULOSIS DEATH RATE per 100,000 people showed a dramatic decline with the introduction of antibiotics and later of the antituberculosis drug isoniazid. As tuberculosis becomes less prevalent, so does its threat to genetically susceptible individuals, who are enabled to survive and reproduce. The chart is based upon information from the U. S. National Office of Vital Statistics.

without quarter. The idea has long been discredited.

We do not deny the reality of competition and combat in nature, but suggest that they do not tell the whole story. Struggle for existence may be won not only by strife but also by mutual help. The surviving fit in human societies may in some circumstances be those with the strongest fists and the greatest readiness to use them. In others they may be those who live in peace with their neighbors and assist them in hour of need. Indeed, co-operation has a long and honorable record. The first human societies, the

hunters of the Old Stone Age, depended on co-operation to kill big game.

Moreover, modern genetics shows that "fitness" has a quite special meaning in connection with evolution. Biologists now speak of Darwinian fitness, or adaptive value, or selective value in a reproductive sense. Consider the condition known as achondroplastic dwarfism, caused by a gene mutation that produces people with normal heads and trunks, but short arms and legs. As adults they may enjoy good health. Nevertheless, E. T. Mørch in Denmark has discovered that achondroplastic dwarfs produce, on

the average, only some 20 surviving children for every 100 children produced by their normal brothers and sisters. In technical terms we say that the Darwinian fitness of achondroplasts is .2 or, alternatively that achondroplastic dwarfism is opposed by a selection-coefficient of .8.

This is a very strong selection, and the reasons for it are only partly understood. What matters from an evolutionary point of view is that achondroplasts are much less efficient in transmitting their genes to the following generations than are nondwarfs. Darwinian fitness is reproductive fitness. Genetically the surviving fittest is neither superman nor conquering hero; he is merely the parent of the largest surviving progeny.

With these definitions in mind, we can answer the question whether natural selection is still active in mankind by considering how such selection might be set aside. If all adults married, and each couple produced exactly the same number of children, all of whom survived to get married in turn and so on,

there would be no selection at all. Alternatively, the number of children, if any, that each person produced might be determined by himself or some outside authority on the basis of the desirability of his hereditary endowment. This would be replacing natural selection by artificial selection. Some day it may come to pass. Meantime natural selection is going on.

It goes on, however, always within the context of environment. As that changes,

the Darwinian fitness of various traits changes with it. Thus by his own efforts man is continually altering the selective pressure for or against certain genes.

The most obvious example, and one with disturbing overtones, is to be found in the advance of medicine and public health. Retinoblastoma, the eye cancer of children, is almost always fatal if it is not treated. Here is "natural" selection at its most rigorous, weeding out virtually all of the harmful mutant genes

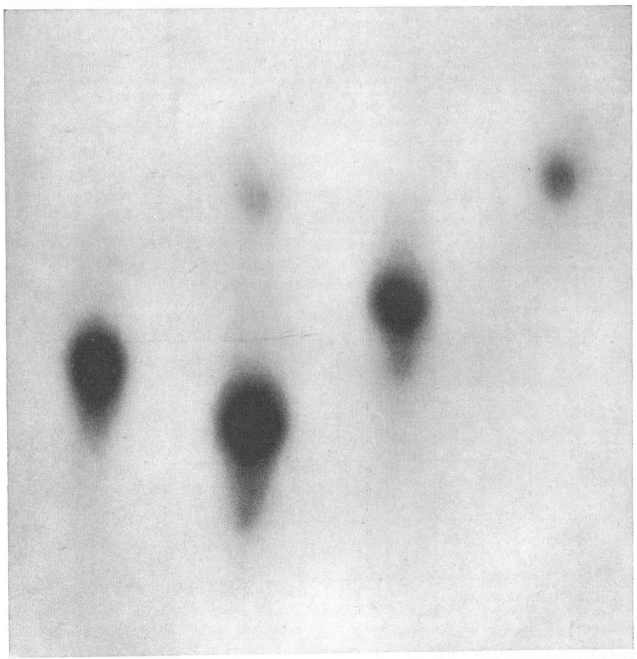

CHEMICAL STRUCTURE OF HEMOGLOBIN in an individual's blood is determined by his genes. Normal and abnormal hemoglobins move at different speeds in an electric field. This photograph, made by Henry G. Kunkel of the Rockefeller Institute, shows surface of a slab of moist starch on which samples of four kinds of human hemoglobin were lined up at top, between a negative electrode at top and a positive electrode at bottom (electrodes are not shown). When current was turned on, the samples migrated toward positive electrode. At right hemoglobin C, the cause of a rare hereditary anemia, has moved down only a short way. Second from right is hemoglobin S, the cause of sickle-cell anemia, which has moved farther in same length of time. Normal hemoglobin, third from right, has separated into its A and A_2 constituents. At left is normal fetal hemoglobin F, obtained from an umbilical cord.

before they can be passed on even once. With proper treatment, however, almost 70 per cent of the carriers of the gene for retinoblastoma survive, become able to reproduce and therefore to transmit the defect to half their children.

More dramatic, if genetically less clear-cut, instances are afforded by advances in the control of tuberculosis and malaria. A century ago the annual death rate from tuberculosis in industrially advanced countries was close to 500 per 100,000. Improvement in living conditions and, more recently, the advent of antibiotic drugs have reduced the death rate to 7.5 per 100,000 in the U. S. today. A similarly steep decline is under way in the mortality from malaria, which used to afflict a seventh of the earth's population.

Being infectious, tuberculosis and malaria are hazards of the environment. There is good evidence, however, that individual susceptibility, both as to contracting the infection and as to the severity of the disease, is genetically conditioned. (We have already mentioned the protective effect of the gene for sickle-cell anemia. This is probably only one of several forms of genetic resistance to malaria.) As the prevalence of these diseases decreases, so does the threat to susceptible individuals. In other words, the Darwinian fitness of such individuals has increased.

It was pointed out earlier that one effect of civilization is to increase mutation rates and hence the supply of harmful genes. A second effect is to decrease the rate of discrimination against such genes, and consequently the rate of their elimination from human populations by natural selection. In thus disturbing

the former genetic equilibrium of inflow and outflow, is man not frustrating natural selection and polluting his genetic pool?

The danger exists and cannot be ignored. But in the present state of knowledge the problem is tremendously complex. If our culture has an ideal, it is the sacredness of human life. A society that refused, on eugenic grounds, to cure children of retinoblastoma would, in our eyes, lose more by moral degradation than it gained genetically. Not so easy, however, is the question whether a person who knows he carries the gene for retinoblastoma, or a similarly deleterious gene, has a right to have children.

Even here the genetic issue is clear, although the moral issue may not be. This is no longer true when we come to genes that are harmful in double dose,

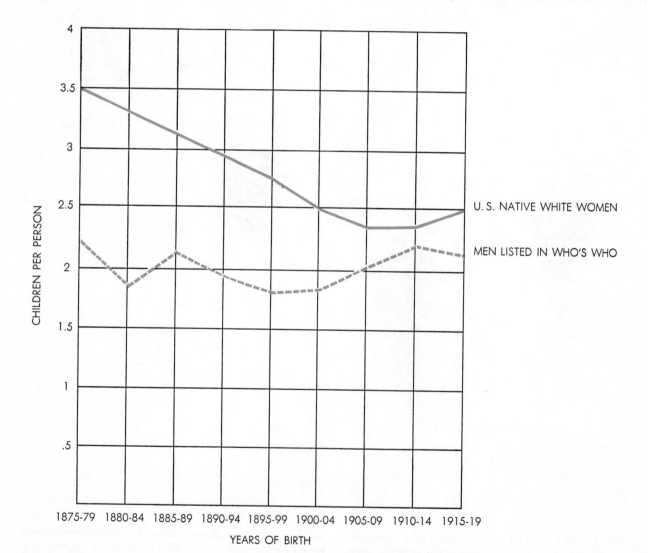

FERTILITY RATE among relatively intelligent people, as represented by a random sample of men listed in *Who's Who in America* for 1956 and 1957, is lower than fertility rate of the U. S. population as a whole, as represented by all native white women. The two fertility rates have recently been moving toward each other. Vertical scale shows average number of children per person; horizontal scale shows approximate birth date of parents. Chart is based upon information collected by Dudley Kirk of the Population Council.

but beneficial in single. If the central African peoples had decided some time ago to breed out the sickle-cell gene, they might have succumbed in much larger numbers to malaria. Fortunately this particular dilemma has been resolved by successful methods of mosquito control. How many other hereditary diseases and malformations are maintained by the advantages their genes confer in heterozygous carriers, we simply do not know.

Conversely, we cannot yet predict the genetic effect of relaxing selection pressure. If, for example, susceptibility to tuberculosis is maintained by recurrent mutations, then the conquest of the disease should increase the concentration of mutant genes as time goes on. On the other hand, if resistance arises from a single dose of genes that make for susceptibility in the double dose, the effects of eradication become much less clear. Other selective forces might then determine the fate of these genes in the population.

In any case, although we cannot see all the consequences, we can be sure that ancient genetic patterns will continue to shift under the shelter of modern medicine. We would not wish it otherwise. It may well be, however, that the social cost of maintaining some genetic variants will be so great that artificial selection against them is ethically, as well as economically, the most acceptable and wisest solution.

If the evolutionary impact of such biological tools as antibiotics and vaccines is still unclear, then computers and rockets, to say nothing of social organizations as a whole, present an even deeper puzzle. There is no doubt that human survival will continue to depend more and more on human intellect and technology. It is idle to argue whether this is good or bad. The point of no return was passed long ago, before anyone knew it was happening.

But to grant that the situation is inevitable is not to ignore the problems it raises. Selection in modern societies does not always encourage characteristics that we regard as desirable. Let us consider one example. Much has been written about the differential fertility that in advanced human societies favors less intelligent over more intelligent people. Studies in several countries have shown that school children from large families tend to score lower on so-called intelligence tests than their classmates with few or no brothers and sisters. Moreover, parents who score lower on these tests have more children on the average

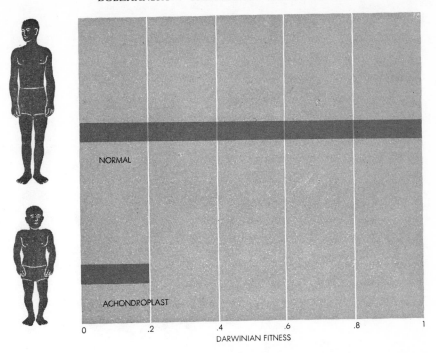

DARWINIAN FITNESS of achondroplastic dwarfs is low. Dwarfs may be healthy, but they have only 20 surviving children to every 100 surviving children of normal parents.

than those who get higher marks.

We cannot put our finger on the forces responsible for this presumed selection against intelligence. As a matter of fact, there is some evidence that matters are changing, in the U. S. at least. People included in *Who's Who in America* (assuming that people listed in this directory are on the average more intelligent than people not listed there) had fewer children than the general population during the period from 1875 to 1904. In the next two decades, however, the difference seemed to be disappearing. L. S. Penrose of University College London, one of the outstanding human geneticists, has pointed out that a negative correlation between intelligence and family size may in part be corrected by the relative infertility of low-grade mental defectives. He suggests that selection may thus be working toward maintaining a constant level of genetic conditioning for intelligence in human populations. The evidence presently available is insufficient either to prove or to contradict this hypothesis.

It must also be recognized that in man and other social animals qualities making for successful individuals are not necessarily those most useful to the society as a whole. If there were a gene for altruism, natural selection might well discriminate against it on the individual level, but favor it on the population level. In that case the fate of the gene would be hard to predict.

If this article has asked many more questions than it has answered, the purpose is to suggest that answers be sought with all possible speed. Natural selection is a very remarkable phenomenon. But it does not even guarantee the survival of a species. Most living forms have become extinct without the "softening" influence of civilization, simply by becoming too narrowly specialized. Natural selection is opportunistic; in shaping an organism to fit its surroundings it may leave the organism unable to cope with a change in environment. In this light, man's explosive ability to change his environment may offer as much threat as promise. Technological evolution may have outstripped biological evolution.

Yet man is the only product of biological evolution who knows that he has evolved and is evolving further. He should be able to replace the blind force of natural selection by conscious direction, based on his knowledge of nature and on his values. It is as certain that such direction will be needed as it is questionable whether man is ready to provide it. He is unready because his knowledge of his own nature and its evolution is insufficient; because a vast majority of people are unaware of the necessity of facing the problem; and because there is so wide a gap between the way people actually live and the values and ideals to which they pay lip service.

BIBLIOGRAPHIES

I NUTRITIONAL INFLUENCES

1 The Physiology of Starvation

THE BIOLOGY OF HUMAN STARVATION. Ancel Keys, Josef Brozek, Austin Henschel, Olaf Mickelsen and Henry Longstreet Taylor. The University of Minnesota Press, 1950.

PROTEIN MALNUTRITION IN YOUNG CHILDREN. Nevin S. Scrimshaw and Moisés Béhar in Science, Vol. 133, No. 3470, pages 2039–2047; June 30, 1961.

MAN UNDER CALORIC DEFICIENCY. Francisco Grande in Handbook of Physiology, Section 4: Adaptation to the Environment. American Physiological Society, 1964.

STARVATION IN MAN. George F. Cahill, Jr., in The New England Journal of Medicine, Vol. 282, No. 12, pages 668–675; March 19, 1970.

2 Endemic Goiter

ETIOLOGY AND PREVENTION OF SIMPLE GOITER. David Marine in Medicine, Vol. 3, No. 4, pages 453–479; November, 1924.

THYROXINE. Edward C. Kendall. The Chemical Catalog Company, Inc., 1929.

ENDEMIC GOITRE. World Health Organization Monograph Series No. 44. World Health Organization, 1960.

THE THYROID GLAND. Rulon W. Rawson in Clinical Symposia, Ciba Pharmaceutical Company, Vol. 17, No. 2, pages 35–63; April-May-June, 1965.

3 Atherosclerosis

ATHEROMA LESIONS. Cardiovascular Pathology: Vol. 1. Reginald E. B. Hudson. The Williams and Wilkins Company, 1965.

EPIDEMIOLOGY OF CARDIOVASCULAR DISEASES: METHODOLOGY. Edited by Herbert Pollack and Dean E. Krueger. Supplement to American Journal of Public Health and the Nation's Health, Vol. 50, No. 10; October, 1960.

THE ETIOLOGY OF MYOCARDIAL INFARCTION. Edited by Thomas N. James and John W. Keyes. Henry Ford Hospital International Symposium. Little, Brown and Company, 1963.

METABOLISM AND STRUCTURE OF THE ARTERIAL WALL IN ATHEROSCLEROSIS. Abel L. Robertson, Jr., in Cleveland Clinic Quarterly, Vol. 32, No. 3, pages 99–117; July, 1965.

PROBLEMS IN THE STUDY OF CORONARY ATHEROSCLEROSIS IN POPULATION GROUPS. David M. Spain in Annals of the New York Academy of Sciences, Vol. 84, Article 17, pages 816–834; December 8, 1960.

4 Nutrition and the Brain

BRAIN SEROTONIN CONTENT: INCREASE FOLLOWING INGESTION OF CARBOHYDRATE DIET. John D. Fernstrom and Richard J. Wurtman in Science, Vol. 174, No. 4013, pages 1023–1025; December 3, 1971.

L-TRYPTOPHAN, L-TYROSINE, AND THE CONTROL OF BRAIN MONOAMINE BIOSYNTHESIS. R. J. Wurtman and J. D. Fernstrom in Perspectives in Neuropharmacology, edited by Solomom H. Snyder. Oxford University Press, 1972.

BRAIN SEROTONIN CONTENT: PHYSIOLOGICAL REGULATION BY PLASMA NEUTRAL AMINO ACIDS. John D. Fernstrom and Richard J. Wurtman in *Science*, Vol. 178, No. 4059, pages 414–416; October 27, 1972.

CONTROL OF BRAIN 5-HT CONTENT BY DIETARY CARBOHYDRATES. J. D. Fernstrom and R. J. Wurtman in *Serotonin and Behavior*, edited by J. Barchas and E. Usdin. Academic Press, 1973.

EFFECT OF PERINATAL UNDERNUTRITION ON THE METABOLISM OF CATECHOLAMINES IN THE RAT BRAIN. W. J. Shoemaker and R. J. Wurtman in *The Journal of Nutrition*, Vol. 103, No. 11, pages 1537–1547; November, 1973.

5 Lactose and Lactase

A RACIAL DIFFERENCE IN INCIDENCE OF LACTASE DEFICIENCY. Theodore M. Bayless and Norton S. Rosensweig in *The Journal of the American Medical Association*, Vol. 197, No. 12, pages 968–972; September 19, 1966.

MILK. Stuart Patton in *Scientific American*, Vol. 221, No. 1, pages 58–68; July, 1969.

PRIMARY ADULT LACTOSE INTOLERANCE AND THE MILKING HABIT: A PROBLEM IN BIOLOGIC AND CULTURAL INTERRELATIONS. II. A CULTURAL HISTORICAL HYPOTHESIS. Frederick J. Simoons in *The American Journal of Digestive Diseases*, Vol. 15, No. 8, pages 695–710; August, 1970.

LACTASE DEFICIENCY: AN EXAMPLE OF DIETARY EVOLUTION. R. D. McCracken in *Current Anthropology*, Vol. 12, No. 4–5, pages 479–517; October–December, 1971.

MEMORIAL LECTURE: LACTOSE AND LACTASE–A HISTORICAL PERSPECTIVE. Norman Kretchmer in *Gastroenterology*, Vol. 61, No. 6, pages 805–813; December, 1971.

II RESPONSES TO ENVIRONMENTAL CHEMICALS

6 How the Liver Metabolizes Foreign Substances

PHARMACOLOGICAL IMPLICATIONS OF MICROSOMAL ENZYME INDUCTION. A. H. Conney in *Pharmacological Reviews*, Vol. 19, pages 317–366; 1967.

MICROSOMES AND DRUG OXIDATIONS. Edited by R. W. Estabrook, J. R. Gillette and K. C. Liebman. The Williams & Wilkins Company, 1972.

MICROSOMAL HYDROXYLASES: INDUCTION AND ROLE IN POLYCYCLIC HYDROCARBON CARCINOGENESIS AND TOXICITY. Harry V. Gelboin, Nadao Kinoshita and Friedrich J. Wiebel in *Federation Proceedings*, Vol. 31, No. 4, pages 1298–1309; July–August, 1972.

METABOLIC INTERACTIONS AMONG ENVIRONMENTAL CHEMICALS AND DRUGS. A. H. Conney and J. J. Burns in *Science*, Vol. 178, No. 4061, pages 576–586; November 10, 1972.

7 Pesticides and the Reproduction of Birds

PESTICIDES AND THE LIVING LANDSCAPE. Robert L. Rudd. The University of Wisconsin Press, 1964.

PESTICIDE-INDUCED ENZYME BREAKDOWN OF STEROIDS IN BIRDS. D. B. Peakall in *Nature*, Vol. 216, No. 5114, pages 505–506; November 4, 1967.

PEREGRINE FALCON POPULATIONS: THEIR BIOLOGY AND DECLINE. Edited by Joseph J. Hickey. The University of Wisconsin Press, 1969.

MARKED DDE IMPAIRMENT OF MALLARD REPRODUCTION IN CONTROLLED STUDIES. Robert G. Heath, James W. Spann and J. F. Kreitzer in *Nature*, Vol. 224, No. 5214, pages 47–48; October 4, 1969.

8 Lead Poisoning

THE EXPOSURE OF CHILDREN TO LEAD. J. Julian Chisolm, Jr., and Harold E. Harrison in *Pediatrics*, Vol. 18, No. 6, pages 943–958; December, 1956.

THE ANAEMIA OF LEAD POISONING: A REVIEW. H. A. Waldron in *British Journal of Industrial Medicine*, Vol. 23, No. 2, pages 83–100; April, 1966.

THE RENAL TUBULE IN LEAD POISONING, I: MITOCHONDRIAL SWELLING AND AMINOACIDURIA. Robert A. Goyer in *Laboratory Investigation*, Vol. 19, No. 1, pages 71–77; July, 1968.

THE RENAL TUBULE IN LEAD POISONING, II: IN VITRO STUDIES OF MITOCHONDRIAL STRUCTURE AND FUNCTION. Robert A. Goyer, Albert Krall and John P. Kimball in *Laboratory Investigation*, Vol. 19, No. 1, pages 78–83; July, 1968.

THE USE OF CHELATING AGENTS IN THE TREAT-
MENT OF ACUTE AND CHRONIC LEAD INTOXI-
CATION IN CHILDHOOD. J. Julian Chisolm in *The
Journal of Pediatrics*, Vol. 73, No. 1, pages 1–
38; July, 1968.

LEAD POISONING IN CHILDHOOD – COMPREHEN-
SIVE MANAGEMENT AND PREVENTION. J. Julian
Chisolm, Jr., and Eugene Kaplan in *The Journal
of Pediatrics*, Vol. 73, No. 6, pages 942–950;
December, 1968.

III RESPONSES TO THE PHYSICAL ENVIRONMENT

9 The Diving Women of Korea and Japan

THE ISLAND OF THE FISHERWOMEN. Fosco
Maraini. Harcourt, Brace & World, Inc., 1962.

KOREAN SEA WOMEN: A STUDY OF THEIR PHYSI-
OLOGY. The departments of physiology, Yonsei
University College of Medicine, Seoul, and the
State University of New York at Buffalo.

THE PHYSIOLOGICAL STRESSES OF THE AMA.
Hermann Rahn in *Physiology of Breath-Hold
Diving and the Ama of Japan*. Publication 1341,
National Academy of Sciences – National Re-
search Council, 1965.

NATURAL AND SYNTHETIC SOURCES OF CIRCU-
LATING 25-HYDROXYVITAMIN D IN MAN. John
G. Haddad and Theodore J. Hahn in *Nature*,
Vol. 244, No. 5417, pages 515–516; August 24,
1973.

THE EFFECTS OF LIGHT ON MAN AND OTHER
MAMMALS. Richard J. Wurtman in *Annual
Review of Physiology*, Vol. 37, pages 467–483;
1975.

DAILY RHYTHM IN HUMAN URINARY MELATONIN.
H. J. Lynch, R. J. Wurtman, M. A. Moskowitz,
M. C. Archer and M. H. Ho in *Science*, Vol. 187,
No. 4172, pages 169–171; January 17, 1975.

10 The Physiology of High Altitude

EFFECTS OF ALTITUDE ON BROWN FAT AND
METABOLISM OF THE DEER MOUSE, PEROMYS-
CUS. Jane C. Roberts, Raymond J. Hock and Robert E. Smith in *Federation Proceedings*, Vol.
28, No. 3, pages 1065–1072; May–June, 1969.

PHYSIOLOGICAL RESPONSES OF DEER MICE TO
VARIOUS NATIVE ALTITUDES. R. J. Hock in
*The Physiological Effects of High Altitude:
Proceedings of a Symposium Held at Interlaken,
September 18–22, 1962*, edited by W. H.
Weihe. Pergamon Press, 1964.

HUMAN ADAPTATION TO HIGH ALTITUDE. Paul
T. Baker in *Science*, Vol. 163, No. 3872, pages
1149–1156; March 14, 1969.

12 Radiation and the Human Cell

THE ACTION OF RADIATION ON MAMMALIAN
CELLS. T. T. Puck in *The American Naturalist*,
Vol. 44, No. 874, pages 95-109; January-Febru-
ary, 1960.

THE SOMATIC CHROMOSOMES OF MAN. J. H. Tjio
and T. T. Puck in *Proceedings of the National
Academy of Sciences*, Vol. 44, No. 12, pages
1229–1236; December, 1958.

GENETIC MECHANISMS: STRUCTURE AND FUNC-
TION. The Biological Laboratory, Cold Spring
Harbor, L. I., New York, 1956.

THE ORGANIZATION AND DUPLICATION OF
CHROMOSOMES AS REVEALED BY AUTORADIO-
GRAPHIC STUDIES USING TRITIUM-LABELED
THYMIDINE. J. Herbert Taylor, Philip S. Woods
and Walter L. Hughes in *Proceedings of the
National Academy of Sciences*, Vol. 43, No. 1,
pages 122–128; January, 1957.

A SYMPOSIUM ON THE CHEMICAL BASIS OF
HEREDITY. William D. McElroy and Bentley
Glass. Johns Hopkins Press, 1957.

11 The Effects of Light on the Human Body

PREVENTION OF HYPERBILIRUBINEMIA OF PRE-
MATURITY BY PHOTOTHERAPY. J. F. Lucey, M.
Ferreiro and J. Hewitt in *Pediatrics*, Vol. 41,
pages 1047–1056; 1968.

IV RESPONSES TO PSYCHOSOCIAL STRESS

13 Stress and Behavior

ADRENOCORTICAL ACTIVITY AND AVOIDANCE LEARNING AS A FUNCTION OF TIME AFTER AVOIDANCE TRAINING. Seymour Levine and F. Robert Brush in *Physiology & Behavior*, Vol. 2, No. 4, pages 385–388; October, 1967.

HORMONES AND CONDITIONING. Seymour Levine in *Nebraska Symposium on Motivation: 1968*, edited by William J. Arnold. University of Nebraska Press, 1968.

EFFECTS OF PEPTIDE HORMONES ON BEHAVIOR. David de Wied in *Frontiers in Neuroendocrinology*, edited by William F. Ganong and Luciano Martini. Oxford University Press, 1969.

THE NEUROENDOCRINE CONTROL OF PERCEPTION. R. I. Henkin in *Perception and Its Disorders: Proceedings of the Association for Research in Nervous Mental Disease*, 32, edited by D. Hamburg. The Williams & Wilkins Co., 1970.

14 Stimulation in Infancy

DIFFERENTIAL MATURATION OF AN ADRENAL RESPONSE TO COLD STRESS IN RATS MANIPULATED IN INFANCY. Seymour Levine, Morton Alpert and George W. Lewis in *The Journal of Comparative and Physiological Psychology*, Vol. 51, No. 6, pages 774–777; December, 1958.

EFFECTS OF EARLY EXPERIENCE UPON THE BEHAVIOR OF ANIMALS. Frank A. Beach and Julian Jaynes in *Psychological Bulletin*, Vol. 51, No. 3, pages 239–263; May, 1954.

A FURTHER STUDY OF INFANTILE HANDLING AND ADULT AVOIDANCE LEARNING. Seymour Levine in *Journal of Personality*, Vol. 25, No. 1, pages 70–80; September, 1956.

INFANTILE EXPERIENCE AND RESISTANCE TO PHYSIOLOGICAL STRESS. Seymour Levine in *Science*, Vol. 126, No. 3270, page 405; August 30, 1957.

15 Deprivation Dwarfism

GROWTH FAILURE IN MATERNAL DEPRIVATION. Robert Gray Patton and Lytt I. Gardner. Charles C. Thomas, Publisher, 1963.

EMOTIONAL DEPRIVATION AND GROWTH RETARDATION SIMULATING IDIOPATHIC HYPOPITUITARISM, II: ENDOCRINOLOGIC EVALUATION OF THE SYNDROME. G. F. Powell, J. A. Brasel, S. Raiti and R. M. Blizzard in *The New England Journal of Medicine*, Vol 276, No. 23, pages 1279–1283; June 8, 1967.

SHORT STATURE ASSOCIATED WITH MATERNAL DEPRIVATION SYNDROME: DISORDERED FAMILY ENVIRONMENT AS CAUSE OF SO-CALLED IDIOPATHIC HYPOPITUITARISM. Robert Gray Patton and Lytt I. Gardner in *Endocrine and Genetic Diseases of Childhood*, edited by L. I. Gardner. W. B. Saunders Company, 1969.

GROWTH HORMONE IN NEWBORN INFANTS DURING SLEEP-WAKE PERIODS. Bennett A. Shaywitz, Jordan Finkelstein, Leon Hellman and Elliot D. Weitzman in *Pediatrics*, Vol. 48, No. 1, pages 103–109; July, 1971.

16 Psychological Factors in Stress and Disease

SOMATIC EFFECTS OF PREDICTABLE AND UNPREDICTABLE SHOCK. Jay M. Weiss in *Psychosomatic Medicine*, Vol. 32, pages 397–408; 1970.

EXPERIMENTALLY INDUCED GASTRIC LESIONS: RESULTS AND IMPLICATIONS OF STUDIES IN ANIMALS. Robert Ader in *Advances in Psychosomatic Medicine*, Vol. 6, pages 1–39; 1971.

EFFECTS OF COPING BEHAVIOR IN DIFFERENT WARNING SIGNAL CONDITIONS ON STRESS PATHOLOGY IN RATS. Jay M. Weiss in *Journal of Comparative and Physiological Psychology*, Vol. 77, No. 1, pages 1–30; October, 1971.

17 Learning in the Autonomic Nervous System

INSTRUMENTAL LEARNING OF HEART RATE CHANGES IN CURARIZED RATS: SHAPING, AND SPECIFICITY TO DISCRIMINATIVE STIMULUS. Neal E. Miller and Leo DiCara in *Journal of Comparative & Physiological Psychology*, Vol. 63, No. 1, pages 12–19; February, 1967.

INSTRUMENTAL LEARNING OF VASOMOTOR RESPONSES BY RATS: LEARNING TO RESPOND DIFFERENTIALLY IN THE TWO EARS. Leo V. DiCara and Neal E. Miller in *Science*, Vol. 159, No. 3822, pages 1485–1486; March 29, 1968.

HOMEOSTASIS AND REWARD: T-MAZE LEARNING INDUCED BY MANIPULATING ANTIDIURETIC HORMONE. Neal E. Miller, Leo V. DiCara and George Wolf in *American Journal of Physiology*, Vol. 215, No. 3, pages 684–686; September, 1968.

LEARNING OF VISCERAL AND GLANDULAR RESPONSES. Neal E. Miller in *Science*, Vol. 163, No. 3866, pages 434–445; January 31, 1969.

18 The Physiology of Meditation

STUDIES ON SHRI RAMANAND YOGI DURING HIS STAY IN AN AIR-TIGHT BOX. B. K. Anand, G. S. Chhina and Baldev Singh in *The Indian Journal of Medical Research*, Vol. 49, No. 1, pages 82–89; January, 1961.

PHYSIOLOGICAL EFFECTS OF TRANSCENDENTAL MEDITATION. Robert Keith Wallace in *Science*, Vol. 167, No. 3926, pages 1751–1754; March 27, 1970.

A WAKEFUL HYPOMETABOLIC PHYSIOLOGIC STATE. Robert Keith Wallace, Herbert Benson and Archie F. Wilson in *American Journal of Physiology*, Vol. 221, No. 3, pages 795–799; September, 1971.

V IMMUNE DEFENSES OF THE BODY

19 Wound Healing

WOUND HEALING. Leslie B. Arey in *Physiological Reviews*, Vol. 16, No. 3, pages 327–406; July, 1936.

SOME BIOCHEMICAL ASPECTS OF FIBROGENESIS AND WOUND HEALING. David S. Jackson in *The New England Journal of Medicine*, Vol. 259, No. 17, pages 814–820; October 23, 1958.

CELL FINE STRUCTURE AND BIOSYNTHESIS OF INTERCELLULAR MACROMOLECULES. Keith R. Porter in *Biophysical Journal*, Vol. 4, No. 1, Part 2, pages 167–196; January, 1964.

THE CONNECTIVE TISSUE FIBER FORMING CELL. Russell Ross in *Treatise on Collagen, Vol. II, Part A: Biology of Collagen*, edited by Bernard S. Gould. Academic Press, 1968.

THE FIBROBLAST AND WOUND REPAIR. Russell Ross in *Biological Reviews*, Vol. 43, No. 1, pages 51–96; February 1968.

20 The Immune System

THE CLONAL SELECTION THEORY OF ACQUIRED IMMUNITY. Sir Macfarlane Burnet. Cambridge University Press, 1959.

ANTIGEN SENSITIVE CELLS: THEIR SOURCE AND DIFFERENTIATION. J. F. A. P. Miller, G. F. Mitchell, A. J. S. Davies, Henry N. Claman, Edward A. Chaperon and R. B. Taylor in *Transplantation Reviews*, Vol. 1, 1969.

INDIVIDUAL ANTIGENIC SPECIFICITY OF IMMUNOGLOBULINS. John E. Hopper and Alfred Nisonoff in *Advances in Immunology*, Vol. 13, pages 57–99; 1971.

THE TAKE-HOME LESSON – 1971. Melvin Cohn in *Annals of the New York Academy of Sciences*, Vol. 190, pages 529–584; December 31, 1971.

THE REGULATORY INFLUENCE OF ACTIVATED T CELLS ON B CELL RESPONSES TO ANTIGEN. David H. Katz and Baruj Benacerraf in *Advances in Immunology*, Vol. 15, pages 1–94; 1972.

ANTIGEN DESIGN AND IMMUNE RESPONSE. Michael Sela in *The Harvey Lectures 1971–1972*, Series 67. Academic Press, 1973.

THE PROBLEM OF MOLECULAR RECOGNITION BY A SELECTIVE SYSTEM. Gerald M. Edelman in *The Problem of Reduction in Biology*, edited by F. Ayala and T. Dobzhansky. The Macmillan Company, in press.

21 Markers of Biological Individuality

THE UNIQUENESS OF THE INDIVIDUAL. P. B. Medawar. Basic Books, Inc., 1957.

THE IMMUNOLOGY OF TRANSPLANTATION. P. B. Medawar in *The Harvey Lecture, Series LII (1956–1957)* Academic Press, 1958.

TRANSPLANTATION OF TISSUES AND CELLS. Edited by R. E. Billingham and Willys K. Silvers. The Wistar Institute Press, 1961.

TRANSPLANTATION ANTIGENS. Barry D. Kahan and Ralph A. Reisfeld in *Science*, Vol. 164, No. 3879, pages 514–521; May 2, 1969.

TRANSPLANTATION ANTIGENS. R. A. Reisfeld and B. D. Kahan in *Advances in Immunology*, Vol. 12, Pages 117–200; 1970.

TRANSPLANTATION ANTIGENS: METHODS OF BIOLOGIC INDIVIDUALITY. Edited by Barry D. Kahan and Ralph A. Reisfeld. Academic Press, in press.

22 How the Immune Response to a Virus Can Cause Disease

DAS LEBEN UND WIRKEN DES WIENER KLINIKERS CLEMENS FREIHERRN V. PIRQUET. Dissertation. E. Hoff. Verlag G. H. Nolte, 1937.

EFFECTS OF VIRUS INFECTIONS ON THE FUNC-

TION OF THE IMMUNE SYSTEM. Abner Louis
Notkins, Stephan E. Mergenhagen and Richard
J. Howard in *Annual Review of Microbiology*,
Vol. 24, pages 525–538; 1970.

DESTRUCTION OF VIRUS-INFECTED CELLS BY
IMMUNOLOGICAL MECHANISMS. David D. Porter
in *Annual Review of Microbiology*, Vol. 25,
pages 283–290; 1971.

IMMUNE COMPLEX DISEASE IN CHRONIC VIRAL
INFECTIONS. Michael B. A. Oldstone and Frank
J. Dixon in *The Journal of Experimental Medi-
cine*, Vol. 134, No. 3, Part 2, pages 32S– 40S;
September, 1971.

INFECTIOUS   VIRUS-ANTIBODY   COMPLEXES:
INTERACTION WITH ANTI-IMMUNOGLOBULINS,
COMPLEMENT, AND RHEUMATOID FACTOR.
Abner Louis Notkins in *The Journal of Experi-
mental Medicine*, Vol. 134, No. 3, Part 2, pages
41S–51S; September, 1971.

IMMUNOPATHOGENESIS OF ACUTE CENTRAL
NERVOUS SYSTEM DISEASE PRODUCED BY
LYMPHOCYTIC   CHORIOMENINGITIS   VIRUS.
Donald H. Gilden, Gerald A. Cole, Andrew A.
Monjan and Neal Nathanson in *The Journal of
Experimental Medicine*, Vol. 135, No. 4, pages
860–869; April, 1972.

VI AGING

23 Getting Old

CHEMICAL ASPECTS OF AGING AND THE EFFECT
OF DIET UPON AGING. C. M. McCay in *Cow-
dry's Problems of Aging*, edited by A. I. Lansing.
Williams & Wilkins Company, 1952.

AGING: THE BIOLOGY OF SENESCENCE. Alexander
Comfort. Holt, Rinehart and Winston, Inc.,
1964.

BEHAVIOR AND ADAPTATION IN LATE LIFE.
Edited by Ewald W. Busse and Eric Pfeiffer.
Little, Brown and Company, 1969.

THE IMMUNOLOGIC THEORY OF AGING. Roy A.
Walford. Munksgaard, Copenhagen, 1969.

SOCIAL IMPLICATIONS OF A PROLONGED LIFE-
SPAN. Bernice L. Neugarten in *The Geron-
tologist*, Vol. 12, No. 4, page 323, pages 438–
440; Winter, 1972.

HANDBOOK OF SOCIAL GERONTOLOGY. Edited by
Clark Tibbitts. The University of Chicago Press,
1960.

MENTAL DISORDERS IN LATER LIFE. Edited by
O. J. Kaplan. Stanford University Press, 1956.

PHYSIOLOGICAL AND PATHOLOGICAL AGEING.
V. Korenchevsky. Hafner Publishing Co.,
1961.

TRENDS IN GERONTOLOGY. Nathan W. Shock.
Standford University Press, 1957.

24 The Physiology of Aging

THE BIOLOGY OF SENESCENCE. Alex Comfort.
Rinehart & Company, Inc., 1956.

GERIATRIC MEDICINE: MEDICAL CARE OF LATER
MATURITY. Edited by Edward J. Stieglitz. J. B.
Lippincott Co., 1954.

25 Human Cells and Aging

SENESCENCE AND CULTURED CELLS. Leonard
Hayflick in *Perspectives in Experimental
Gerontology: A Festschrift for Doctor F. Ver-
zár*. Charles C Thomas, Inc., 1966.

THE SERIAL CULTIVATION OF HUMAN DIPLOID
CELL STRAINS. L. Hayflick and P. S. Moorhead
in *Experimental Cell Research*, Vol. 25, No. 3,
pages 585–621; December, 1961.

TOPICS IN THE BIOLOGY OF AGING. Edited by
Peter L. Krohn. Interscience Publishers,
1966.

VII EPILOGUE

26 The Present Evolution of Man

EVOLUTION, GENETICS AND MAN. Theodosius
Dobzhansky. John Wiley & Sons, Inc., 1955.

MIRROR FOR MAN. Clyde Kluckhohn. McGraw-
Hill Book Co., Inc., 1949.

RADIATION, GENES AND MAN. Bruce Wallace and
Theodosius Dobzhansky. Henry Holt & Co.,
Inc., 1959.

INDEX

Aaron, Jean, 118
Abercrombie, Michael, 197
Accidents, 235
Acclimatization
  genetics of, 1-5, 89-91
  high altitude and, 103-104, 106
Acetylcholine, 168, 180
Ada, Gordon L., 206
Adaptation
  aging body and, 243-251
  environment and, 1-5, 89-91
  genetics of, 260-265
  goiter as, 21
  to high altitude, 103-106, 111
  lactase tolerance and, 50-51
  physiological, 177, 185-187
  starvation and, 12-19
Adenosine triphosphate (ATP), 76
Adrenal glands, 141, 143-145
  ACTH and, 137-138
  high altitude and, 108, 110
  hormones, 249
Adrenocorticotrophic hormone
    (ACTH)
  cortisol and, 132
  retardation and, 152, 154-155
  stress behavior and, 136-141,
    143-148
Aged
  groups of, 236, 238, 239
  sexual activity of, 241
Aging, 235-242
  cell degeneration in, 232, 252-257
  individual differences in, 243-251
Agricultural revolution,
    malnutrition and, 8-9
Akutsu, K., 177
Albumin and tryptophan, 43-44
Alcohol metabolism, 67
Allison, Anthony C., 261
Allotype, 206, 207, 215
Alpha waves, 177, 178, 180, 181
Altitude acclimatization, 89,
    103-106, 111
Ama, adaptation by, 89, 92-101
Amino acid, 15, 38, 39, 41, 192, 194
Amino acid sequence
  antibody molecules, 199, 200,
    201, 205

transplant marker, 214-215
Anand, B. K., 177
Andrew, Warren, 248
Anemia
  lead and, 81, 84
  sickle-cell, 261, 264
Aneurysm, 33
Anitschkow, N., 33
Anorexia nervosa, 153-154
Anson, George, 198
Antibiotic, metabolism of, 63
Antibody
  epitope recognition, 199-200,
    205-206
  genetic code, 204, 206
  immune system, 186, 199, 203,
    204, 218-227, 242
  lymphatic system and, 203, 204,
    205-207
  structure of, 199, 201, 203, 205
  variability, 199, 201
Anticoagulant therapy, 65-68
Antigen
  antibody and, 186, 199, 302, 204
  cell and, 211-213
  protein marker, 211-215, 217
  viral, 218, 220-221, 222, 227
Antigen-antibody complex, 203-207,
    221-222, 224-227
Antigenic determinant. See Epitope
Antiserum, 216, 217
Antithyroid compounds, 25, 27
Anxiety
  lactate and, 180
  measurements of, 178, 179, 180
  physiology of, 136-141
Aorta, atherosclerosis and, 30, 31,
    32, 33
Appetitive responses, 138, 139
Appley, Mortimer K., 136-137
Arteries, atherosclerosis and, 29, 30,
    31, 32, 33
Arteriosclerosis, 236, 239
  See also Atherosclerosis
Arteritis, 225, 226
Atheroma, 29, 239
Atherosclerosis, 29-37, 231, 238, 239
  See also Arteriosclerosis

Autoimmune disease, 187, 226-227
Autonomic nervous system (ANS),
    146, 167-175
  meditation and, 180, 181
Autonomic response, 172-173
Avoidance learning, 136-141,
    159-165

Bagchi, B., 177
Baker, Sir George, 83
Baker, Paul T., 103, 105
Bakwin, Harry, 149
Baldspot, child's, 152
Banuazizi, Ali, 170
Baranska, Wanda, 227
Barcroft, Joseph, 103
Barger, George, 23
Basal metabolism, 100-101, 106,
    242, 249, 251
Baumann, Eugen, 23
Bayless, Theodore M., 49, 51
Behavior
  coping, 157-165
  regulation of, 136-141
  stimulus conditioned, 143-148
Behring, Emil von, 201
Benacerraf, Barnj, 204
Benditt, Earl P., 191, 192, 194
Bends, 98
Benedict, F. G., 12
Berger, Daniel D., 72
Bilirubin, 62-63, 113, 118
Billingham, Rupert E., 210
Biological individuality, 208-217,
    243, 249
Biological rhythm, 113, 118,
    119-121
Biotransformation, 58, 60-69
Birds, 71-76
Black, Paul H., 227
Blood
  aging and, 241, 248
  clotting of, 32
  function of, 38-42, 43-44
  high altitude and, 104-106,
    108, 110
Blood-group genetics, 260